T0141422

PLANNING,
GEOMETRY,
and
COMPLEXITY
of
ROBOT MOTION

Jerry R. Hobbs, Editor

PLANNING, GEOMETRY, AND COMPLEXITY OF ROBOT MOTION

edited by

Jacob T. Schwartz
Courant Institute of Mathematical Sciences
New York University

Micha Sharir
Department of Mathematical Sciences
Tel Aviv University

John Hopcroft
Department of Computer Science
Cornell University

ABLEX SERIES IN ARTIFICIAL INTELLIGENCE

ABLEX PUBLISHING CORPORATION
NORWOOD, NEW JERSEY

Copyright © 1987 by Ablex Publishing Corporation

All rights reserved. No part of this publication may be reproduced, stored in a retrieval system, or transmitted, in any form or by any means, electronic, mechanical, photocopying, microfilming, recording, or otherwise, without permission of the publisher.

Printed in the United States of America.

CHAPTER 1: From *Communications on Pure and Applied Mathematics,* vol. 36, pp. 345–398, 1983. Copyright © 1983, John Wiley & Sons, Inc.; reprinted by permission.

CHAPTER 2: Reprinted by permission from *Advances in Applied Mathematics,* vol. 4, pp. 298–351, 1983. Copyright © 1983 by Academic Press, Inc.

CHAPTER 3: A shorter version appeared in *The International Journal of Robotics Research,* vol. 2, no. 3, pp. 46–75, 1983, copyright © 1983 by Massachusetts Institute of Technology. By permission of The MIT Press.

CHAPTER 4: From *Communications on Pure and Applied Mathematics,* vol. 37, pp. 479–493, 1984. Copyright © 1984, John Wiley and Sons, Inc.; reprinted by permission.

CHAPTER 5: From *Communications on Pure and Applied Mathematics,* vol. 37, pp. 815–848, 1984. Copyright © 1984, John Wiley & Sons, Inc.; reprinted by permission.

CHAPTER 6: Reprinted by permission from *Journal of Algorithms,* March 1985. Copyright © by Academic Press, Inc.

CHAPTER 7: From *Proceedings of the 15th Symposium on the Theory of Computing,* copyright © 1983, Association for Computing Machinery, Inc.; reprinted by permission.

CHAPTER 8: Reprinted by permission from *SIAM Journal on Computing,* vol. 14, no. 2, May 1985.

CHAPTER 9: From *International Journal of Robotics Research,* vol. 2, no. 4, pp. 77–80, 1983. Copyright © 1983 by Massachusetts Institute of Technology. By permission of the MIT Press.

CHAPTER 11: An earlier version appeared in the *Proceedings of the 20th IEEE Symposium on Foundations of Computer Science* (San Juan, Puerto Rico), pp. 421–427, 1979.

CHAPTER 12: Reprinted by permission from SIAM Journal on Computing, vol. 13, pp. 610–629, August 1984.

CHAPTER 13: Reprinted by permission from SIAM Journal on Computing, vol. 14, no. 2, May 1985.

Library of Congress Cataloging in Publication Data
Planning, geometry, and complexity of robot motion.

(Ablex series in artificial intelligence)
Bibliography: p.
Includes index.
1. Robots—Motion.   I. Schwartz, Jacob T.   II. Sharir,
Micha.   III. Hopcroft, John E., 1939–   .   IV. Series.
TJ211.4.P57   1986   629.8'92   86-20623
ISBN 0-89391-361-8

Ablex Publishing Corporation
355 Chestnut Street
Norwood, New Jersey 07648

Contents

Preface

During the last few years, robotics, whose earlier history lies principally in the fields of mechanical engineering and artificial intelligence, has come to attract the attention of mathematicians and theoretical computer scientists in rapidly increasing degree. Initial investigations have shown that robotics is a rich source of deep theoretical problems, which range over computational geometry, control theory, and many aspects of physics, and whose solutions draw upon methods developed in subjects as diverse as automata theory, algebraic topology, and Fourier analysis. The present volume presents some of this new theoretical robotics research, emphasizing papers relating to the geometric aspects of robot motion planning.

The first group of papers in the volume study various cases of the motion planning problem. Chapter 1 treats the relatively simple case of a rigid polygonal body which must be moved from a specified starting position to a specified final position in two dimensions, amidst polygonal barriers which the body must avoid during its motion. The problem of rigorously deciding whether or not there exists a continuous motion of the body satisfying these conditions is solved by a "decomposition" method: the space of all obstacle-avoiding positions of the body is cut up into relatively simple connected "cells" and the relationships of adjacency among those cells is determined, thereby reducing the problem of finding a *continuous* motion of the body to that of finding a *finite combinatorial* path through a simple "adjacency graph" whose nodes are the aforementioned cells. Chapter 3 extends this same "decomposition" approach to the case of several independent circular bodies moving amidst polygonal obstacles in two dimensions, and shows how the decomposition method can be applied recursively to treat problems in which the "configuration" space of obstacle-avoiding positions of the movable bodies has relatively high dimensionality. The same method is applied in Chapter 5 to a geometrically less simple three-dimensional case: a rod moving amidst polyhedral obstacles.

The computational cost of the algorithms developed in these three chapters grows rapidly as one increases the number of degrees of freedom of the system whose motion must be planned. The theoretical results derived in later papers in this volume show that, for sufficiently general systems, such rapidly rising computational costs are inevitable. However, Chapter 4 shows this inherent difficulty need not affect systems whose geometric structure is sufficiently simple, even if their number of degrees of freedom is large. More specifically, Chapter 4 considers the problem of planning the motion of a "spider"-like planar object consisting of n rods of various lengths linked at a single common point around which the arms are free to rotate; the object moves in an environment of polygonal obstacles. It is shown that the problem of planning motions of such a "spider" grows relatively slowly (although still exponentially) with its number of "arms."

Chapter 2 sums up and generalizes the decomposition approach common to the four chapters mentioned above, by showing that it can be extended (at a high but nevertheless polynomial computational cost) to curved bodies consisting of any number of jointed members, curved obstacles, and any number of dimensions, provided that all the geometric constraints defining the problem can be written algebraically. To obtain this general result it uses the systematic algebraic cell decomposition introduced by George Collins to prove a famous result of Tarski on the decidability of arbitrary statements in the elementary algebraic theory of real numbers.

Collectively, the first five chapters in the present volume show that the motion planning problem, for a fixed number of degrees of freedom, is solvable "in polynomial time"; but the high exponents involved indicate that the approach followed in these chapters is not likely to be usable in practical situations. Subsequent work has therefore sought substantially more efficient approaches. A "retraction" technique which appears promising in this regard is introduced in Chapter 6 and continued in Chapter 7. In this alternative approach, instead of decomposing the whole space F of obstacle-avoiding positions into cells, one first "standardizes" the motions that will be used, by finding a smaller subset of F (e.g., all positions in which the body lies at the center of a "corridor" defined by two or more walls, i.e. at equal minimum distance from several walls), and shows that any motion of the body between specified initial and final position can be reduced to a short segment which moves the body to some "standardized" position, an intermediate motion which moves the body along a curve consisting of standardized positions only, and a short final segment which moves the body from its last standardized position to its specified target position. The reduction in dimensionality achieved by passing from the full space of all obstacle-avoiding body positions to the smaller space of standardized positions is seen to allow substantial improvement in the asymptotic efficiency of the motion planning procedure that results.

This latter approach to motion planning connects with an interesting theme belonging to the rapidly developing field of computational geometry, namely the special properties of the so-called "Voronoi diagram," which as originally defined consists of all points lying at equal distance from at least two of the points belonging to a finite set S of points in the plane. Particularly efficient ways for calculating this diagram from the set S have been developed, and it has been shown that use of this diagram can accelerate many important two-dimensional geometric calculations. The retraction approach to motion planning suggests new forms of Voronoi diagram, which are subsets of curved spaces of object positions rather than of Euclidean spaces; investigations of the geometric properties of these "generalized Voronoi diagrams" was initiated in papers by O'Dunlaing, Sharir, and Yap too recent to be included in the present collection. However, Chapter 8 illustrates the use of a slightly generalized planar Voronoi diagram to obtain an efficient solution of a collision detection problem in two dimensions: how to tell when any two of a set of circles used to cover a collection of supposedly disjoint planar bodies intersect. Here the circles covering each body represent a simplified overestimate of its geometric extent, and we are only interested in intersections between circles covering different bodies—differently 'colored' circles in the terminology of Chapter 8.

As seems to be typical in computational geometry, the corresponding three-dimensional problem is substantially harder; however, Chapter 9 gives an efficient solution for a simplified sphere intersection problem. The theme common to the two last-mentioned chapters in the present collection, namely techniques for accomplishing some of the many geometric calculations which robotics is likely to require, is connected in Chapter 10, which discusses a particularly tractable class of surfaces, namely general quadrics, and describes techniques for calculating curves and points of intersection of two and three such surfaces respectively.

The final group of papers in this volume are more theoretical, and establish *lower* bounds for the computational cost of motion planning problems. Chapter 11, which is one of the earliest papers in theoretical robotics, opened this theme in 1979 by showing that the problem of planning the motion through a series of narrow branching tunnels of a three-dimensional robot consisting of numerous arms able to revolve about a common axis is PSPACE-complete, i.e. that (as is generally believed) its solution inherently requires a number of steps exponential in the size of the input data needed to define the problem. Subsequent work on this theme has simplified the structure of the moving objects for which a statement of this kind can be established. Chapter 12 establishes such a result for the case of simple planar "linkages" consisting of rigid straight arms meeting at points ("hinges") which may either be fixed or free to move in the plane. The proof makes use of an interesting late 19th century result of Kempe which asserts that a mechanical linkage of this sort can be constructed to calculate any polynomial function of n variables $x_1, \ldots, x_n$ (represented mechanically by the position of hinge points which are free to move along a straight line in the plane). The related Chapter 13 shows that certain motion planning problems for simpler linked structures, resembling folding carpenter's rulers, can still be difficult (NP-complete) if the structures must move in the presence of obstacles, but remain tractable if the free space has a very simple form, namely the interior of a circular region.

New York City and Ithaca
June 1984

CHAPTER 1

On the Piano Movers' Problem: I. The Case of a Two-Dimensional Rigid Polygonal Body Moving Amidst Polygonal Barriers

JACOB T. SCHWARTZ

Courant Institute of Mathematical Sciences
New York University

MICHA SHARIR

Department of Mathematical Sciences
Tel Aviv University

We present an algorithm that solves a two-dimensional case of the following problem which arises in robotics: Given a body B, and a region bounded by a collection of "walls", either find a continuous motion connecting two given positions and orientations of B during which B avoids collision with the walls, or else establish that no such motion exists. The algorithm is polynomial in the number of walls ($O(n^5)$ if n is the number of walls), but for typical wall configurations can run more efficiently. It is somewhat related to a technique outlined by Reif.

0. INTRODUCTION

The piano mover's problem (see Reif, 1979; Lozano-Perez & Wesley, 1979; Ignat'yev, Kulakov, & Pokrovskiy, 1973; Udupa, 1977) is that of finding a continuous motion that will take a given body from a given initial position to a desired final position, but which is subject to certain geometric constraints during the motion. These constraints forbid the body to come in contact with certain obstacles or "walls." In this paper, the first in a series, we focus on a simplified two-dimensional version of this problem: Let B be a given two-dimensional bounded polygonal body, and let V be a two-dimensional open region bounded by a collection of convex closed polygonal curves; each line segment on this boundary is considered to be a "wall." The vertices of the polygonal curves are called "wall corners," and the full collection of walls is not required to be connected (see Figure 0.1). The problem that we set out to solve is: Given two positions and orientations in which B does not touch any walls, find a continuous wall-avoiding motion of B between these two positions, or establish that no such motion exists.

Variants of this problem have been studied by various authors (see Schwartz,

This work has been supported in part by ONR Grant N00014-75-C-0571, by NSF Grant MCS-80-04349, and by U.S. DOE Office of Energy Research Contract EY-76-C-02-3077.

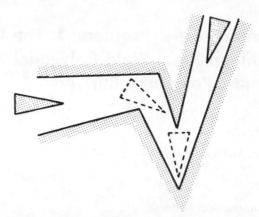

Figure 0.1.   An instance of our case of the
piano movers' problem. The positions drawn
in full are the initial and final positions of B;
the intermediate dotted positions describe a
possible motion of B between the initial and
final positions.

1968; Lozano-Perez & Wesley, 1979; Udupa, 1977; Ignat'yev et al., 1973), some
of whom simplify the general problem by imposing various restrictions on the
allowed motions, e.g., insisting that B must move in a fixed orientation, or at any
rate that B may change its orientation at most once during its motion (see Lozano-
Perez & Wesley, 1979). Our solution allows B to move in a general manner, both
translating and rotating. The problem has also been discussed in an interesting paper
by Reif (1979), who sketches an approach quite different from ours. Reif's approach, to be presented more fully in a forthcoming paper, involves direct computation of the "forbidden" algebraic manifolds of various dimensions traced out in the
space of positions of a moving body as points on the body move in contact with a
given set of obstacles.

Like Reif's suggested approach, our algorithm is based on a study of the space of
all collision-free positions of B. We show that this space can be partitioned into a
collection of connected components as follows. Let P be some fixed corner of B
around which B can rotate, and let V be the complement of the set of walls. We
partition V into subregions separated from each other by "critical curves." A
"critical curve" is a locus of points X such that, if B is placed in V with P at X, a
"critical" contact between B and the walls can occur at some orientation of B.
More specifically, either B can touch some wall perpendicularly, or a corner of B
can touch a corner point between two walls, or B can touch two walls simultaneously, etc. Having thus divided V into subregions R, we first state necessary and
sufficient conditions for two positions, for which P lies in the same subregion R, to
be reachable from each other via a continuous motion during which P remains

within R (see Lemmas 1.8 and 2.7). Then we study possible crossings from one subregion R to another R', and show that we have only to concern ourselves with finitely many possible types of crossings, all of which do actually occur at each crossing point between R and R' (see Lemmas 1.11, 1.12, 2.9, 2.10). Taken together, these results enable us to reduce our original continuous problem to a finite combinatorial problem, described by a finite "connectivity graph" CG which characterizes all possible inter-region crossings, and we show that two given positions of B are reachable from one another by a continuous wall-avoiding motion if and only if two associated vertices in the connectivity graph CG are reachable from one another in CG.

Having thus reduced the initially given topological problem to a combinatorial one, we go on to examine the reduction in more detail, studying the geometric properties of the critical curves that arise in our case of the movers' problem. These curves are seen to consist of line segments, circular arcs, ellipses, and certain kinds of fourth-degree curves. This makes straightforward construction of the connectivity graph possible, as seen in the final version of the algorithm presented below. The complexity of the algorithm is then analyzed, and is shown to be $O(n^5)$, where n is the number of distinct convex walls appearing in the original problem. (The complexity of the body itself is assumed constant in this analysis.)

From the point of view of algebraic topology, our approach can be interpreted as an explicit triangulation of a three-dimensional bounded manifold defined by various geometric constraints. Our approach is elementary, but must deal with an irregular collection of geometric constraints which deeper topological investigations normally avoid.

A much more general problem could be formulated as follows: given a pair V_1, V_2 of real algebraic manifolds and a pair x, y of points lying on V_1-V_2, decide whether or not x and y belong to the same component of V_1-V_2. As will be shown in a later paper, this problem is amenable to treatment by a method closely related to that of Tarski (1951), but of course such very general methods do not lead to algorithms of polynomial complexity. The more limited and detail-ridden approach followed in this paper turns out to involve the definition and examination of various kinds of "critical" configurations at which the range of positions or orientations available to the body as it moves in restricted ways changes abruptly. Similar approaches can be used to develop algorithms of polynomial complexity for other related moving-body problems, for example the problem of a ladder moving in three-space among polyhedral obstacles, the problem of a "hinged" two-dimensional polygonal body, and the problem of two circular bodies moving among polygonal obstacles. We hope to discuss these problems in later reports.

The problem we study in this paper can be regarded as a highly simplified model of the problem of planning a motion of a robot through a confined space. The more complex problems outlined in the preceding paragraph use more realistic and more sophisticated models of this robotic motion-planning problem. Each such generalization yields a manifold of free positions having high dimensionality, and the methods developed in this paper do not extend immediately to these cases. Nevertheless, the general approach of detecting and analyzing "critical" configurations, of which this paper is only a limited special case, may prove useful in attacking those more formidable problems.

The paper is organized as follows. In Section 1 we describe our solution to the movers' problem in the special case where B is just a line segment, or, in our terminology, a "ladder." To avoid duplication, most of the results in Section 1 are stated without proof. These results are later generalized and their generalizations proved in Section 2, yielding a solution to the case in which B is a general polygonal body. An algorithm based on the theory developed in these sections is sketched and analyzed in Section 3. A simple example of motion planning using our technique is presented in the final Section 4.

1. THE CASE OF A "LADDER"

1.1. Topological and Geometric Relationships. The Connectivity Graph.

Rather than tackling the general case immediately, we begin by assuming that B is a line segment PQ. We consider P to be the "marked end" and Q the "unmarked end" of B. We assume that the region V in which B is free to move is a two-dimensional open region bounded by finitely many polygonal walls which can be partitioned into a disjoint collection of simple polygonal closed curves. We assume also that the complement of V' of V is a two-dimensional region (called the wall region) having the same boundary as V. This excludes cases in which the complement of V contains one-dimensional "slits" or isolated points. Furthermore, the assumption that the walls separating V' from V fall into a disjoint union of closed polygonal curves excludes cases in which a boundary point of V is an inner point of two distinct boundary curves. Assume that B is to be placed in V with P at some point X. X is called an *admissible point* if there exists at least one orientation angle for which B does not touch any wall. We shall call a position of B at which it does not touch any wall a *free position,* and a position of B at which it does not cross through any wall a *semi-free position.* (Thus, a semi-free position is either a free position or a position at which B touches a point of some wall, but does not enter the interior of the wall region.) The set FP of all free positions of B is plainly an open three-dimensional manifold, and the set SFP of all semi-free positions is closed.

Since, by assumption, the boundary of the two-dimensional region V in which the line B is free to move consists of a finite collection of walls, the complement V' of V can be divided into convex subregions by finitely many chords (see Figure 1.1). For technical reasons, we require that each of the convex subregions R into which the wall region V' is divided must consist of a continuous sequence of (zero or more) walls W, followed by a sequence of (zero or more) chords of V' (this partitioning may require splitting some wall edges into two or more subedges, which we continue to call "walls"). Moreover, we also require that none of the

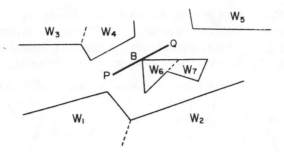

Figure 1.1. Walls divided into seven convex wall sections. The ladder B is shown touching a wall.

chords bounding R pass through a convex corner of V'. The first of these conditions is easily satisfied. To satisfy the second condition, we can simply begin the process of partitioning V' by finding all convex corners Z of V' and cutting off a small triangle containing each such Z; then we partition the remainder of V' into convex subregions by chords as before. Since we want V' to be partitioned into as few subregions as possible, we can improve this procedure by grouping the individual convex corners of V' into disjoint sequences of contiguous convex corners whose convex hull H is contained in V', and such that the chord connecting the two extreme corners in the sequence is contained in the interior of V'. Then for each such group of points we can cut from V' a convex polygon which contains H by drawing a chord that connects two internal points of wall boundary segments but otherwise lies entirely in the interior of V'. After this, we can complete the partitioning as before. (See Figure 1.1 for such a partitioning.)

DEFINITION 1.1. Suppose that V' has been partitioned in the manner just described. We call each (maximal connected) sequence of one or more walls constituting part of the boundary of a convex subsection of V' a *(convex) wall section*. When no confusion arises, we shall also use this term to denote the entire corresponding convex subsection of V'.

Note that, if two wall sections touch each other, they meet at an exterior angle less than or equal to 180°. (In Section 2, where we treat the case in which B is a general polygonal body, this partitioning of the walls into convex sections is not used, for technical reasons discussed there. The partitioning described above is useful in reducing the complexity of our procedure, though, in general, not its order of magnitude.)

To analyse the irregularly shaped three-dimensional manifold *SFP*, we project it into a more easily graspable space of fewer dimensions. Two possibilities suggest themselves: (a) Project *SFP* onto the two-dimensional region in which the point P is free to move, and then consider the set of orientations available to B for each fixed position of P; or (b) project *SFP* onto the one-dimensional space of all possible orientations of B, and then consider the set of all positions available to P for each fixed orientation. Although both approaches are feasible, we prefer in this paper to follow approach (a) (however, approach (b) and its generalizations will be discussed in a forthcoming paper), which leads us at once to the following initial

LEMMA 1.1. *For each fixed admissible position X of P, the set $O(X)$ of all orientations representing free positions of B is a finite collection of open angular sectors, whose endpoints represent semi-free orientations of B.*
(Note that orientations are represented by angles on the unit circle.)

Proof: That $O(X)$ is open is obvious from the fact that FP is open (and that the projection from FP onto the orientation of B is an open mapping.) Since SFP is closed, it is clear that each boundary point of $O(X)$ represents a semi-free orientation of B. In each such orientation, B either touches a wall at its end Q, or touches one or more wall corners at some points between P and Q. Hence, for a given X there can be only finitely many boundary points of $O(X)$, making it plain that $O(X)$ consists of finitely many intervals. (In fact, since each convex wall section can be touched by B in at most two orientations (with P fixed at X), it follows that the number of endpoints of these intervals is at most $2n$, where n is the number of convex wall sections.)

We shall use $O^*(X)$ to denote the set of all orientations representing semi-free positions of B with center at X. Clearly $T(X)=O^*(X)-O(X)$ consists of finitely many points, some of which are endpoints of intervals of $O(X)$, and others of which may be isolated.

LEMMA 1.2. *Let X be an admissible point. Then for each $\varepsilon>0$ there exists a neighborhood N of X such that, for all $Y \in N$, each point of $O^*(Y)$ lies within a distance ε of some point of $O^*(X)$, and conversely each nonisolated point of $O^*(X)$ lies within a distance ε of some point of $O^*(Y)$.*

Proof: Suppose that there exists a sequence $Y_n \rightarrow X$ and orientations θ_n in $O^*(Y_n)$ such that dist $(\theta_n, O^*(X))>\varepsilon$. Passing to a subsequence, we can assume that $\theta_n \rightarrow \theta$ for some θ. Since SFP is closed, it follows that $[X, \theta]\in SFP$, i.e., $\theta\in O^*(X)$, a contradiction proving the first part of our lemma.

To prove the second part, we argue as follows. Since $O^*(X)-O(X)$ is finite, we can find a compact subset K of $O(X)$ such that each nonisolated point of $O^*(X)$ lies within a distance ε of K. Each $\theta\in K$ represents a position in FP and, hence (since FP is open and K is compact), there exists a neighborhood N of X such that if $Y \in N$, then $O(Y)$ contains K, clearly completing the proof of the lemma.

COROLLARY. *If X is an admissible point, then for each compact subinterval K of $O(X)$, there exists a neighborhood N of X such that, for each $Y \in N$, $O(Y)$ contains K.*

DEFINITION 1.2. Let X be an admissible position of X, and let $\theta\in T(X)$. Then θ is an orientation at which PQ touches at least one wall section W, either at its extremity Q or at a point intermediate between P and Q. We denote by $s(\theta, X)$ the set of wall sections W which B touches when placed in the orientation θ with P at X. Since, in most cases that will arise, $s(\theta, X)$ will contain a single wall section, we shall also use $s(\theta, X)$ also to denote that unique wall section.

DEFINITION 1.3. (a) Let $[X, \theta] \in SFP$. Give B the position/orientation $[X, \theta]$, and extend the segment B to a line L. Let S be the (closed) side (half-space) of L that B will enter if rotated slightly in the clockwise (respectively counterclockwise) direction with X fixed. Suppose that, when B is given the location/orientation $[X, \theta]$, it touches wall section W. If the intersection of W with the disc of center X and radius equal to the length of B lies entirely within S, then $[X, \theta]$ is called a *clockwise* (respectively *counterclockwise*) stop of B against W.

(b) The set of pairs

$$\{[s(\theta, X), s(\theta', X)]: \theta, \theta' \in T(X) | \theta \text{ is a clockwise stop, } \theta' \text{ is a}$$
counterclockwise stop, and the whole interval between θ and θ' belongs to
$$O(X)\}$$

is called the *characteristic* of X and will be written as $\sigma(X)$. (If each of the sets $s(\theta, X)$ and $s(\theta', X)$ contains a single wall section W, W', respectively, then, following our earlier convention, we shall consider the pair $[W, W']$, rather than the pair $[s(\theta, X), \sigma(\theta', X)]$, to be an element of $\sigma(X)$.) If $O(X)$ consists of the full angular space, we put $\sigma(X) = \{[\Omega, \Omega]\}$, where Ω designates a "nonexistent" stop.

With the exception of Lemma 1.7 and Theorem 1.1 below, the rest of the theory presented in this section is given with proofs omitted. These proofs can be found in the technical report (Schwartz & Sharir, 1983a), which is a fuller version of this chapter.

LEMMA 1.3. *Let $X_n \to X$ and suppose that a wall section W belongs to all the sets $s(\theta_n, X)$ for all n and for some sequence $\theta_n \in T(X_n)$. Then the limit orientation tj of any convergent subsequence of the θ_n belongs to $T(X)$, and $W \in s(\theta, X)$. Furthermore, if, for all n, $[X_n, \theta_n]$ is a clockwise (respectively a counterclockwise) stop of B against W, then $[X, \theta]$ is also a clockwise (respectively counterclockwise) stop of B against W.*

LEMMA 1.4. (a) *If θ is the clockwise (respectively counterclockwise) endpoint of an interval of $O^*(X)$, it is a clockwise (respectively counterclockwise) stop of B.*

(b) *If θ and θ' are both clockwise (respectively counterclockwise) stops of B against the same wall section W, then $\theta = \theta'$.*

(c) *If $[X, \theta]$ is a location/orientation of B such that $\theta \in T(X)$ with B touching a convex wall section W, then it is either a clockwise stop or counterclockwise stop of B against W, or both.*

COROLLARY. *For each wall section W and each admissible point X, there are at most two orientations $\theta, \theta' \in T(X)$ at which B touches W.*

DEFINITION 1.4. Let X be an admissible point. Then, for each wall section W which labels a clockwise stop (respectively a counterclockwise stop) in $T(X)$, we let $\psi(X, W)$ (respectively $\psi'(X, W)$) denote the unique clockwise (respectively counterclockwise) stop θ such that $W \in s(\theta, X)$.

If $\sigma(X) = \{[\Omega, \Omega]\}$ we shall find it technically convenient in what follows to define $\psi(X, \Omega)$ to be the angle $0°$, and $\psi'(X, \Omega)$ to be the angle $360°$.

LEMMA 1.5. *Suppose that P lies at an admissible point X of V which has the following properties:*

(i) There does not exist a semi-free orientation of B (with P at X) such that B touches the boundary of V' in more than one point.

(ii) There does not exist a semi-free orientation of B (with P at X) in which B touches a wall perpendicularly at Q, or in which Q touches a corner or end of a wall section.

Let $\theta \in T(X)$. Then θ is not an isolated point of $O^*(X)$, and is not the endpoint of more than one arc of $O^*(X)$. Moreover, $s(\theta, X)$ contains just one wall section W.

LEMMA 1.6. *The set of points satisfying the hypotheses (i) and (ii) of Lemma 1.5 is open.*

LEMMA 1.7. *Let X satisfy the hypotheses (i) and (ii) of Lemma 1.5. Then there exists an open neighborhood N of X such that for $Y \in N$ we have $\sigma(Y) = \sigma(X)$, and $O^*(Y)$ consists of exactly as many arcs as $O^*(X)$. Moreover, if $[W, W'] \in \sigma(X) = \sigma(Y)$, then $\psi(Y, W)$ (respectively $\psi'(Y, W')$) depends continuously on Y for $Y \in N$.*

Proof: It follows readily from Lemmas 1.2 and 1.5 that, for Y sufficiently close to X, the set $O^*(Y)$ consists of at least as many arcs as $O^*(X)$. (Indeed, for Y sufficiently close to X each arc of $O^*(Y)$ is contained in an ε-expansion of some arc of $O^*(X)$, and these expansions will be disjoint if ε is small enough.) Suppose next that there exists a sequence Y_n of points converging to X such that each $O^*(Y_n)$ consists of more arcs than $O^*(X)$. By Lemma 1.6, we can suppose without loss of generality that each Y_n satisfies the hypotheses of Lemma 1.5, so that each set $O^*(Y_n)$ consists of finitely many arcs whose endpoints are (clockwise and counterclockwise) stops, and no $O^*(Y_n)$ contains any isolated point. Let the arcs of $O(Y_n)$ be $A_1(n), \cdots, A_k(n)$, and let the clockwise and counterclockwise ends of these arcs be $\theta_1(n), \cdots, \theta_k(n)$, and $\theta'_1(n), \cdots, \theta'_k(n)$. Passing to a subsequence if necessary, we can suppose without loss of generality that all these quantities converge as $n \to \infty$, to quantities $\theta_1, \cdots, \theta_k, \theta'_1, \cdots, \theta'_k$. By the corollary to Lemma 1.2, every ε-contraction of an arc I of $O^*(X)$ is contained in $O^*(Y_n)$ for sufficiently large n; hence it is contained in one of the circular intervals $[\theta_j, \theta'_j]$. Since this holds for all $\varepsilon > 0$, the arc I is itself contained in such an interval. On the other hand, it follows by Lemma 1.2 that each of these intervals is contained in $O^*(X)$. It is also easy to check that these intervals are pairwise disjoint (except for their endpoints). Since

$O^*(X)$ is assumed to have fewer than k arcs, it follows that at least one of the intervals $[\theta_j, \theta_j']$ must consist of a single point which also coincides with an endpoint of a neighboring interval.

It follows therefore that some clockwise or counterclockwise stop $\theta \in O^*(X)$ must be the limit as $n \to \infty$ of at least two different sequences $\theta_i(n)$, $\theta_j(n)$ of (both clockwise or both counterclockwise) stops. By Lemma 1.4 and its corollary, since $\theta_i(n) \neq \theta_j(n)$, it follows that these are stops against two different wall sections W_1, W_2, and, passing to a subsequence if necessary, we can conclude that, in the limiting orientation $[X, \theta]$, B must touch at least two different wall sections, contradicting the assumption that X satisfies the hypotheses of Lemma 1.5. This proves that, for all Y sufficiently close to X, the set $O^*(Y)$ consists of exactly as many arcs as $O^*(X)$, and that the clockwise and counterclockwise ends of the arcs of $O^*(Y)$ converge to the clockwise and counterclockwise ends of corresponding arcs (in the sense of Lemma 1.2) of $O^*(X)$ as Y approaches X.

Suppose next that $Y_n \to X$, that θ_n is the clockwise end of an arc of $O^*(Y_n)$, and that $\theta_n \to \theta$, so that θ is the clockwise end of an arc of $O^*(X)$. Then, if each θ_n is a stop of B on a wall section W, it is plain that, when B is given the limiting orientation $[X, \theta]$, it must also touch W. Hence $\sigma(Y_n) = \sigma(X)$ for each sufficiently large n, i.e., $\sigma(Y) = \sigma(X)$ for Y sufficiently close to X. Moreover, as $Y_n \to X$, no subsequence of $\psi(Y_n, W)$ can converge to any angle but $\psi(X, W)$. The same argument holds if X is replaced by any Y in N, since the points in N satisfy the hypotheses of Lemma 1.5. This proves that $\psi(Y, W)$ depends continuously on $Y \in N$, and similarly for $\psi'(Y, N')$.

Let K be the locus of all admissible points violating one of the conditions (i), (ii) of Lemma 1.5. We shall see below that K is the union of a finite collection of curves, which we call the *critical curves* of our case of the mover's problem. Removal of these critical curves divides the two-dimensional space of all admissible points into a finite collection of connected open regions R, which we call the *noncritical regions* of our problem. The following corollary is an immediate consequence of what has already been shown:

COROLLARY. *The set $\sigma(X)$ is constant on each connected subregion R of the set A of admissible positions for which R contains no critical curve.*

DEFINITION 1.5. For each such R, we put $\sigma(R) = \sigma(X)$, where X is a point chosen arbitrarily from R.

Let d be the length of B. Then the critical curves of our problem fall into the four following categories.

Type I: For each convex wall section W, the locus of all points at distance d from w (where this distance is attained at an interior point of W).

Type II: For each common endpoint Z of a pair of neighboring wall sections, the circle of radius d about Z.

Type III: For each wall section W_1 and each corner C of a different wall section

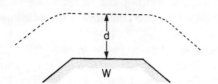

Figure 1.2.   A type I critical curve.   **Figure 1.3.   A type II critical curve.**

W_2, the set of all points traced by P as B moves touching W_1 and C (see Figures 1.4(a), 1.4(b)).

Type IV: For each wall section W, the set of all points traced by P as B slides along an edge of W (see Figure 1.4(c)).

These four types of critical curves will be analyzed in greater detail in the next section, where various degenerate cases will also be noted.

DEFINITION 1.6.   Let X be an admissible point, and let $\theta \in O(X)$. We denote by $\gamma(\theta,X)$ (respectively $\gamma'(\theta,X)$) the clockwise (respectively counterclockwise) endpoint of the arc of $O^*(X)$ to which θ belongs.

LEMMA 1.8.   *Let R be a connected component of the noncritical subregion of the set A of all admissible points. Then one can move continuously through FP from a given free position $[X, \theta]$ to another free position $[X', \theta']$, where both $X, X' \in R$, via a motion during which P remains in R, if and only if $s(\gamma(\theta,X),\hat{X})= s(\gamma(\theta',X'),X')$ and $s(\gamma'(\theta,X),X)=s(\gamma'(\theta',X'),X')$.*

DEFINITION 1.7.   Let R be a connected open noncritical region. Then
(a) $C(R)$ is the set of all free positions $[X, \theta]$ such that $X \in R$;

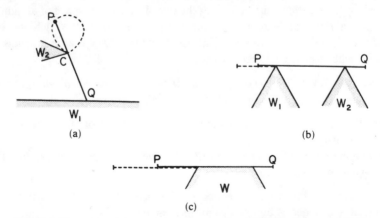

(a)

(b)

(c)

Figure 1.4.   (a) B **in contact with two walls.** Q **lies along a wall. (b)** B **in contact with two walls.** Q **does not lie along a wall. (c)** B **in contact with two points of the same wall.**

(b) if $[W,W']\in\sigma(R)$, and $[W, W']\neq[\Omega,\Omega]$, then $C([W, W'], R)$ is the set of all $[X, \theta]\in C(R)$ such that θ belongs to the open interval $(\psi(X, W), \psi'(X, W'))$;

(c) if $\sigma(R) = \{[\Omega,\Omega]\}$, then we put $C([\Omega,\Omega], R) = C(R)$.

It is obvious from Lemma 1.8 that $C(R) = C([\Omega,\Omega], R)$ is connected if $\sigma(R)$ contains this element; similarly, if $\sigma(R)$ does not contain that element, it follows that the connected components of $C(R)$ are the sets $C([W, W'], R)$, $[W, W']\in\sigma(R)$.

Next we consider what happens when B crosses between regions R_1, R_2 separated by a critical curve. The following simple lemma rules out extreme cases that would otherwise be troublesome.

LEMMA 1.9. *Let $p(t)=[x(t),\theta(t)]$ be a continuous curve in the open three-dimensional manifold FP of free positions of B. Suppose that the endpoints $[X, \theta]$, $[X',\theta']$ of p are specified. Let $\{X_1, \cdots, X_n\}$ be any finite collection of points in the two-dimensional space V not containing either X or X'. Then, by moving p slightly, we can assume that, during the motion described by p, P never passes through any of the points $X_1, \cdots, X_n$.*

Proof: The subset of *FP* for which *P* lies at one of the points $X_1, \cdots, X_n$ is a finite union of submanifolds of dimension 1, and this can never disconnect the three-dimensional manifold *FP*, even locally (see Schwartz, 1968, pp. 5ff.).

Remark. A similar argument, based on Sard's lemma (see Schwartz, 1968, p. 7) shows that, by modifying any given free motion very slightly, we can always ensure that the curve $x(t)$ traced out by *P* during the motion *p* has a nonvanishing tangent everywhere along its length, and that, given any finite set $\beta_1, \cdots, \beta_n$ of smooth curves in two-dimensional space, we can assume that the tangent to $x(t)$ lies transversal to β_j at any point in which $x(t)$ intersects β_j (see Schwartz, 1968). Moreover, we can assume that $\theta(t)$ is constant and $x(t)$ is linear in t for all points along p lying in a sufficiently small neighborhood of each such intersection.

It will be seen in the next section that the critical curves of the case of the mover's problem considered here are always smooth, and that two critical curves can have only finitely many intersections. Thus, to characterize the connected components of the three-dimensional manifold *FP*, it is sufficient to analyze what happens as $P = x(t)$ crosses between regions R_1, R_2 along a line L transversal to a critical curve β separating these two regions, such that L does not pass through any point common to two critical curves. Moreover, we can suppose that θ maintains a constant orientation in the neighborhood of each such crossing.

LEMMA 1.10. *Suppose that (a portion of) the critical curve β forms part of the boundary of a noncritical region R, and that $[W,W']\in\sigma(R)$. Let $X\in\beta$, and let $Y_n\in R$, and $Y_n \to X$. Then $\psi(Y_n, W)$ (respectively $\psi'(Y_n, W')$) converges to $\psi(X, W)$ (respectively $\psi'(X, W')$). Moreover, the function $\psi(X, W)$ (respectively $\psi'(X, W')$) depends continuously on X for $X\in\beta$. Finally, the entire (counterclockwise) open angular sector $(\psi(X,W),\psi'(X,W'))$ belongs to $O^*(X)$.*

LEMMA 1.11.   *Suppose that (a portion of) a smooth critical curve β separates two connected noncritical regions R_1, R_2 and that $R_1 \cup R_2 \cup \beta$ is open. Let $[W_1, W_1'] \in \sigma(R_1)$ (respectively $[W_2, W_2'] \in \sigma(R_2)$), and let $C_1 = C([W_1, W_1'], R_1)$, and $C_2 = C([W_2, W_2'], R_2)$. Then the following two sets of conditions are equivalent:*

Condition A. There exists a point $X \in \beta$ such that the open angular intervals $(\psi(X, W_1), \psi'(X, W_1'))$ and $(\psi(X, W_2), \psi'(X, W_2'))$ are subsets of $O^*(X)$ and have a non-null intersection.

Condition B. There exists a smooth path $c(t) = [x(t), \theta(t)] \in FP$ which has the three following properties:

   (i) *$c(0) \in C_1, c(1) \in C_2$,*

   (ii) *$x(t) \in R_1 \cup R_2 \cup \beta$ for all $0 \leq t \leq 1$,*

   (iii) *$x(t)$ crosses β just once, transversally, when $t = t_o$, $0 < t_o < 1$, and $\theta(t)$ is constant for t in the vicinity of t_o.*

Note that condition B amounts to saying that C_1 and C_2 lie in the same arcwise connected (and hence connected) component of FP (see also the remarks made in Definition 1.7).

Next, we show that if β is an (open) critical curve section not interested by any other critical curve, and if condition A of Lemma 1.11 holds for one point X along β, then it holds for all X along β. This makes it easy to calculate the relationships of connectivity in which we are interested by applying Lemma 1.11 to an arbitrarily selected point of β.

LEMMA 1.12.   *Let the smooth critical curve β separate the two noncritical regions R_1, R_2. Let β' be a connected open segment of β not intersecting any other critical curve, and suppose that $\beta' \cup R_1 \cup R_2$ is open. Let $[W_1, W_1'] \in \sigma(R_1)$ (respectively $[W_2, W_2'] \in \sigma(R_2)$). Suppose that for some $X \in \beta'$ the open clockwise angular sectors $(\psi(X, W_1), \psi'(X, W_1'))$ and $(\psi(X, W_2), \psi'(X, W_2'))$ are subsets of $O^*(X)$ having nonempty intersection. Then this property holds for every $X \in \beta'$.*

The chain of lemmas described so far enables us to reduce the case of the movers' problem considered here to a finite combinatorial search.

DEFINITION 1.8.   The *connectivity graph CG* of an instance of our case of the movers' problem is an undirected graph whose nodes are all the pairs $[R, T]$, where R is a connected open noncritical subregion of V and where $T \in \sigma(R)$. An edge connects two nodes $[R_1, T_1]$ and $[R_2, T_2]$ in CG if and only if the following conditions hold:

   (i) R_1 and R_2 are adjacent and meet along a critical curve β.

   (ii) There exists a (maximal) open portion β' of β contained in the common boundary of R_1 and R_2 and not intersecting any other critical curve, such that for some (hence every) point X on β' the open angular sectors $(\psi(X, W_1), \psi(X, W_1'))$ and $\psi'(X, W'_2), \psi'(X, W_2'))$ overlap, where $T_1 = [W_1, W_1']$ and $T_2 = [W_2, W_2']$.

We are now in a position to prove our main theorem:

THEOREM 1.1. *There exists a motion c of B through the space FP of free positions from an initial* $[X_1, \theta_1]$ *to a final* $[X_2, \theta_2]$ *if and only if the vertices* $[R_1, S_1]$ *and* $[R_2, S_2]$ *of the connectivity graph CG introduced above can be connected by a path of CG, where* R_1 *(respectively* R_2*) is the noncritical region containing* X_1 *(respectively* X_2*), and where*

$$S_1 = [s(\gamma(\theta_1, X_1), X_1), s(\gamma'(\theta_1, X_1), X_1)],$$

and

$$S_2 = [s(\gamma(\theta_2, X_2), X_2), s(\gamma'(\theta_2, X_2), X_2)].$$

Remark. For this theorem to apply, X_1 and X_2 must not lie on a critical curve. If, say, X_1 lies on a critical curve, we first move B a little so as to change X_1 to a point inside some noncritical region, and then apply the theorem as stated above.

Proof: Assume that there exists a continuous path c from $[X_1, \theta_1]$ to $[X_2, \theta_2]$ in *FP*. By the remark preceding Lemma 1.7, we can assume without loss of generality that the curve $c(t)=[x(t), \theta(t)]$ is smooth, that $x(t)$ avoids all intersections of critical curves and crosses all critical curves transversally, and that $\theta(t)$ is constant for t in the neighborhood of each point t_0 at which $x(t)$ intersects a critical curve. Let $R'_1, \cdots,$ R'_k be the sequence of noncritical regions traversed by $x(t)$, and let $C_1 = C(S'_1, R'_1),$ $\cdots, C_k = C(S'_k, R'_k)$ be the corresponding sequence of connected subcomponents of *FP* (cf. Definition 1.7) traversed by $c(t)$. Then it follows from Lemma 1.11 that $[R'_1, S'_1], \cdots, [R'_k, S'_k]$ is a path through the connectivity graph *CG*, proving the ''only if'' part of the present theorem. It also follows from Lemma 1.10 that $C(S'_j, R'_j)$ and $C(S'_j, R'_j)$ belong to the same connected component of *FP* if $[R'_i, S'_i]$ and $[R'_j, S'_j]$ are connected by an edge in *CG*, and this clearly implies the ''if'' part of our theorem.

A closer analysis of the critical curves, to be given below, will show that as we cross any critical curve β from a region R_1 to a region R_2 (and assuming that β does not represent the coincidence of two or more separate critical curves, a possibility that will be analyzed in more detail below), precisely one interval (or a pair of related intervals) of $O(X)$ or its ''marking'' changes discontinuously as X crosses β, in one of the three following ways:

(i) One interval of $O(X)$ may shrink to a point and then disappear (or vice-versa one new interval may appear).

(ii) As we leave R_1 a new stop (which is both clockwise and counterclockwise) may appear within some interval I of $O(X)$, dividing I into two parts which then pull apart as we move into R_2.

(iii) As we cross β the ''marking'' S' of one endpoint of some interval may change to S''.

Suppose for specificity that $\sigma(R_1)$ always contains at least as many elements as $\sigma(R_2)$. Then, in case (i), $\sigma(R_1)$ contains $\sigma(R_2)$, and $\sigma(R_1)-\sigma(R_2)$ contains exactly one element S'_1. In this case, we connect each node $[R_2, S]$, $S \in \sigma(R_2)$, of the connectivity graph *CG* to $[R_1, S]$, but leave $[R_1, S'_1]$ unconnected to any node $[R_2, S]$.

In case (ii), $\sigma(R_2)-\sigma(R_1)$ has exactly one element $S_2' = [W_1, W_2]$, and $\sigma(R_1)-\sigma(R_2)$ has exactly two elements $S_1' = [W_1, W]$ and $S_1'' = [W, W_2]$. Here we connect each node $[R_1, S]$ to $[R_2, S]$ for $S \neq S_1',S_1'',S_2'$, but connect $[R_2,S_2']$ to both $[R_1,S_1']$ and $[R_1,S_1'']$.

In case (iii), $\sigma(R_1)-\sigma(R_2)$ and $\sigma(R_2)-\sigma(R_1)$ are both singletons, the former containing an element S_1', and the latter containing an element S_2'. In this case, we connect $[R_1,S]$ to $[R_2,S]$ for $S \neq S_1, S_2$ but connect $[R_1,S]$ to $[R_2,S_2']$.

All these rules can be summarized as follows (assuming that β is not the coincidence of several critical curves): connect $[R_1,S]$ to $[R_2,S]$ for each $S \in \sigma(R_1) \cap \sigma(R_2)$, and connect each $[R_1,S_1]$, $S_1 \in \sigma(R_1)-\sigma(R_2)$, to each $[R_2,S_2]$ $S_2 \in \sigma(R_2)-\sigma(R_1)$.

1.2. Geometric and Algorithmic Details.

In this subsection we shall describe the geometric structure of the critical curves introduced in the preceding subsection, and go on to give additional algorithmic detail concerning calculation of the connectivity graph CG. We begin by associating appropriate "crossing rules" with the critical curves, which, as has been noted, fall into four main categories.

Type 1 curves: These, the simplest of the critical curves, consist of concatenated sequences of line segments and circular arcs at distance d (equal to the length of B) from a convex wall section W. (In more precise terms, except for the endpoints of a type I curve, all its other points are at distance d from *interior* points of the sequence of wall edges and convex wall corners constituting a convex wall section.) Ignoring the exceptional case (treated below) in which there exist two parallel walls exactly $2d$ apart, we can easily see what happens as we cross such a curve β at a point x not lying on any other critical curve.

When we cross at X from R_1 to R_2 in Figure 1.5, there appears exactly one new (clockwise) stop at which B touches W, at an orientation θ which is an interior point of some open angular sector $[\theta_1, \theta_2]$ of $O^*(X)$ which corresponds to a pair $[W_1, W_2] \in \sigma(R_1)$. Then, as we penetrate into R_2, the clockwise sectors $[\theta_1, \theta]$ and $[\theta, \theta_2]$ separate from each other, and in $\sigma(R_2)$ the pair $[W_1, W_2]$ is replaced by the two pairs $[W_1, W]$ and $[W, W_2]$. (However, if $\sigma(R_1)=\{[\Omega,\Omega]\}$, then the new stop(s) against W which appear as we cross into R_2 do not split $O(X)$ into two components; in this case the pair $[\Omega,\Omega]$ in $\sigma(R_1)$ is replaced by the pair $[W,W]$ in $\sigma(R_2)$.)

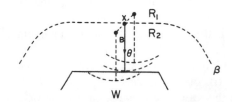

Figure 1.5.   Crossing a type I critical curve.

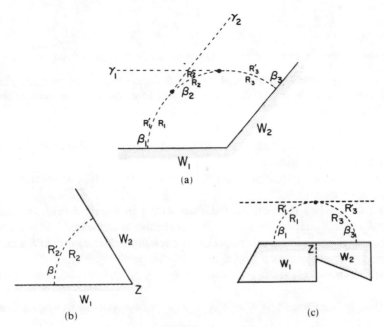

Figure 1.6. (a) Crossing a type II critical curve (obtuse case). (b) Crossing a type II critical curve (acute case). (c) Type II critical curve, two wall sections meeting at 180°.

Type II curves: These require slightly more complicated treatment than those of type I. Recall that a type II curve is a circular arc β centered at a corner point Z at which two wall sections W_1, W_2 intersect. The radius of β is d. We distinguish between three subcases:

(a) The interior angle at Z is less than 180° but greater than 90°, as shown in Figure 1.6(a).

In this case, β is partitioned into three segments β_1, β_2, β_3 by the type I critical curves γ_1, γ_2 that intersect β (see Figure 1.6(a)). As we cross the section β_3 of β from R'_3 to R_3, a counterclockwise stop against W_2 becomes a counterclockwise step against W_1. Hence one pair $[W, W_2]$ of $\sigma(R'_3)$ becomes $[W, W_1]$ in $\sigma(R_3)$.

As we cross the curve β_1, a symmetric situation occurs, namely a clockwise stop against W_1 becomes a clockwise stop against W_2. Thus one pair $[W_1, W]$ in $\sigma(R'_1)$ becomes $[W_2, W]$ in $\sigma(R_1)$.

Next consider the critical arc β_2. As X crosses from R'_2 to R_2 along β_2, the angular sector of $O^*(X)$ labeled by $[W_1, W_2] \in \sigma(R'_2)$ diminishes to a point and then disappears as we cross into R_2. Thus $\sigma(R_2)$ has one less pair than $\sigma(R'_2)$; the pair which disappears is $[W_1, W_2]$.

(b) Next suppose that the interior angle α at Z is less than or equal to 90°; see Figure 1.6(b).

(a)   (b)

Figure 1.7. (a) Critical curve traced by P as B slides along a pair of walls. (b) Critical curve traced by P as B slides along a pair of walls. The two walls lie on opposite sides of B.

In this case, β_1 and β_3 disappear, and the whole curve β behaves exactly like the section β_2 in the previous case.

(c) Finally, we have the case in which two wall sections W_1 and W_2 make contact at an angle of exactly 180°.

This behaves like the limit of the case shown in Figure 1.6(a), but here the regions R_2 and R_2' appearing in Figure 1.6(a) have disappeared. As we cross β_1 (respectively β_3) the clockwise (respectively counterclockwise) stop of B against W_1 (respectively W_2) becomes a stop against W_2 (respectively W_1).

Type III *curves:* These are the most interesting critical curves. They are of two kinds:

(i) Straight lines traced by P as B rests against two wall corners or sections (see Figures 1.7(a),(b)).

(ii) Curves traced by P as B rests against a corner C of one wall W_1 and Q slides along another wall W_2. (See Figure 1.8; the complementary case in which W_1 is placed on the other side of PQ is discussed below.)

A curve β of this second type III kind is the upper loop of a "conchoid of Nicomedes" (see [2]). Its equation is readily calculated (see Figure 1.9):

Since $y/x = h/(-z)$ and $|PQ| = d$, we have

$$d^2 = (y + h)^2 + (z - x)^2 = (y + h)^2 + (x + hx/y)^2,$$

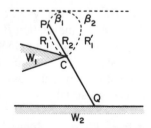

Figure 1.8. "Conchoid" critical curve traced by P as β rests against a corner C of W_1 and Q slides along a wall W_2.

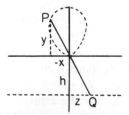

Figure 1.9. Analysis of the conchoid curve generated by the line $y = -h$ and the origin $(0, 0)$.

i.e.,

$$x^2 = y^2\left(\frac{d^2}{(y + h)^2} - 1\right)$$

As seen in Figure 1.8, the type I curve meeting β at distance d from W_2 divides β into two sections β_1, β_2, for which distinct crossing rules must be stated. As we cross over β_1 from R_1 to R_2 a clockwise stop against wall W_2, nonexistent in R_1, appears, and the corresponding stop W_1 disappears. Thus one pair $[W_1, W] \in \sigma(R_1)$ becomes $[W_2, W] \in \sigma(R_2)$.

As we cross over β_2 from R_2 to R_1', an isolated clockwise stop against W_1 (respectively a counterclockwise stop against W_2) appears and grows into an angular sector in R_1', in which region stops against W_1 and W_2 are both possible. Hence $\sigma(R_1')$ has exactly one more element than $\sigma(R_2)$, namely $[W_1, W_2]$. The symmetrical case, in which the wall W_1 appears on the opposite side of B, is analyzed in much the same manner, and yields similar but symmetric crossing rules.

Next we examine cases 1.7(a)–(b). In case 1.7(a), as we cross over β from R_1 to R_2, a pair $[W_2, W]$ disappears from $\sigma(R_1)$ just as $[W_1, W]$ appears in $\sigma(R_2)$. (Note that R_1 and R_2 cannot be connected to each other through noncritical points, because as β ends, another type III critical curve immediately begins, namely a conchoid traced by P as Q slides along the left edge of W_2, and B touches the corner of W_1.)

In case 1.7(b), as we cross β from R_1 to R_2, exactly one pair $[W, W_1]$ of $\sigma(R_1)$ is replaced in $\sigma(R_2)$ by $[W, W_2]$. (Note that here too, R_1 and R_2 are separated from each other by additional critical curves appearing after β ends.)

The symmetric cases to those shown in Figure 1.7(a)–(b), i.e., those in which W_1 and W_2 appear on the other sides of B, can be treated in a completely symmetric fashion.

Type IV *curves:* These curves turn out to be the most trivial among the critical curves. Recall that such a curve β is the locus traced by P as B lies along an edge of some wall section W.

As we cross β from R_1 to R_2 in Figure 1.10, a stop against the corner of this edge near Q changes into a stop against the opposite corner. Since both these corners belong to W only, the pattern of stops remains unchanged as β is crossed. It is thus preferable to discard these curves altogether from our considerations, and so merge R_1 and R_2 in Figure 1.12 into one connected noncritical subregion.

Finally, we need to consider various extreme cases ignored in the preceding discussion. Since all our critical curves are algebraic, they can have nonisolated intersections only if they coincide. Two circles can coincide only if they have the

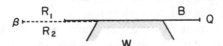

Figure 1.10.   Crossing a type IV critical curve.

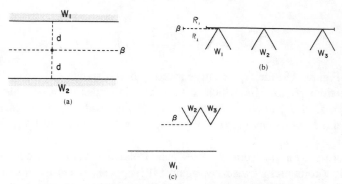

Figure 1.11. (a) Two type I lines become coincident. (b) Two type III lines become coincident. (c) Coincidence between type I and type III lines.

same center and radius, and all such cases have already been considered above. Two different conchoids cannot coincide, since this would imply that their double points, i.e., the corner which generates them, and their lines of symmetry, would have to be the same, so the two would have to be generated by the same corner and wall. Hence the only cases of coincident critical curves we need to consider are

(a) coincident type I lines, which arise in the case of two parallel walls separated by the distance $2d$ (see Figure 1.11(a));

(b) coincident type III lines, which arise in the case of three or more collinear corners (see Figure 1.11(b));

(c) a coincident pair of lines, one of type I, the other of type III, as shown in Figure 1.11(c).

The crossing rules applicable in all these cases are easily derived by imagining certain of the walls to be shifted by some randomly chosen infinitesimal amount. This splits the coincident critical lines into several lines separated by infinitesimal distances; and then any crossing of the original line can be considered as a crossing of all these infinitely close split lines. Several infinitely thin strip regions will appear between the various groups of lines split in this way from single lines. After splitting, and in the presence of these thin regions, all coincidences of critical curves will have been removed, and then the crossing rules stated above will apply.

Consider, for example, the case shown in Figure 1.11(b). In this case, it is convenient to treat each of the segments of β lying between successive wall corners separately. Assume therefore that, for P placed anywhere along such a segment of β, B rests against corners of the wall sections $W_1, \cdots, W_k$, with W_1 between P and W_2, and with the remaining corners lying along the (half) line PW_1W_2 in the sequence $W_3, \cdots, W_k$. Then, if the ladder B is not rigidly "clamped" by three or more walls touching it from opposite sides, only two configurations are possible:

(i) All the stops against $W_1, \cdots, W_k$ are clockwise (as in Figure 1.13(b)). Then a stop against W_1 is possible in R_2, but changes to a stop against W_k as we cross β from R_2 to R_1, replacing a pair $[W_1, W]$ in $\sigma(R_2)$ by a pair $[W_k, W]$ in $\sigma(R_1)$.

Similarly, if all the stops against $W_1, \cdots, W_k$ were counterclockwise, we would have replaced a pair $[W, W_k]$ in $\sigma(R_2)$ by $[W, W_1]$ in $\sigma(R_1)$.

(ii) For some $1 \leq i < k$ all the stops against W_j, $j \leq i$, are clockwise, while all stops against W_j, $j > i$, are counterclockwise. (This case is analogous to that shown in Figure 1.7(c), so we continue to use the labeling of regions shown in that figure.) Here $\sigma(R_1) - \sigma(R_2)$ is the single pair $[W_i, W_{i+1}]$. The symmetric case in which for some $1 \leq i < k$ all the stops against W_j, $j \leq i$, are counterclockwise, while all stops against W_j, $j > i$, are clockwise, behaves similarly, and we leave it to the reader to analyze.

In the case shown in Figure 1.13(a) (which, after splitting, would involve two type I lines infinitely near to one another), it is convenient to introduce additional nodes $[R', T]$ corresponding to the stops that occur for P in the "infinitely thin" region R' separating these two split lines. Then we can simply apply the normal type I crossing rules to the crossing from R_1 to R' and to the crossing from R_2 to R'.

Similarly, the case shown in Figures 1.13(c) can be split, and yields a type I line and a type III line separated by an infinitely thin strip R'. Again, we can introduce additional nodes $[R', W]$ corresponding to the stops that occur for P in R'. Crossings from R_1 to R' are then governed by the standard type I crossing rules, and crossings from R' to R_2 by the appropriate type III crossing rules.

Remark. The infinitesimally thin noncritical regions R' appearing between coincident critical curves after splitting, should be crossed only transversally. If such a strip R' is intersected transversally by another critical curve, then crossing that curve from one portion of R' to another is not allowed.

2. THE CASE OF A GENERAL POLYGONAL BODY

2.1. Topological and Geometric Relationships

Now that the case in which B is a line segment has been treated, we begin to analyze the case in which B is a compact connected two-dimensional polygonal region whose boundary consists of finitely many simply polygonal curves. We assume that B is allowed to move in a region V of the form described in the preceding section. However certain technical modifications of the terminology pertaining to the regions V and V' established in Section 1 are needed. Rather than combining several contiguous wall edges into a convex wall section, we divide each wall into its separate edges $W;$ these edges replace the "wall sections" of Section 1. To describe the positions of B inside V, we choose a reference coordinate system on $B;$ call P the origin and P_x the positive direction of the x-axis of these coordinates (in what follows, P will be taken to be a vertex of B). We can then describe each position of B by a pair $[X, \theta]$, where X is the position of P and where θ is the orientation of P_x in V (relative to some fixed coordinate system in V). The set A of all *admissible points* in V, the set FP of all *free positions*, and the set SFP of all *semi-free positions* of B are then defined as in Section 1. As before, FP is a three-dimensional open manifold, and SFP is a closed manifold.

Lemmas 1.1 and 1.2 and the corollary to Lemma 1.2 of Section 1 remain true for a general polygonal body, because (for a particular position X of P) each edge of the boundary of B can touch a wall edge in only finitely many orientations, and because in our new problem there are still only finitely many body edges and wall edges.

We shall also continue to use the notations $O(X)$ (respectively $O^*(X)$) for the set of all free (respectively semi-free) orientations of B for which P lies at an admissible point X in V. As in Section 1, the set $T(X)$ of all *touches* of B for P at X is defined to be $O^*(X) - O(X)$. However, we shall now use the term *clockwise stop* (respectively *counterclockwise stop*) to refer to clockwise (respectively counterclockwise) endpoints of open intervals in $O(X)$, rather than adopting the somewhat more general notion of clockwise and counterclockwise stops appearing in Definition 1.3(a).

As before, we wish to label each $\theta \in T(X)$ with information characterizing the manner in which B touches the walls at the position/orientation $[X, \theta]$. We do this as follows. First we partition the boundary of B into a collection of segments which we call "half-edges." A half-edge is a maximal subsegment UV of a boundary edge of B having the property that the point on UV nearest to P is either U or V. Obviously, each boundary edge, say, EF, of B consists of either one or two half-edges, namely one half-edge in case one of the angles PEF, PFE is greater than or equal to $90°$, but two half-edges in case both of these angles are acute. In this latter case, EF is partitioned into two half-edges by the foot of the perpendicular from P to EF. (See Figure 2.1 for such a partitioning.)

We then define the labeling $s(\theta, X)$ as follows:

DEFINITION 2.1.   (a) Let X be an admissible point in V, and let $\theta \in T(X)$. Define $s(\theta, X)$ to be the set of all pairs $[L, W]$, where L is a body half-edge, and W is a wall edge such that B touches W at a point on L when placed with P at X and with orientation θ.

(b) Define the *characteristic* $\sigma(X)$ of X to be the set

$$\{[s(\theta,X), s(\theta',X)]: \theta, \theta' \in T(X) | \theta \text{ is a clockwise stop, } \theta' \text{ is a}$$
counterclockwise stop, and the whole open angular interval (θ, θ') belongs to
$$O(X)\}.$$

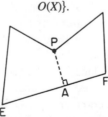

Figure   2.1.   The edge EF of the body B partitioned into two half-edges EA and AF by the foot A of the perpendicular from P to EF.

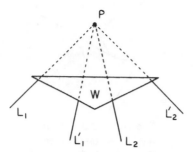

Figure 2.2. A ladder having two clockwise and two counterclockwise stops against the same wall section W.

If $O(X)$ is the full angular space, we put $\sigma(X)=\{[\Omega,\Omega]\}$, for some nominal element Ω, as in the case of a ladder.

The following variant of Lemma 1.3 holds in the polygonal case; its proof is an immediate generalization of the proof of Lemma 1.3, and is left to the reader.

LEMMA 2.1. *Let $X_n \to X$ and suppose that there exists a wall edge W and a body half-edge L such that $[L, W] \in s(\theta_n,X)$ for all n and for some sequence $\theta_n \in T(X_n)$. Then the limit orientation θ of any convergent subsequence of the θ_n belongs to $T(X)$, and $[L, W]\in s(\theta,X)$.*

To understand why using convex wall sections (as defined in Section 1) to label stops in $T(X)$ is not appropriate in the case of a general polygonal body B, we note that the direct generalization of Lemma 1.4 is not true in the general polygonal case if convex wall sections containing more than one wall edge are allowed. In fact, even in the case where B is a ladder this lemma becomes false if one allows the reference point P to lie outside the ladder (but to remain in a fixed position in a coordinate system attached to the ladder). Indeed, Figure 2.2 displays a case in which, for a fixed position of P, the ladder B has two clockwise and two counterclockwise stops against the same wall section W, when the coordinate system attached to B rotates about P. An obvious generalization of the situation shown in Figure 2.2 can yield examples in which B has arbitrarily many clockwise and counterclockwise stops against the same wall section W. However, by distinguishing between different wall edges when we label stops, we can attach different labels to the different clockwise stops which arise as B is rotated about P. (See Lemma 2.5 for a precise statement of this remark.)

The following simple geometric lemma gives initial justification to our use of half-edges.

LEMMA 2.2. *Let L be a half-edge of the body B. Suppose that L has the position U_iV_i when B occupies the position/orientation $[X,\theta_i],i = 1,2$, with $\theta_1 \neq \theta_2$. Then the segments U_1V_1 and U_2V_2 are disjoint.*

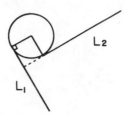

Figure 2.3. Two half-lines tangent to the same circle.

Proof: Any half-edge $L = UV$ is contained in a semi-infinite half-line L' which ends at the foot U' of the perpendicular from P to L'. If the two rotated positions U_1V_1 and U_2V_2 of L were not disjoint, then the two corresponding positions L_1', L_2' of L' would not be disjoint either. But it is easily seen (cf. Figure 2.3) that the lines determined by L_1' and L_2' have their sole intersection at a point lying outside either L_1' or L_2', proving our assertion.

LEMMA 2.3. *Suppose that P lies at an admissible point X of V which has the following properties:*

(I) *There does not exist a semi-free orientation of B (with P at X) such that B touches the boundary of V' in more than one point.*

(II) *There does not exist a semi-free orientation of B (with P at X) in which a convex corner Q of B touches a wall W in such a way that PQ is perpendicular to W at the point of contact, or in which the endpoint Q of a body half-edge touches an endpoint of a wall edge W.*

Let $\theta \in T(X)$. Then θ is not an isolated point of $O(X)$, and is not the endpoint of more than one arc of $O^*(X)$. Moreover, $s(\theta,X)$ contains exactly two pairs. These pairs are either of the form

(a) $[L_1, W], [L_2, W]$, *where L_1 and L_2 are body half-edges coming together at a convex corner Q, and W is a wall edge, in which case Q touches W at an internal point when B is placed at position $[X, \theta]$; or of the form*

(b) $[L, W_1], [L, W_2]$, *where W_1 and W_2 are two wall edges coming together at a convex corner Z and L is a body half-edge, in which case Z touches an internal point of L when B is at position $[X,\theta]$.*

Proof: Place B in the semi-free position $[X, \theta]$. Suppose first that θ is an isolated point of $O^*(X)$. By hypothesis, only one point of B, call it Q, touches the wall region V' at a single point Z. Moreover, either Q is not an endpoint of a body half-edge, or Z is not a corner of any wall edge. Assume first that Q is not an endpoint of a body half-edge. Then it must be an internal point of some half-edge UV of B. In this case, Z must be a convex corner of V' at which two wall edges W_1, W_2 come together.

Consider Figure 2.4. Since Z is the only intersection of B with the wall region V',

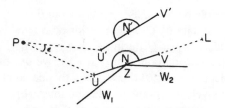

Figure 2.4. An internal point Q of the half-edge UV touches a convex corner of W, and a small rotation of B about this position.

if we remove a small semi-circular neighborhood N of Z from B, $B - N$ will be a compact set disjoint from V'. Therefore, after any sufficiently small rotation $B - N$ will remain disjoint from V'. Thus, to show that θ is not isolated in $O^*(X)$, we have only to find a direction of rotation such that any sufficiently small rotation of N, in this direction, from its original orientation θ, will move it into a position disjoint from W, which is the convex hull of two sufficiently small subsegments of W_1 and W_2 coming together at Z. Since W is convex, it must lie entirely on one side (say, for the sake of definiteness, the counterclockwise side) of L, and since N must be entirely on the other side of L, it is clear that a sufficiently small rotation of N in the appropriate direction (in our assumed case, this is the clockwise direction) will leave N disjoint from W. Thus, θ is not isolated in $O^*(X)$; rather, it is an endpoint of some open interval of $O(X)$. (Note that this argument depends crucially on the fact that L is a half-edge. If not, and Z happened to be the foot of the perpendicular from P to L, then it would no longer be the case that all small counterclockwise rotations of the small semi-circular neighborhood N of Z left N entirely on the counterclockwise side of L.)

Next assume that Z is an internal point of a wall edge W. Then Q must be a convex corner of B, at which two body half-edges L_1 and L_2 come together. Since we have assumed that PQ is not perpendicular to W, there must exist a direction in which B can be rotated from its initial orientation θ, in such a way as to move Q out of W. Arguing as in the preceding paragraph we can show that small rotations in this direction will put B into a position disjoint from W; therefore θ is not isolated in $O^*(X)$, but rather is the endpoint of some open interval of $O(X)$.

This proves that θ cannot be an isolated point at $O^*(X)$.

Next suppose that θ is a common endpoint of two arcs in $O(X)$. Then there exists $\varepsilon > 0$ such that as B is rotated with P at X from orientation $\theta - \varepsilon$ to orientation $\theta + \varepsilon$ it does not meet any wall, except at a unique point Q at orientation θ. Suppose first that Q is an endpoint of some body half-edge, so that it must meet some wall edge W at an interior point. It follows that W is tangent to the arc of radius $|PQ|$ about X, so that B touches W perpendicularly at orientation θ, contrary to our assumptions. Similarly, the other case, in which Q is an inner point of some body half-edge L, is also seen to be impossible because, as B rotates from $\theta - \varepsilon$ to $\theta + \varepsilon$, L traces out a portion of an annulus containing Q as an interior point, and it is plainly impossible for L to touch the walls only at Q during such a rotation.

To prove our final assertion concerning $s(\theta,X)$, we use the fact that when B is placed at position/orientation $[X, \theta]$ there is exactly one point Q of contact between B and the walls. Moreover, this point is either an internal point of a body half-edge L, in which case Q must also be a convex corner of V', which must clearly belong to two wall edges W_1, W_2, or else Q is an internal point of a wall edge W, in which case Q must be a convex body corner at which two body half-edges L_1 and L_2 come together. In the first of these two cases we have $s(\theta,X)=\{[L, W_1],[L, W_2]\}$; in the second case, $s(\theta,X)=\{[L_1, W],[L_2, W]\}$.

LEMMA 2.4. *The set of points X satisfying the hypotheses* (I) *and* (II) *of Lemma* 2.3 *is open.*

Proof: Suppose the contrary. Then there exists a sequence of points X_n converging to a point X, such that X satisfies (I) and (II) but X_n does not. Passing to a subsequence if necessary, we can assume that there exists a sequence $[X_n, \theta_n]$ of semi-free positions/orientations of B, such that $\theta_n \rightarrow \theta \in T(X)$ and such that one of the following conditions holds uniformly in n: either

(a) B touches the boundary of V' at two different points;

or

(b) a convex corner Q of B touches a wall edge W in such a way that PQ is perpendicular to W;

or

(c) an endpoint Q of a body half-edge touches a corner or an endpoint of some wall edge.

It is clear that, if (b) or (c) holds for all $[X_n,\theta_n]$, it must also hold for $[X, \theta]$, which contradicts our assumption. Next, suppose that (a) holds for all $[X_n, \theta_n]$. By passing to a subsequence if necessary we can assume that either

(i) in all positions/orientations $[X_n, \theta_n]$, B touches the wall region V' in two points U_n, U'_n which are contained in different wall edges W, W' and definitely not contained in any single wall edge;

or

(ii) in all these orientations, B touches two points of a single wall edge W.

In discussing the various intersections which can arise in this situation, it will be convenient to speak of two line segments as "overlapping" if their intersection includes any nonempty line segment. In case (i), if the points U_n, U'_n stay a fixed positive distance apart from each other as $n \rightarrow \infty$, then the limiting orientation $[X, \theta]$ would have two points of contact with the walls, which is impossible by hypothesis (I). Hence it is clear that U_n and U'_n must both converge to the same wall corner U, each along a different wall edge. Then for sufficiently large n the corners U_n and U'_n are both internal points of wall edges, and so each of them is either touched by a convex body corner, or else by a body half-edge which overlaps the wall edge (W or

W') containing the point U_n or U'_n. Passing to subsequences of U_n, U'_n if necessary, we have three subcases to consider:

(i.1) For all n, U_n and U'_n are both touched by body corners. These must clearly be distinct corners of the body B (since only one orientation θ_n is involved). This is impossible, since U_n and U'_n approach each other while, of course, different corners of B cannot.

(i.2) For all n, U_n is touched by a body corner Q and U'_n is touched by a body half-edge L which overlaps the wall edge W'. Since L and W' overlap, they belong to the same line; and, clearly, U_n does not belong to that line. Hence the corner Q of the body lies outside the body half-edge L. Thus, in any orientation of the body, Q and L maintain the same nonzero separation. Since, at the orientation $[X_n, \theta_n]$, Q lies at U_n and U'_n lies on L, it follows that U_n and U'_n cannot converge to the same point, a contradiction.

(i.3) Finally, it may be that for all n the corners U_n and U'_n are both touched by overlapping body half-edges L_1, L_2, respectively. Then if L_1 and L_2 are disjoint (closed) half-edges, we can argue as in the preceding two subcases. Otherwise, L_1 and L_2 meet at a corner Q of B, which must clearly coincide with the corner U of W_1 and W_2. But this configuration fixes the position X_n of P, so that $X_n = X$ for all n, which implies that at position $[X, \theta]$ a corner of B touches a wall corner, contradicting assumption (II) for X.

In case (ii), several subcases need to be considered: Either

(ii.1) for all n, there is a body half-edge UV overlapping W, in which case either such an overlap will also exist in the limiting position $[X, \theta]$, or else an endpoint of UV will touch an endpoint or a corner of W in this limiting position; or

(ii.2) for all n, the two distinct points U_n, V_n on B making contact with W when B has position $[X_n, \theta_n]$ are endpoints of body half-edges, in which case it is clear that, at position $[X, \theta]$, B would also make contacts with W, an impossibility by (I); or

(ii.3) for all n, the two distinct points C_n, D_n on W making contact with B at position $[X_n, \theta_n]$ are both endpoints of W, in which case it is also clear that B would also make two contacts with W in the limiting position $[X, \theta]$, again impossible by (I); or, finally,

(ii.4) for all n, at position/orientation $[X_n, \theta_n]$, there is a convex corner U_n of B making contact with a point C_n internal to W, and there is a point V_n on B internal to some half-edge of B making contact with a convex corner D_n of W. If we exclude subcase (ii.1), which has already been taken care of, it follows that U_n and V_n cannot both belong to the same half-edge of B, and so their distance remains bounded away from 0. This implies that, in this case, B would also make two contacts with W in the limiting position $[X, \theta]$, again contradicting (I).

We have thus shown that in all possible cases X must violate one of the hypotheses (I) or (II), concluding the proof of the present lemma.

Next we prove a variant of Lemma 1.4 to the general polygonal case.

LEMMA 2.5. *Let X be an admissible position of P satisfying the hypotheses of Lemma 2.3, and suppose that θ_1 and θ_2 are both clockwise (respectively counterclockwise) stops in $T(X)$ for which $s(\theta_1, X) = s(\theta_2, X)$. Then $\theta_1 = \theta_2$.*

Proof: Suppose that $\theta_1 \neq \theta_2$. Then there are two possible cases, according to Lemma 2.3: either

(a) $s(\theta_1,X)=s(\theta_2,X)=\{[L_1, W],[L_2, W]\}$, where L_1 and L_2 are two body half-edges coming together at a convex body corner Q, and W is a wall edge;
or

(b) $s(\theta_1,X)=s(\theta_2,X)=\{[L, W_1],[L, W_2]\}$, where L is a body half-edge, and where W_1 and W_2 are two wall edges coming together at a corner Q.

In case (a), there would exist two clockwise (respectively counterclockwise) stops of B in which Q touches an internal point of W. Suppose that the orientations θ_1, θ_2 at which these stops occur are such that θ_2 lies clockwise from θ_1. If P lies on the outward side of W (i.e., in the half-plane bounded by the line containing W which contains points lying near W outside the wall region), then the arc swept by Q as B rotates about P from orientation θ_1 to θ_2 lies on the opposite (inner) side of W from P. Thus, as B approaches θ_2 moving clockwise, it would have to penetrate the wall region V', which is impossible, since the stop at θ_2 is a clockwise stop. If P lies on the inward side of W (i.e., in the complementary half-plane bounded by the line containing W, note that this can happen only for body shapes and wall configurations of special form; see also the following description of the critical curves involved in this case of the movers' problem), then a symmetric argument shows that θ_1 cannot be a clockwise stop. Hence case (a) is impossible.

Case (b) is also impossible, for it implies that the two positions assumed by L when B is given the two distinct orientations θ_1 and θ_2 both contain the point Q, which contradicts Lemma 2.2. Hence we must have $\theta_1 = \theta_2$.

Lemma 2.5 justifies the introduction of the following variants of the functions ψ, ψ' (of Section 1) as functional inverses of the mapping s.

DEFINITION 2.2. Let X be an admissible point satisfying the hypotheses of Lemma 2.3. Then for each $[S,S'] \in \sigma(X)$ we let $\psi(X,S)$ (respectively $\psi'(X,S')$) denote the unique clockwise (respectively counterclockwise) stop $\theta \in T(X)$ for which $s(\theta,X)=S$ (respectively $s(\theta,X)=S'$). We also put $\psi(X,\Omega)=0°$ and $\psi'(X,\Omega)=360°$, as in the case of a ladder.

LEMMA 2.6. *Let X satisfy the hypotheses of Lemma 2.3. Then there exists an open neighborhood N of X such that for $Y \in N$ we have $\sigma(Y)=\sigma(X)$, and $O^*(Y)$ consists of exactly as many arcs as $O^*(X)$. Moreover, if $[S,S']\in\sigma(X)=\sigma(Y)$, then $\psi(Y,S)$ (respectively $\psi'(Y,S')$) depends continuously on Y for $Y \in N$.*

Proof: Using arguments similar to those appearing in the proof of Lemma 1.7, we can show that $O^*(Y)$ consists of at least as many arcs as $O^*(X)$ for Y sufficiently close to X. Suppose that there exists a sequence $Y_n \to X$ such that, for each n, $O^*(Y_n)$ consists of k arcs, but that $O^*(X)$ consists of fewer than k arcs. Then, as in Lemma 1.7, there must exist two sequences $\theta_n \neq \theta'_n$ of stops in $O^*(Y_n)$, such that either all these stops are clockwise or all are counterclockwise, and such that θ_n, θ'_n both converge to the same limit θ. By Lemma 2.4 we can assume, with no loss of generality, that for all n the point Y_n satisfies the hypotheses (I) and (II) of Lemma 2.3. Hence, by Lemma 2.5, $s(\theta_n, Y_n) \neq s(\theta'_n, Y_n)$, so that by Lemma 2.3 the union $S = s(\theta_n, Y_n) \cup s(\theta'_n, Y_n)$ must contain at least three pairs. However, Lemma 2.1 implies that S is contained in $s(\theta, X)$, and this contradicts Lemma 2.3.

It follows that $O^*(Y)$ contains exactly as many arcs as $O^*(X)$, and an argument similar to that given in Lemma 1.7 shows that the endpoints of these arcs converge to the endpoints of the corresponding arcs (in the sense of Lemma 1.2) of $O^*(X)$ as Y approaches X. Next suppose that $Y_n \to X$, and that, for all n, $\theta_n \in T(Y_n)$ and that $\theta_n \to \theta$. Then Lemma 2.1 implies that, for all sufficiently large n, $s(\theta_n, Y_n)$ is a subset of $s(\theta, X)$. But since all the sets $s(\theta, X)$, $s(\theta_n, Y_n)$ have precisely two elements by Lemma 2.3, it follows that, for sufficiently large n, $s(\theta_n, Y_n) = s(\theta, X)$. This of course implies that $\sigma(Y_n) = \sigma(X)$. Moreover, it follows by compactness that if $Y_n \to X$ and $[S, S'] \in \sigma(X)$, then $\psi(Y_n, S)$ (respectively $\psi'(Y_n, S')$) is well defined for sufficiently large n, and $\psi(Y_n, S) \to \psi(X, S)$ (respectively $\psi'(Y_n, S') \to \psi'(X, S')$). This completes our proof.

As in Section 1, we now go on to give a more detailed description of the locus C of admissible points X violating one of the conditions (I) or (II) of Lemma 2.3. We shall see shortly that C consists of a finite collection of critical curves whose removal partitions the space A of admissible points into finitely many connected open subregions R, called the noncritical regions of our problem. Lemma 2.6 then gives us the following corollary:

COROLLARY. *The set $\sigma(X)$ is constant on each connected noncritical subregion R of A. Furthermore, if we denote this constant set by $\sigma(R)$, then for each $[S, S'] \in \sigma(R)$ the function $\psi(X, S)$ (respectively $\psi'(X, S')$) appearing in Definition 2.2 is a continuous function on R.*

To classify the critical curves that arise in the case of a general polygonal body, note that the condition (II) of Lemma 2.3 is only violated in two cases, namely in the case of a "perpendicular touch" or in the case of a "corner touching corner" (see below for more precise statements). These two critical conditions lead to the first two categories (type I and type II) of critical curves listed below. Condition (I) is violated by any orientation of B for which there exist (at least) two distinct contact points of B with the walls. If no body half-edge overlaps a wall edge and condition (II) is not violated, then each of these two contact points corresponds either to a touch of a convex body corner against an inner point of a wall edge, or to a touch of a convex wall corner against an inner point of a body half-edge. Taking all combinations of such touches into account, we get five additional classes of critical curves, which correspond to the following conditions: either two body corners touch

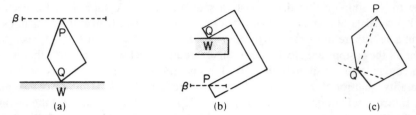

Figure 2.5. (a) A type I critical curve β lying above W. (b) A type I critical curve β lying below W. (c) No type I critical curve corresponding to Q can exist.

the same wall edge (type III), or two body corners touch two different wall edges (type IV), or one body corner touches a wall edge and one wall corner touches a body half-edge (type V), or two wall corners touch the same body half-edge (type VI), or two wall corners touch different body half-edges (type VII). Finally, when a body half-edge overlaps a wall edge, the final class of critical curves (type VIII) is generated. (If more than two distincf contacts occur beteeen B and the walls at the same orientation, we have a coincidence between several of these critical curves; see a discussion concerning treatment of these cases at the end of subsection 2.2.)

Thus in the case of a general polygonal body the critical curves fall into eight categories, which we review in more detail below. Note that in all these cases the shape of the body and of the wall region can constrain the critical curves to avoid particular regions of the space A of all admissible points, and sometimes will even imply that certain critical curves whose presence might be inferred from isolated "local" consideration of the geometric relationship of PQ to a given wall cannot in fact exist, because the critical orientations corresponding to points lying on these curves must be physically permissible.

Details of the eight classes of critical curves are as follows:

Type I. For each wall edge W and each convex corner Q of B, the locus β of points at distance PQ from W (cf. Figures 2.5(a)–(c)). The relationship of P and Q to the rest of the body B dictates the side of W on which β should be drawn, and may even indicate that β cannot be traversed by P at all (see Figures 2.5(a), 2.5(b), 2.5(c)).

Type II. For each endpoint Q of a body half-edge, and each corner Z of the wall region, the locus of points at distance PQ from Z. This is a circular arc I about Z, and the set of points which P can actually traverse is constrained by the size of the angle α_1 formed by the two wall edges meeting at Z and by the size of the angle α_2 formed by the two body half-edges meeting at Q. The angular size of I is easily seen to be

$$360° - \alpha_1 - \alpha_2.$$

If this difference is negative, I is empty (see Figure 2.6).

Type III. For each wall edge W and each pair of convex body corners Q_1 and Q_2,

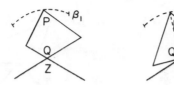

Figure 2.6. Type II critical curves β_1, β_2.

the locus β of P as both Q_1 and Q_2 touch W. (Again, this locus, which is a line segment parallel to W, may or may not be traversable by P; a necessary condition that β should be traversable is that the segment connecting Q_1 and Q_2 does not intersect any interior point of B and that the length of this segment is less than the length of W.) (See Figure 2.7.)

Type IV. For each pair of wall edges W_1, W_2, and each pair of convex body corners Q_1, Q_2, the locus of P as Q_1 slides along W_1 and Q_2 slides along W_2 (see Figure 2.8).

Type V. For each convex corner Q of B, each half-edge L of B (which may or may not contain Q), each convex corner Z of V' and each wall edge W, the locus of P as Q slides along W and Z slides along L (see Figure 2.9).

Type VI. For each pair of convex corners Z_1, Z_2 of V' and each body half-edge L, the locus of P as L slides touching both Z_1 and Z_2 (this locus is a line segment; see Figure 2.10).

Type VII. For each pair of convex corners Z_1, Z_2 of V' and each pair of body half-edges L_1, L_2, the locus of P as Z_1 slides along L_1 and Z_2 slides along L_2 (see Figure 2.11).

Type VIII. For each body half-edge L and each wall edge W, the locus of P as B moves with L and W overlapping each other (this locus is a line segment; see Figure 2.12).

The geometrically nontrivial critical curves, namely those of types IV, V, and VII, belong to the general family of curves known as *glissettes* (cf. Lockwood & Prag, 1961). The detailed algebraic and geometric properties of these critical curves will be discussed in subsection 2.2. It will be shown there that the type IV curves are ellipses or straight lines; type V curves are quartic curves that degenerate into

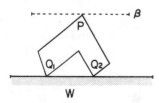

Figure 2.7. A type III critical curve β.

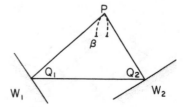

Figure 2.8. A type IV critical curve β.

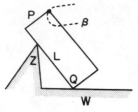

Figure 2.9.   A type V critical curve β.

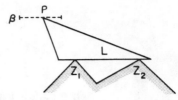

Figure 2.10.   A type VI critical curve β.

conchoids in the special case of a ladder discussed in subsection 1.2; type VII curves are also quartic curves. It therefore follows that all these curves are algebraic (and therefore smooth except at their singular points.).

LEMMA 2.7.   *Let R be a connected open noncritical subregion of the set A of all admissible points. Then one can move continuously through FP from a given free position and orientation $[X_1, \theta_1]$ for which $X_1 \in R$, to another such $[X_1, \theta_2]$, where $X_2 \in R$ also, via a motion during which P remains in R if and only if*

$$s(\gamma(\theta_1,X_1),X_1)=s(\gamma(\theta_2,X_2),X_2),$$

and

$$s(\gamma'(\theta_1,X_1),X_1)=s(\gamma'(\theta_2,X_2),X_2).$$

(See Definition 1.6 for the definition of γ,γ'.)

Proof: First, note that if $O^*(X)$ is a full circle for all points X in R, there is nothing to be proved. Otherwise, using Lemma 2.6 it follows that $\gamma(\theta,Y)$ and γ' (θ,Y) are continuous functions of $[Y,\theta]\in FP$, and that these angles retain constant markings $s(\gamma(\theta,Y),Y)$ and $s(\gamma'(\theta,Y),Y)$ as long as Y remains in R. This proves the "only if" part of the present lemma.

For the converse, we take a curve $x(t)$, $0\leq t\leq 1$, that connects X_1 to X_2 in R, and put

$$S=s(\gamma(\theta_1,X_1),X_1)=s(\gamma(\theta_2,X_2),X_2).$$

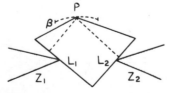

Figure 2.11.   A type VII critical curve β.

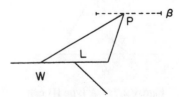

Figure 2.12.   A type VIII critical curve β.

It follows from Lemma 2.6 that the markings of the counterclockwise endpoints of the arcs containing θ_1 and θ_2, respectively, are also identical, i.e., that

$$s(\gamma'(\theta_1,X_1),X_1)=s(\gamma'(\theta_2,x_2),X_2)=S'.$$

It is plain that θ_1 (respectively θ_2) belongs to the open angular arc $(\psi(X,S),$ $\psi'(X,S'))$ (respectively $(\psi(X',S),\psi'(X',S')))$. By Lemma 2.6 and its corollary, the mapping

$$t\mapsto[x(t),(\psi(x(t),S)+\psi'(x(t),S'))]$$

defines a continuous free motion of B for which $x(t)$ remains in R (note that $\psi(x(t),S)$ and $\psi(x(t),S')$ are defined for all $0\leq t\leq1$ because $[S,S']\in\sigma(R)$ and $x(t)\in R$), and the desired motion from $[X_1,\theta_1]$ to $[X_2,\theta_2]$ is then constructed from this motion augmented by an appropriate initial and final rotation from θ_1 (respectively θ_2) to the midpoint of the arc of $O(X_1)$ (respectively $O(X_2)$) containing that orientation.

Definition 1.4 can be extended to the polygonal case presently under consideration as follows.

DEFINITION 2.3. Let R be a connected noncritical region. Then
(a) define $C(R)$ as the set of all free positions $[X,\theta]$ such that $X \in R$;
(b) for each pair $\xi = [R,S]$, where $[S,S']\in\sigma(R)$, define $C(\xi)$ to be the set of all $[X,\theta]\in C(R)$ such that $S = s(\gamma(\theta,X),X)$ (and $S' = s(\gamma'(\theta,X),X)$). (If $\sigma(R)=\{[\Omega,\Omega]\}$, we put $C([R,\Omega]) = C(R)$, as in section 1.)

Next we begin to analyze the way in which the functions ψ and ψ' introduced above behave in the immediate neighborhood of a critical curve. Lemma 1.9 and the remark following it remain true in the general polygonal case. Hence it is sufficient to study motions of B for which the point P of B crosses all critical curves β transversally, always at a point which does not lie on any other critical curve.

LEMMA 2.8. *Suppose that (a portion of) the critical curve β forms part of the boundary of a noncritical region R, and that $[S,S']\in\sigma(R)$. Put $\xi=[R,S]$, $\xi'=[R,S']$. Let $X \in \beta$, and let $Y_n \in R$, $Y_n \to X$. Then $\psi(Y_n, S)$ (respectively $\psi'(Y_n,S'))$ converges to a unique angle, which we denote by $\phi(X, \xi)$ (respectively $\phi'(X, \xi'))$. The function $\phi(X,\xi)$ (respectively $\phi'(X, \xi'))$ depends continuously on X for $X \in \beta$. Moreover, the entire open angular arc $(\phi(X, \xi),\phi'(X,\xi'))$ belongs to $O^*(X)$.*

Proof: Let θ be a limit orientation of any convergent subsequence of $\psi(Y_n, S)$. By Lemma 2.1, S is contained in $s(\theta,X)$, so that θ is in $T(X)$. We must show that θ is unique. For this, note that we have already remarked that all critical curves are smooth, so that the region R is locally arcwise connected in the neighborhood of X. Thus we can construct a continuous curve $Y(t)$, $0\leq t\leq1$, which lies in R for $t<1$ and approaches X as $t \to 1$. Moreover, this curve can be made to pass through all the points Y_n. For each $t_0 < 1$, the range $\varrho(t_0)$ of the restriction of the continuous function (cf. Lemma 2.6) $h(t) = \psi(Y(t), S)$ to $t_0<t<1$ is an interval. If there are two distinct limit orientations of subsequences of $\psi(Y_n,S)$, then plainly there exists an open angular sector (θ_1, θ_2) which belongs to $\varrho(t_0)$ for all $t_0 < 1$. Thus each orientation in (θ_1, θ_2) is a limit orientation, and thus in $T(X)$, which is impossible, since $T(X)$ is finite. Hence θ is unique. We write this limit as $\phi(X,\xi)$. The corresponding "counterclockwise" function $\phi'(X,\xi')$ is defined, and its definition justified, in a similar manner.

Next we show that the functions ϕ and ϕ' are continuous on β. Let X_n be a sequence of points on β converging to some $X \in \beta$, and suppose that the distance between $\phi(X_n,\xi)$ and $\phi(X,\xi)$ remains greater than some $\varepsilon > 0$. For each n take Y_n, $Z_n \in R$ such that the distance between Y_n and X_n and between Z_n and X is less than $1/n$, and such that the distance between $\psi(Y_n,S)$ and $\phi(X_n,\xi)$ and between $\psi(Z_n,S)$ and $\phi(X,\xi)$ is also less than $1/n$. Then the two sequences Y_n,Z_n both converge to X but $\psi(Y_n,S) - \psi(Z_n,S)$ does not converge to 0, which is impossible by what has already been shown. Hence ϕ, and similarly ϕ', are continuous in $X \in \beta$. The final assertion of the present lemma follows from Lemma 1.2.

LEMMA 2.9.   *Suppose that (a portion of) a smooth critical curve β separates two connected noncritical regions R_1 and R_2 and that $R_1 \cup R_2 \cup \beta$ is open. Let $[S_1,S_1'] \in \sigma(R_1)$ and $[S_2,S_2'] \in \sigma(R_2)$. Put $\xi_1 = [R_1, S_1]$, $\xi_2 = [R_2, S_2]$, and let $C_1 = C(\xi_1)$, $C_2 = C(\xi_2)$, $\xi'_1 = [R_1, S_1']$ and $\xi'_2 = [R_2, S_2']$. Then the following conditions are equivalent:*

Condition A: There exists a point $X \in \beta$ such that the open angular intervals $(\phi(X,\xi_1),\phi'(X,\xi_1'))$ and $(\phi(X,\xi_2),\phi'(X,\xi_2'))$ (which are subsets of $O^*(X)$) have non-null intersection.

Condition B: There exists a smooth path $c(t)=[x(t),\theta(t)] \in FP$ which has the following properties:

   (i) $c(0) \in C_1, C(1) \in C_2$;
   (ii) $x(t) \in R_1 \cup R_2 \cup \beta$ for all $0 \le t \le 1$;
   (iii) $x(t)$ crosses β just once, transversally, when $t = t_0$, $0 < t_0 < 1$, and $\theta(t)$ is constant for t in the vicinity of t_0.

Proof: Suppose first that there exists a path $[x(t), \theta]$ in the open three-dimensional manifold FP of free positions of B satisfying (i)–(iii) of Condition B. (By Lemma 2.7 there is no loss of generality in assuming that θ is constant throughout this whole path.) As in Lemma 2.7, let $\gamma(\theta,x(t))$ (respectively $\gamma'(\theta,x(t))$) denote the clockwise (respectively counterclockwise) endpoint of the arc in $O^*(x(t))$ containing θ. Since $c(0) \in C_1$, it follows from Lemma 2.8 that for $t < t_0$ we have $\gamma(\theta,x(t)) = \psi(x(t),S_1)$ and $\gamma'(\theta,x(t)) = \psi'(x(t),S_1')$. Similarly, for $t > t_0$, $\gamma(\theta, x(t)) = \psi(x(t),S_2)$ and $\gamma'(\theta,x(t)) = \psi'(x(t),S_2')$ (where $\xi_2' = [R_2,S_2']$). Moreover, since for $t > t_0$ we have $\theta \in (\psi(x(t),S_1),\psi'(x(t),S_1'))$, it follows from Lemma 2.8 that $\phi(X,\xi_1) \le \theta \le \phi'(X,\xi_1')$. Since we cannot have equality here (θ is a free orientation and the other two orientations are touches), we conclude that θ lies in the open interval $(\phi(X,\xi_1),\phi'(X,\xi_1'))$. Similar reasoning for $t > t_0$ shows that θ also lies in the open interval $(\phi(X,\xi_2),\phi'(X,\xi2'))$, and hence these intervals have a non-null intersection, thus establishing Condition A.

Next suppose that Condition A holds. Since $T(X)$ is finite, there exists an orientation $\theta \in O(X)$ belonging to both the open angular arcs $(\phi(X,\xi_1),\phi'(X,\xi_1'))$ and $(\phi(X,\xi_2),\phi'(X,\xi_2'))$. Since $[X, \theta] \in FP$, there exists a sufficiently short curve $x(t)$ crossing β at X (for some $t=t_0$) from R_1 to R_2, such that $x(t)$ satisfies (ii) and (iii) of Condition B, and such that, for all t, $[x(t), \theta] \in FP$. It follows from the definition of ϕ and ϕ' that $\psi(x(t),S_j) \mapsto \phi(X,\xi_j)$ and $\psi'(x(t),S_j') \mapsto \phi'(X,\xi_j')$ as $x(t) \to X$ from the R_j part of the curve, $j = 1, 2$. Hence, assuming once more that the curve $x(t)$ is short enough, we have $\theta \in (\psi(x(t), S_j), \psi'(x(t),S_j'))$, for $t<t_0$, $j = 1$, and also for $t>t_0$, $j = 2$. Hence $c(t) \in C(\xi_1)$ for $t<t_0$ and $c(t) \in C(\xi_2)$ for $t>t_0$, showing that Condition B holds.

Next we must prove an appropriate generalization of Lemma 1.12, that is, we must show that if Condition A of the preceding lemma holds for one point lying on a portion of β' of β not intersected by any other critical curve, this same condition holds for all points of β'.

LEMMA 2.10. *Let the smooth critical curve β separate the two noncritical regions R_1 and R_2. Let β' be a connected open segment of β not intersecting any other critical curve, and suppose that $\beta' \cup R_1 \cup R_2$ is open. Let S_1, S'_1, S_2, S'_2, ξ_1, ξ'_1, ξ_2, ξ'_2 be defined as in Lemma 2.9. Then the set of $X \in \beta'$ for which the open angular sectors $(\phi(X,\xi_1),\phi'(X,\xi_1'))$ and $(\phi(X,\xi_2),\phi'(X,\xi_2'))$ are subsets of $O^*(X)$ and have a non-null intersection is either all of β' or is empty.*

Proof: We shall suppose that S_1 and S_2 are not Ω, leaving it to the reader to consider the case in which either S_1 or S_2 is Ω.

Our first step is to establish the following auxiliary

CLAIM. *Suppose that for some point $X \in \beta'$ we have $\phi(X,\xi)=\phi'(X,\xi')$, where ξ is either ξ_1 or ξ_2, and where ξ' is either ξ_1' or ξ_2'. Then $\phi(X,\xi)=\phi'(X,\xi')$ holds for all $X \in \beta'$.*

Proof of the claim: Consider the set of all $Y \in \beta'$ for which $\phi(Y,\xi)=\phi'(Y,\xi')$. By the continuity of ϕ, ϕ' this set is closed, and hence we have only to show that it is also open. Suppose the contrary; then there exists an $X \in \beta'$ such that $\phi(X,\xi)=\phi'(X, \xi')$ but for which there also exists a sequence $Y_n \in \beta'$, $Y_n \to X$ such that, for all n, $\phi(Y_n,\xi) \neq \phi'(Y_n,\xi')$. We let $\xi = [R,S]$ and $\xi' = [R',S']$, and proceed by cases.

(a) First suppose that $S = S'$. Then, in the two different positions/orientations $[Y_n,\phi(Y_n,\xi)]$ and $[Y_n,\phi'(Y_n,\xi')]$, B makes the stop indicated by S. It follows by Lemma 2.2 that S cannot designate a wall-corner against body-edge stop, since no half-edge L of B can pass through the same point for a fixed location of P and two different orientations of B. Hence S must designate a stop of some fixed corner Q of B against a wall edge W. The two points Z_n, Z'_n at which B makes contact with W, when given positions/orientations $[Y_n,\phi(Y_n,\xi)]$, $[Y_n,\phi'(Y_n,\xi')]$, form the sides of an isosceles triangle. When $Y_n \to X$ we have $|Z_n - Z'_n| \to 0$ and hence this triangle converges to an isosceles right triangle. Thus PQ must be perpendicular to W at Q, and X must lie along the type I curve determined by W and Q. Hence all of β' must be included in this curve. It is plain that for each Y along this curve the point Q

makes only one stop against W. Hence $\phi(Y,\xi)=\phi'(Y,\xi')$ for all $Y \in \beta'$, a contradiction which proves that $S \neq S'$.

(b) Since $S \neq S'$ and B makes both these stops when given the position/orientation $[X,\phi(X,\xi)]$, it follows that β' is entirely contained within the critical curve defined by the touches of the corners, wall edges and body half-edges designated by S and S'. Since this curve is critical, it follows that everywhere along β' there exists an angle $\theta(Y)$ such that B makes both stops S, S' if given the position/orientation $[Y, \theta(Y)]$. Since $\phi(Y_n, \xi) \neq \phi'(Y_n, \xi')$, there certainly exists either a sequence of points $Z_n \in \beta'$ approaching X for which $\phi(Z_n, \xi) \neq \theta(Z_n)$, or a sequence of points $Z'_n \in \beta'$ approaching X for which $\phi'(Z'_n, \xi') \neq \theta(Z'_n)$.

First suppose that just one such sequence exists, e.g., that $\phi(Z_n, \xi) \neq \theta(Z_n)$ but $\phi'(Y, \xi') = \theta(Y)$ for all $Y \in \beta'$ sufficiently near X. Then, as in (a), it follows that S must designate a stop of a body corner Q against a wall edge W, and moreover in the limiting position X the segment PQ must be perpendicular to W, so that β' is included in the type I curve defined by W and Q. But along this curve there is only one stop of Q against W, so that $\phi(Z_n, \xi) = \theta(Z_n)$, contrary to assumption.

(c) Hence both the sequences Z_n and Z'_n must exist. Again arguing as in (a), it follows that S and S' must both designate stops of body corners Q, Q' against wall edges W, W'. That is, β' must be part of the critical curve of type II, III, or IV defined by Q, Q', W, W'. A detailed analysis of these curves, which can be found in the technical report (Schwartz & Sharir, 1983a), will show that for each Y in such a curve there either exists just one critical orientation $\theta_1(Y)$ at which B makes the critical stop marked by S and S' (this is the case for type II and type III curves, and for the nondegenerate type IV curves), or that there exist two such stops $\theta_1(Y)$, $\theta_2(Y)$ (in this case it is possible that one of these stops does not materialize due to the interference of other walls and body edges). Moreover, in the former case, $\theta_1(Y)$ varies continuously with Y. If only θ_1 exists, then $|\theta_1(Z_n)-\phi(Z_n, \xi)|$ is positive but converges to 0 as $n \to \infty$, so that we can derive a contradiction just as in (b) above. Finally, suppose that both θ_1 and θ_2 exist, so that $\theta_1(Y) \neq \theta_2(Y)$ for Y near X. Since for a given position Y of P there cannot exist three separate stops of a body corner q against the same wall W, it follows that for each Y either $\theta_1(Y)=\phi(Y, \xi)$ or $\theta_2(Y)=\phi(Y,\xi)$. Hence, for all Y near X, B makes both stops S, S' when given position/orientation $[Y,\phi(Y,\xi)]$. This allows us to argue just as in case (a) above, leading to a final contradiction and proving our claim.

It is now easy to finish the proof of the present lemma. Suppose that there exists $X \in \beta'$ for which the open angular arcs $(\phi(X, \xi_1),\phi'(X, \xi'_1))$ and $(\phi(X, \xi_2),\phi'(X, \xi'_2))$ overlap. Since the set K of all such points is open by Lemma 2.8, it suffices to show that it is also closed. Let $X_n \in K$ approach a point X which does not belong to K. Then $\phi(X, \xi)=\phi'(X, \xi')$ for some $\xi \in \{\xi_1, \xi_2\}$ and some $\xi' \in \{\xi'_1, \xi'_2\}$. By what we have proved this equality must hold throughout β', which in turn would imply that K is empty. Hence K is closed, so that, by the connectivity of β', K is all of β'.

We can now define the connectivity graph for the case of a general polygonal body.

DEFINITION 2.4. The *connectivity graph CG* of an instance of our case of the movers' problem is an undirected graph whose nodes are of the form $[R, T]$, where R is some connected noncritical region (bounded by critical curves) and *where* $T \in \sigma(R)$ (or $T = \Omega$ if $\sigma(R)$ is null). Let $T_1 = [S_1, S_1'] \in \sigma(R_1)$, $T_2 = [S_2, S_2'] \in \sigma(R_2)$, and put $\xi_j = [R_j, S_j]$, $\xi_j' = [R_j, S_j']$ for $j = 1, 2$. The graph contains an edge connecting $[R_1, T_1]$ and $[R_2, T_2]$ if and only if the following conditions hold:

(I) R_1 and R_2 are adjacent and meet along a critical curve β.

(II) For some one of the open connected portions β' of β contained in the common boundary of R_1 and R_2, and not intersecting any other critical curve, and for some (hence every) point $X \in \beta'$ the open angular arcs $(\phi(X,\xi_1),\phi'(X,\xi_1'))$ and $(\phi(X,\xi_2),\phi'(X,\xi_2'))$ have a non-null intersection.

We can now state the main result of this section.

THEOREM 2.1. *There exists a motion* $c(t)=[X(t), \theta(t)]$ *of B through the space FP of free positions from an initial position* $[X_1, \theta_1]$ *to a final position* $[X_2, \theta_2]$ *if and only if the vertices* $[R_1, T_1]$ *and* $[R_2, T_2]$ *of the connectivity graph CG introduced above can be connected by a path* π *in CG, where* R_1, R_2 *are the noncritical regions containing* X_1, X_2, *respectively, and where* $T_1 = [S_1, S_1']$ (*respectively* $T_2 = [S_2, S_2']$) *is the "marking" of the interval of* $O(X_1)$ (*respectively* $O(X_2)$) *containing* θ_1 (*respectively* θ_2), *i.e., the interval for which* $S_1 = s(\gamma(\theta_1, X_1),X_1)$ (*respectively* $S_2 = s(\gamma(\theta_2,X_2),X_2)$). *Moreover, the motion* $c(t)$ *can be effectively calculated from the path* π *in CG.*

Remark. We assume here that neither X nor X' lies on a critical curve. If either X or X' lies on such a curve, we first move X (or X') slightly into a noncritical region, and then apply the above theorem.

The proof is identical to the proof of the main theorem of subsection 1.1, and is therefore omitted.

2.2. Geometric and Algorithmic Details

In this subsection we shall study the critical curves and their crossing rules in more detail. As will be shown below, the crossing patterns that can arise are completely analogous to those obtained in the ladder case. Specifically, assuming that no two critical curves coincide, exactly one of the three following crossing patterns can arise as we cross a critical curve β at a point X not lying on any other critical curve (in what follows R_1 and R_2 are the regions lying on the two sides of β near X, and for specificity we assume that $\sigma(R_1)$ contains at least as many pairs as $\sigma(R_2)$):

(i) One interval of $O(X)$ may shrink to a point, and then disappear, in which case $\sigma(R_1)$ consists of all pairs in $\sigma(R_2)$ plus an extra pair $[S_1, S_1']$ marking this interval.

(ii) A new stop may appear inside one interval I in $O(X)$, dividing it into two subintervals which then pull away from each other. Let $[S_1, S_1']$ be the marking of I, and let S_2 (respectively S_2') be the marking of the new clockwise (respectively counterclockwise) stop which appears inside I. Then the pair $[S_1, S_1']$ in $\sigma(R_2)$ is replaced in $\sigma(R_1)$ by the two pairs $[S_1, S_2']$ and $[S_2, S_1']$.

(iii) The marking S_1 of one end of some interval of $O(X)$ may change to another marking S_2. In this case one pair having the form $[S_1, S]$ (respectively $[S,S_1]$) in $\sigma(R_1)$ will be replaced by the pair $[S_2, S]$ (respectively $[S, S_2]$) in $\sigma(R_2)$.

Thus our crossing rules can be stated in the following sample and general form: connect $[R_1, T]$ to $[R_2, T]$ for each $T \in \sigma(R_1) \cap \sigma(R_2)$, and connect each $[R_1, T_1]$, $T_1 T \in \sigma(R_1) - \sigma(R_2)$, to each $[R_2, T_2]$, $T_2 \in \sigma(R_2) - \sigma(R_1)$.

In order to develop the theory presented at the preceding subsection into an algorithm which solves the motion-planning problem for a polygonal body, we need to specify in more detail the structure of the various critical curves described in subsection 2.1, state "crossing rules" for each type of curve as special cases of the general crossing rules stated above, and explain how to deal with degenerate cases in which several critical curves become coincident. This information is required in order to build the connectivity graph by a procedure which, given $\sigma(R)$ for some noncritical region R, computes $\sigma(R')$ for adjacent regions R', and calculates the edges which connect nodes $[R, T]$ to nodes $[R', T']$ in CG.

The geometrically nontrivial critical curves are those of types IV, V, and VII. It is easily shown that these curves are all algebraic and that their degree is at most quartic (critical curves of type IV are actually elliptic or straight arcs). Details are given in Schwartz and Sharir (1983a). It therefore follows that the number of intersection points of each pair of critical curves is bounded by some fixed constant number.

The crossing rule which applies to a given curve section can be derived from the algebraic parameters of the curve section in the following manner. Suppose that the point X lies on a critical curve β. Assume that X does not belong to any other critical curve. The curve β is characterized by the fact that when B is placed with P at $X \in \beta$ and is given a certain critical orientation θ, it touches the walls in more than one point, or it touches a wall perpendicularly, or a corner of B touches a wall corner. To illustrate the way in which the crossing rules can be derived, consider the case in which β is a type IV curve, so that when B has position/orientation $[X, \theta]$ two convex body corners Q_1 and Q_2 touch internal points of two wall edges W_1 and W_2, respectively. To obtain the crossing rule at X, move X an infinitesimal distance into R_1 along some line transversal to β, and then an infinitesimal distance into R_2 in the reverse direction. For each of these displacements, compute the new position U_2 that Q_2 would have if Q_1 is constrained to touch W_1, and similarly compute the new position U_1 that Q_1 would have if Q_2 remained in touch with W_2. The positions of U_2 with respect to W_2 and of U_1 with respect to W_1 then determine the pattern of stops that will occur near the critical position/orientation $[X, \theta]$ for X displaced into R_1 and R_2, respectively, allowing us to find the way in which the pattern of stops changes as we cross from one side to the other.

Note. In the next few pages, we assume that no two critical curves coincide, so that the point X allows only one critical orientation, i.e., at most one orientation at which the pattern of stops changes across β. The special case of coincident critical curves is discussed at the end of this subsection.

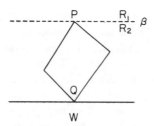

Figure 2.13. Crossing a type I curve β.

We shall describe now the way in which the general approach outlined above applies to each category of critical curves.

Type I curves: Recall that a type I curve β is a line segment at distance PQ from a given wall edge W, where Q is a convex body corner. As noted in subsection 2.1, the side of W on which β lies is determined by the relationship between the segment PQ and the body corner subtended at Q. That is, if the angles formed between PQ and the corner edges are both acute, β lies on the "outer" side of W, while if these two angles are both obtuse, β lies on the "inner" side of W, and if one of these angles is acute and the other is obtuse, β cannot exist. Let R_1 be the side of β such that if P moves slightly from β into R_1, Q moves out of the wall region bounded by W, and let R_2 be the other side of β (see Figure 2.13).

Suppose first that β lies in the "outer" side of W (as in Figures 2.5(a) and 2.13). Then when we cross β at a point X not lying on any other critical curve, the "stopping pattern" changes in much the same way as in the case of a "ladder." That is, as we cross from R_1 to R_2, exactly one new clockwise and one new counterclockwise stop, at both of which Q touches W, appears. This stop occurs at an interior angle θ of some open angular interval $[\theta_1, \theta_2]$ of $O^*(X)$, corresponding to an element $[S_1, S_2]$ of $\sigma(R_1)$. As R_2 is penetrated, the intervals $[\theta_1, \theta]$ and $[\theta, \theta_2]$ separate from each other, so that $\sigma(R_2)$ contains exactly one more element than $\sigma(R_1)$. If we put $S = \{[L_1, W],[L_2, W]\}$, where L_1 and L_2 are the body half-edges coming together at Q, the pair $[S_1, S_2]$ of $\sigma(R_1)$ is replaced in $\sigma(R_2)$ by the two pairs $[S_1, S]$ and $[S, S_2]$.

On the other hand, if β lies in the "inner" side of W then as we cross β from R_1 to R_2 exactly one interval $[\theta_1, \theta_2]$ of $O(X)$ shrinks to a point and then disappears. Then $\sigma(R_1)$ contains one more element than $\sigma(R_2)$, which labels the small interval $[\theta_1, \theta_2]$, and which is not connected to any label in $\sigma(R_2)$.

Type III and type IV curves: We postpone the detailed treatment of type II curves temporarily, as they turn out to have the most complex crossing rules, and go on to consider critical curves of types III and IV. Recall that such a curve β is the locus of P as two convex body corners Q_1, Q_2 slide along (the interior) of two wall edges W_1, W_2 (which are identical in case of a type III curve, but different in case of a type IV curve). Let n be any unit vector in a direction transversal to β at X. (To compute such an n, we first compute the tangent vector t to β at X from the algebraic equation defining β; then we can let n be the perpendicular to t, or any other vector lying close to this perpendicular.) Consider an infinitesimal displacement of X along n, which we write as $X' = X + \varepsilon n$. Let Y, Z denote the points touched by Q_1, Q_2, respectively, and let k, l be unit vectors in the directions of W_1, W_2, respectively (drawn so that the wall region lies to the left if we face these directions). First constrain the infinitesimal motion of B so that Q_1 remains on W_1, i.e., so that the new position of Q_1 is $Y' = Y + \delta k$. To find δ, use the fact that $X'Y' = XY$ to obtain the following equality (note that we ignore terms of higher order):

$$|Y + \delta k - X - \varepsilon n|^2 = |Y - X|^2.$$

In what follows we shall find it convenient to represent vectors as complex numbers. In this notation, the scalar product of two vectors a and b is written as $\mathfrak{Re}(a\bar{b})$, where $\bar{b}$ denotes the complex conjugate of b. Using this notation, the above equation expands into

$$\mathfrak{Re}(Y - X)(\overline{\delta k - \varepsilon n}) = 0.$$

Hence

$$\delta = \varepsilon \cdot \frac{\mathfrak{Re}(Y - X)\bar{n}}{\mathfrak{Re}(Y - X)\bar{k}}. \tag{2.1}$$

(Note that the denominator in (2.1) cannot be 0, since this would imply that X lies on the type I curve corresponding to Q_1, and W_1, contrary to our assumption that X lies only on one critical curve. Moreover, we can avoid choosing n orthogonal to $Y - X$, so that the numerator in (2.1) is not zero either.) Given the new positions X', Y' of P, Q_1, respectively, we can easily compute the new position of Q_2. This is best achieved using complex numbers. Indeed, since the triangle PQ_1Q_2 remains rigid, the position Z of Q_2 can be expressed as a "convex complex" combination of X and Y which does not change as PQ_1Q_2 rotates. That is, for some complex number ζ,

$$Z = \zeta Y + (1 - \zeta)X,$$

and similarly the displaced position Z' of Q_2 is

$$Z' = \zeta Y' + (1 - \zeta)X'.$$

Hence

$$Z' - Z = \zeta \delta k + (1 - \zeta)\varepsilon n.$$

The point Z' lies inside W_2 (respectively outside W_2) if $\mathfrak{Im}(Z' - Z)\bar{l} < 0$ (respectively > 0). If Z' lies outside W_2, then infinitesimal displacement of X in the direction of n results in a stop of Q_1 against W_1. (It is also easy to check whether the stop of Q_1 against W_1 is a clockwise or a counterclockwise stop; specifically we will have a clockwise (respectively counterclockwise) stop if $\mathfrak{Re}(Y - X)\bar{k} < 0$ (respectively >0).) If Z' lies inside W_2, no stop of Q against W_1 is possible after X is displaced in this way. By reversing the roles of Q_1 and Q_2 we can determine whether it is possible for Q_2 to stop against W_2 after X has been displaced in the n direction. Then we can change the sign of ε, to obtain an infinitesimal displacement of X into the opposite region. Clearly, if for one sign of ε we obtain a stop of Q_1 against W_1 (respectively Q_2 against W_2), then if ε is given the opposite sign this stop becomes impossible, and vice versa. It follows clearly that type III and IV curves are always governed by one of two possible rules, which are summarized in the following table:

Case 1

	stops in R_1	stops in R_2
Q_1 against W_1	clockwise	impossible
Q_2 against W_2	counterclockwise	impossible

Case 2

Q_1 against W_1	clockwise	impossible
Q_2 against W_2	impossible	clockwise

(Note that certain configurations symmetric to those given in this table are not shown; these are obtained by interchanging R_1 and R_2 in the tables above, or by interchanging the stop of Q_1 against W_1 with that of Q_2 against W_2. The treatment of these symmetric configurations is completely analogous to that of the configurations given in this table.) The first case describes a situation in which an interval in $O(Y)$ shrinks to a point and then disappears as Y crosses from R_1 to R_2 at X. The second case describes a situation in which the structure of $O(Y)$ remains unchanged at the crossing, but the marking of one of its clockwise (respectively counterclockwise) stops changes (from that denoting a stop of Q_1 against W_1 to that denoting a stop of Q_2 against W_2). In case 1, $\sigma(R_1)-\sigma(R_2)$ consists of a single pair $[S_1,S_1']$ and $\sigma(R_2)-\sigma(R_1)$ is null. In case 2, a single pair $[S_1,S_1']$ in $\sigma(R_1)$ is replaced in $\sigma(R_2)$ by a pair $[S_2,S_1']$ (respectively $[S_1,S'_2]$).

Type VI *and type* VII *curves:* Next we analyze crossings of type VI curves and type VII curves, postponing the discussion of type V curves, since their crossing rules are obtained by using a mixture of the techniques described above for type III and IV curves, and of the techniques to be established shortly for type VI and VIII curves. Recall that a curve β of type VI or VII is the locus of P as two half-edges L_1 and L_2 of B slide in contact with two convex wall corners Z_1 and Z_2, respectively. The crossing rules applicable to these curves are most easily obtained by making use of a simple logical symmetry which allows us to interchange wall edges with body edges. If we fix a center of rotation P, then we can either think of the body rotating around P or consider that the walls rotate about P in the opposite direction. The set $O(X)$ retains its structure irrespective of whether we think of free "body positions" or free "wall positions," except that clockwise and counterclockwise directions are reversed. Moreover, a small translation εn of the body can be thought of as a small translation of the walls in the opposite direction. There is one problem with this "symmetrization": To obtain the crossing rules in the preceding analysis we have first translated the body to a new position X' and then rotated it around the new center X'. If we reversed the roles of body and walls, this would correspond to first performing a small rotation of the walls around the old center, and then translating the walls in an opposite direction to the original displacement. It would therefore appear that the techniques used above to analyze type III and IV curves cannot be directly extended to the reversed situation which appears when we cross type VI and VII curves, because the order of translation and rotation is reversed when the roles of the body and of the walls are interchanged. However, since both translation and rotation are infinitesimal, it is not hard to check that when we reverse their order the terms of first order appearing in the analysis above do not change. Hence it follows that the crossing phenomena which appear in the type VI and VII cases are exactly like those which appear in the type III and IV cases, and can be found by making exactly the prior computations, but this time in the "body," rather than in the "space" coordinate system, and by treating clockwise stops as counterclockwise stops and vice versa. Again, $\sigma(R_1)$ either contains exactly as many pairs as $\sigma(R_2)$, in which case just one pair of $\sigma(R_1)$ is replaced by some other pair in $\sigma(R_2)$; or $\sigma(R_1)$ contains exactly one extra pair which disappears in $\sigma(R_2)$.

Type V curves: Recall that a type V curve β is the locus traced by P as a convex corner Q 'of B slides along a wall edge W and a body half-edge L slides about a convex wall corner Z. The wall/body symmetry principle stated just above makes it easy to extend the technique used for type III and IV curves to this case. Performing an analysis in space coordinates tells us what happens to the body-corner/wall-edge contact as we cross β; a corresponding computation in body coordinates shows what happens to the body-edge/wall-corner contact. If these both appear on one side of β and disappear on the other, then one is clockwise and the other is counterclockwise. In this case exactly one pair in $\sigma(R_1)$ disappears in $\sigma(R_2)$. Otherwise one will have two stops, both clockwise or both counterclockwise, one appearing just as the other disappears. In this case the single pair $[S_1, S_1']$ of $\sigma(R_1)$ is replaced in $\sigma(R_2)$ by some other pair, either of the form $[S_2, S_1']$ or of the form $[S_1, S_2']$.

Type II curves: Recall that a type II curve is a circular arc β with radius PQ about some wall corner Z, where Q is an endpoint of some body half-edge. As noted in subsection 2.1, the angular size of β is 360° minus the sum of the angles $\delta 1$ and $\delta 2$ subtended by the body half-edges L_1, L_2 meeting at Q and by the wall edges W_1, W_2 meeting at Z. Assume that W_2 lies counterclockwise from W_1 when going around Z in the free space V, and that L_2 lies clockwise from L_1 when going around Q outside B. Thus, when Q is placed against Z, W_1 is near L_1 and W_2 is near L_2.

As previously, we analyze this case by displacing X an infinitesimal amount along the direction n normal to the critical curve β, while either constraining the point Q to remain on (the line containing) one of the wall edges W_1 or W_2, or else constraining the point Z to remain on (the line containing) one of the body half-

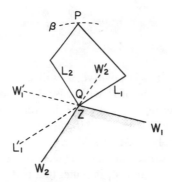

**Figure 2.14. Contact between
convex body and wall corners.**

edges L_1 or L_2. (However, in any given configuration of these body and wall edges, only certain of these constraints will be applicable.) This is shown in Figure 2.14.

First we assume that Q is a convex body corner and that Z is a convex wall corner. Let W_1' (respectively W_2', L_1', L_2') be the extension of W_1 (respectively W_2, L_1, L_2) past Z (or Q), and consider a small perturbation of the body corner Q. Then

(a) If L_2 lies clockwise of W_1', Q can touch W_1 or W_1' after a slight perturbation, but if L_2 lies counterclockwise of W_1', Z can touch L_2 or L'_2 after perturbation;

(b) If L_1 lies counterclockwise of W_2', Q can touch W_2 or W_2' after perturbation, but if L_1 lies clockwise of W_2', Z can touch L_1 or L_1' after perturbation.

Thus depending on the position of L_1 relative to W_2' and of L_2 relative to W'_1, we establish an appropriate pair of constraints.

Let k_1 and k_2 be unit vectors drawn along W_1 and W_2, respectively, drawn so that the wall region lies to the right if we go in the direction k_1 from the point Z, and to the left if we go in the direction k_2 from Z. Similarly, let l_1, l_2 be unit vectors drawn "away" from Q along the body half-edges L_1, L_2, respectively. We can use complex notation, i.e., regard all these vectors as complex numbers. Calculating as previously, we find that an infinitesimal displacement εn of P leads to the following displacement of Q, depending on the constraint (cf. (a), (b) above) that applies:

(i) If Q is constrained to lie along the line containing W_j, then it is displaced by the amount

$$\delta = \varepsilon \cdot \frac{\mathcal{R}e(Q - P)\bar{n}}{\mathcal{R}e(Q - P)\bar{k}_j}, \qquad\qquad j \neq 1, 2,$$

in the k_j direction. Thus, if δ is positive, Q will continue to touch W_j; but if δ is negative, the perturbation of P will cause Q to pull away from W_j. In either case, the reverse perturbation $-\varepsilon n$ of P will cause Q to stay on the other half of the line containing W_j and W'_j.

(ii) If Z is constrained to lie the line containing L_j, then Q is displaced by the amount

$$\delta = \varepsilon \cdot \frac{\Re e(Q - P)\bar{n}}{\Re e(Q - P)\bar{l}_j}, \qquad\qquad j \neq 1, 2,$$

in the l_j direction. Here, if δ is positive, Z will continue to touch L_j, but if δ is negative, then the perturbation of P will cause L_j to pull away from Z. Again, in either case the reverse perturbation $-\varepsilon n$ of P interchanges these two cases.

Note also that L_2 lies clockwise of W'_1 if and only if $\Im m(l_2\bar{k}_1) > 0$, and that L_1 lies clockwise of W'_2 if and only if $\Im m(l_1\bar{k}_2) > 0$.

When it exists, the perturbed contact of Q with W_1 represents a counterclockwise stop if the angle between PQ and W_1 is acute, that is if $\Re e((P-Q)\bar{k}_1) > 0$, otherwise a clockwise stop; and similarly, a perturbed contact of Q with W_2 represents a clockwise stop if $\Re e((P-Q)\bar{k}_2) > 0$, otherwise a counterclockwise stop. Also, when it exists, the perturbed contact between L_1 and Z is a counterclockwise stop if the angle between PQ and L_1 is acute, that is if $\Re e((P-Q)\bar{l}_1) > 0$, otherwise clockwise. Similarly, when it exists, the perturbed contact between L_2 and Z is a clockwise stop if $\Re e((P-Q)\bar{l}_2) > 0$, otherwise counterclockwise. (Here we have used the principle of symmetry between body rotations and wall rotations, as explained above.)

To select the applicable crossing rule, we determine the applicable constraints according to (a) and (b) above. Then, for an infinitesimal displacement εn of P we determine which stops occur at the perturbed position, and whether they are clockwise or counterclockwise. A careful examination of all the cases that can arise will show that only two essentially different cases are possible: either (a) as we cross β from one side to the other, a stop (of Q against Z) appears inside an interval of $O(X)$, dividing it into two intervals which then pull apart on the other side of β; or (b) the interval structure of $O(X)$ remains unchanged as β is crossed, but an endpoint of one of these intervals changes its labeling. In case (a), the node whose label changes can either be a clockwise or a counterclockwise stop. Hence exactly one pair $[S_1, S'_1]$ of $\sigma(R_1)$ will be replaced in $\sigma(R_2)$, by a pair of the form $[S_2, S'_1]$ in the former case, and by a pair $[S_1, S'_2]$ in the latter case. In case (a) let the pair in $\sigma(R_1)$ corresponding to the arc I be $[S_1, S'_1]$, and let the markings of the new clockwise and counterclockwise stops that appear as we cross into R_2 be S_2, S'_2, respectively. Then the pair $[S_1, S'_1]$ in $\sigma(R_1)$ is replaced by the two pairs $[S_1, S'_2]$ and $[S_2, S'_1]$ in $\sigma(R_2)$.

Next we must consider the cases in which either the body corner Q or the wall corner Z touching it is concave. First suppose that the wall corner is concave, so that the body corner touching it must be convex. Draw a circular arc C that the corner Q would sweep if B were rotated around P. Then two cases can arise:

(a) If the portion of C near Z is entirely contained within the wall region, then as we cross β an interval of $O(X)$ corresponding to a pair $[S_1, S'_1]$ of $\sigma(R_1)$ shrinks to a point and then disappears. In this case, $\sigma(R_2)$ is obtained simply by dropping this point from $\sigma(R_1)$.

(b) If this portion of C is not entirely contained within the wall region, then the part of C on one side of Z will lie inside the wall region, and the other side will lie outside that region. In this case, as we cross β one pair $[S_1, S'_1]$ of $\sigma(R_1)$ is replaced by a new pair, having either the form $[S_2, S'_1]$ or $[S_1, {}'S'_2]$, in $\sigma(R_2)$.

By the symmetry principle noted earlier, the case in which the body corner is concave and the wall corner is convex can be treated in essentially the same way, and will exhibit exactly the same crossing pattern.

Finally, we must consider the case in which Q is the common endpoint of two body half-edges constituting a single body edge. This requires a separate analysis, which however is quite easy to perform in body coordinates. Here it is always the case that an interval in $O(X)$ shrinks to a point and then disappears as β is crossed. Hence one extra pair $[S_1, S'_1]$ in $\sigma(R_1)$ is missing in $\sigma(R_2)$. Moreover, this pair will always designate a clockwise stop of Z against L_1 and a counterclockwise stop of A against L_2.

This completes our analysis of type II curves.

Type VIII *curves:* Recall that a type VIII curve β is the locus of P as a half-edge L of B overlaps a wall edge W. The crossing rules for such curves are rather simple. Suppose that when P is placed at a point $X \in \beta$ such that X does not lie on any other critical curve, the edges L and W overlap in an interval CD, where C is an internal point of L and also a (necessarily convex) corner of W, and where D is an internal point of W and also a convex endpoint of L. Then as Y crosses β through X from one side to the other a stop of C against L changes into a stop of D against W, and thus we must include corresponding connections in CG. One case deserving special comment is that in which two collinear half-edges L_1 and L_2 of B rest along the same wall edge W. Here it is best not to treat β as two coincident type VIII curves, but rather as one simple curve. In this case direct analysis shows that as Y crosses β from one side to the other, a small interval of $O(Y)$ shrinks to a point and then disappears.

Finally, we must consider the case in which the critical curves determined by several pairs of walls become coincident. As in the case of a ladder, we resolve these cases by imagining certain of the walls to be shifted by some randomly chosen infinitesimal amount, thereby splitting the coincident critical curves into several lines separated by infinitesimal distances. Several infinitely thin regions will then appear between the various curves split in this way. Each of these strip regions is considered to make contact only with the two infinitely separated lines between which it lies (so that one should not attempt to slide the body a noninfinitesimal distance with P remaining in such a strip, but use this strip only for immediate crossing across the coincident critical curves generating it). After splitting, and in the presence of these regions, the crossing rules stated above will apply. Note that by choosing the reference point P to lie at a different distance from every corner of B, one can ensure that no two type I or type II curves are coincident; since these are the only kinds of critical curves across which two nodes of the connectivity graph connect to a single node, these are the only coincidences that are troublesome to deal with. Detailed examples, applying to the somewhat simpler ladder case, are given at the end of subsection 1.2.

3. SKETCH OF THE ALGORITHM FOR THE LADDER CASE, AND ITS ANALYSIS

In this section we sum up the results of the preceding analysis by sketching an algorithm which could be used to determine the existence of a path between given initial and final positions in the movers' problem that we have considered. For simplicity, we consider the "ladder" case only, and assume that we do not need to deal either with coincident critical curves or with any situation in which the inherent imprecision of floating point arithmetic might lead us to an incorrect identification of two closely neighboring intersection points of pairs of critical curves.

ALGORITHM LADDER: The algorithm proceeds through the following steps:

Phase A. Read in the wall edges, test for consistency (e.g., that distinct walls do not intersect each other), and partition the wall region into a collection of convex wall sections. Additional details, e.g., classification of wall corners as convex or concave, are also worked out in this phase.

Phase B. Find intersection points of all the critical curves with each other and with the walls.

Each critical curve or wall is assumed to be available here as a sequence of subsegments, each of which can either be a full line, a half-line, a compact line segment, a circular arc, or a conchoidal arc. We form all intersections of these segments with all segments of other critical curves and with wall edges. The output of this phase is a map from each curve segment to the set of all intersection points lying on it.

Note that the common endpoint of two adjacent segments of a single critical curve is treated as an intersection point; also note that walls are considered as a special kind of critical curve. This latter convention simplifies the identification of noncritical regions, part of whose boundary can consist of wall edges or portions thereof.

Phase C. Our next step is to find all "critical curve sections," each of which is a section β of some critical curve lying between two intersection points but not containing any internal intersection point. To find these curve sections, we sort the intersection points lying along each critical curve according to their order on this curve. This is easily done for each of the few types of critical curve that can arise. Once we have sorted the intersection points in this manner, each pair of adjacent points on a critical curve defines a curve section. With each such section β we associate two "sides," denoted as "left(β)" and "right(β)." It is also convenient to treat each curve section as a pair of oppositely oriented sections β_1, β_2, which of course satisfy the relationships right(β_1)=left(β_2) and left(β_1)=right(β_2). An additional output of this phase is a map sending each intersection point to the set of all oriented curve sections emerging from it.

Technical Note. Since this algorithm works with floating-point arithmetic, identification of two separately calculated intersection points as being the same point is problematical. In the present sketch we ignore this delicate issue, which must of course be addressed carefully in the full version of the algorithm.

Phase D. Once the curve sections have been constructed and their "sides" designated, we associate a "crossing rule" with each curve section β. This is done by building a map CROSSING which describes all the "nontrivial" connections that arise as some critical curve β is crossed from right(β) to left(β). As shown by the preceding analysis, only the following crossing rules are possible:

(i) If β is the type I curve determined by a wall W, then there will ordinarily exist two other walls W_1, W_2 such that on one side R_1 of β an interval of free orientations whose endpoints touch W_1 and W_2 splits on the other side R_2 of β into two intervals having extreme stops W_1, W and W, W_2, respectively. We record this information in the map CROSSING by mapping $[R_1,[W_1, W_2]]$ into $[R_2,[W_1, W]]$ and into $[R_2,[W, W_2]]$, and vice versa. (An extreme case occurs when $\sigma(R_1)$ is null, in which case we simply map $[R_1,\Omega]$ into $[R_2,[W, W]]$.)

(ii) If β is a type II or a type III curve, there exist two walls W_1, W_2 such that either

(ii.1) on one side R_1 of β an interval of free orientations bounded by stops against W_1 and W_2 shrinks to a point and then disappears on the other side R_2 of β; or

(ii.2) there exists a third wall W_3 such that an interval of free orientations bounded by stops against W_1 and W_3 on R_1 becomes an interval bounded by stops against W_2 and W_3 on R_2.

In the first case, CROSSING will map $[R_1,[W_1, W_2]]$ to a special flag DEAD-END, indicating that no connection across β is possible from $[R_1,[W_1, W_2]]$. In the second case, CROSSING maps $[R_1,[W_1, W_3]]$ into $[R_2,[W_2, W_3]]$ and vice versa, if W_1 and W_2 are clockwise stops, or maps $[R_1,[W_3, W_1]]$ to $[R_2,[W_3, W_2]]$, if W_1 and W_2 are counterclockwise stops.

To obtain a crossing rule applying to the two sides of a critical arc β we proceed as follows: If β is a type I curve, pick any point X interior to β, and compute $\sigma(X)$ (the method used to compute this signature is shown below). We shall then find that W labels a common endpoint of two arcs of $O(X)$, whose full labeling is $[W_1, W]$ and $[W, W_2]$; this gives us the desired crossing rule, as explained in (i) above. (The extreme case in which $O(X)$ contains just one arc, labeled $[W, W]$, is also treated in the manner explained in (i).)

If W does not label any stop in $O^*(X)$, then β is redundant, in the sense that no change in the structure of $O(X)$ occurs as β is crossed over. In this case we can either discard β altogether, or can simply not define any nontrivial crossing rule for β.

Next suppose that β is of type II or type III. Although we could compute $\sigma(X)$ for a point X on β as before, this is not required, since the following simpler technique can be used instead. The type of β and its relative position within the full critical curve containing it determine which of the subcases (ii.1) or (ii.2) applies to β. Moreover, the walls W_1, W_2 appearing in the crossing rules are simply the walls defining β (cf. subsection 1.2), and hence are known *a priori*. Thus in subcase (ii.1) we can at once build the relevant pairs of the CROSSING map in the manner explained above. Subcase (ii.2) is a bit harder, since we do not know what wall W_3 must be paired with W_1 on one side of β and with W_2 on the other. To handle this case, we simply let CROSSING map $[R_1,[W_1, W]]$ into $[R_2,[W_2, W]]$, and vice versa, for each wall section W. (Here we have assumed that W_1 and W_2 are clockwise stops.) By Lemma 1.4, there will be only one arc in $O(X)$ whose clockwise end is a stop against W_1 (or W_2). Thus the extra pairs put into CROSSING will never be used and will not mislead the algorithm.

Phase E. Once having found nonredundant curve sections, we sort the set of all oriented critical curve sections emerging from a common endpoint X in clockwise circular order. To this end we first compute the outgoing tangential direction of each curve section at X, and then sort these directions in cyclic order. If two such directions are equal (as can happen, e.g., when a type I and a type II curve meet (cf. Figure 1.8(a)), or when a type III conchoid meets a type I line at the apex of its loop (cf. Figure 1.10)), then finer techniques are required to determine which of these curves is clockwise to the other at X, i.e., we must use second derivatives, etc.

The output of this phase is simply a map that sends each intersection point to a tuple of critical curve sections sorted in clockwise circular order.

Phase F. This is the last "geometric" phase of our algorithm. Here we read in the initial and final positions $[x_1, \theta_1]$ and $[X_2, \theta_2]$ of B. Then we find a curve section β_1 (respectively β_2) such that X_1 (respectively X_2) belongs to one of its sides R_1 (respectively R_2). To do this, we connect X_1 and X_2 by a line segment, and find the intersection points of this segment with all critical curve sections and wall edges. Then we sort these points into their order along X_1X_2 and find the curve sections β_1, β_2 whose intersections with X_1X_2 lie nearest to X_1, X_2, respectively. After this it is easy to find the sides of β_1, β_2 that contain X_1, X_2, respectively.

Next we compute $\sigma(X_1)$, $\sigma(X_2)$ and find the label $[W_1, W_1']$ (respectively $[W_2, W_2']$) of the arc in $O(X_1)$ (respectively $O(X_2)$) containing θ_1 (respectively θ_2).

Phase G. Step F concludes the geometric part of our algorithm, and we go on at once to the remaining simpler, purely combinatorial, work. Our first step is to "glue" together sides of adjacent critical curve sections, thus establishing the fact that they belong to the same actual noncritical region. To do this, we proceed as follows: Let X be an intersection point, and let $\beta_1, \cdots, \beta_n$ denote all the oriented curve sections having X as their initial endpoint. These curves have been arranged in their clockwise circular order. We build a map "nextside" that maps left(β_j) to right(β_j') and vice versa, where β_j' is the curve following X in clockwise order. Note that this procedure will generally make "nextside" into a two-valued map, since each side of a curve section will be mapped into two adjacent sides of other curve sections, one at each endpoint of the curve.

Technical Note. A slight technical problem arises when a critical curve section β is unbounded at one or both of its extremities. We then have to connect a side of such a section to a side of another unbounded curve section which also belongs to the boundary of the same noncritical region. However, this problem can easily be overcome, i.e., by bounding the free space by some large square, so that all walls (and consequently also critical curves) are bounded.

Phase H. In this final phase we search for a path that connects $[R_1, S_1]$ to $[R_2, S_2]$ in the connectivity graph. To this end, we search through CG, starting from $[R_1, S_1]$. In each step in the search we examine a node $[R, S]$ that we want to connect to adjacent nodes in the graph. Two connection modes are possible:

(a) Connect $[R, S]$ to $[R', S]$, if $[R, R'] \in$ "nextside."

(b) Let R' be the other side of a critical curve whose first side is R. If CROSSING$\{[R, S]\}$ is not empty, connect $[R, S]$ to each element in this image, but never to the special flag DEADEND which can occur. If CROSSING$\{[R, S]\}$ is empty, connect $[R, S]$ to $[R', S]$.

This process continues until either no new points are found, in which case no path from $[R_1, S_1]$ to $[R_2, S_2]$ is possible, or until $[R_2, S_2]$ is encountered during the search. In the latter case, we can reconstruct a path from $[R_1, S_1]$ to $[R_2, S_2]$ through CG, by keeping a "father" mapping during our search; this mapping will map each newly visited node N to the node from which N has been reached. By tracing this map from $[R_2, S_2]$ to $[R_1, S_1]$ we find the desired path.

3.1. Analysis of the Complexity of the Algorithm

Let us next analyze the complexity of the algorithm just sketched. Assume that the number of wall edges is n, and that the number of convex wall sections is also of order n. Note that the number of type I and type I curves is then $O(n)$, but that the number of type III curves can be $O(n^2)$, since each such curve is determined by a wall edge and a wall corner. Moreover, since all the critical curves arising in our problem are algebraic of degree at most 4, the number of intersection points of a pair of critical curves is bounded by some fixed constant k independent of n. Hence we have

LEMMA 3.1. (a) *The total number of intersection points of critical curves with each other and with the walls is* $O(n^4)$.

(b) *The total number of critical curve sections is* $O(n^4)$.

(c) *For any admissible point X, the maximum number of arcs in* $O(X)$ *is* $O(n)$.

Proof: (a) There are $O(n^4)$ pairs of critical curves, and the curves of each such pair can intersect in at most k points, so that there are at most $O(n^4)$ intersection points in all.

(b) Each critical curve can contain at most $O(n^2)$ intersection points, which partition it into $O(n^2)$ curve sections. Hence the total number of such sections is at most $O(n^4)$.

(c) Each stop in $O^*(X)$ is at least against one wall, and by Lemma 1.4 each wall can participate in at most two such stops. Hence there are at most $2n$ stops, and so $O(X)$ contains $O(n)$ intervals.

Lemma 3.1 gives us the estimates required to analyze the efficiency of our algorithm. Phase A requires time at most $O(n^4)$, by Lemma 3.1(a). (We assume that each geometric operation, such as finding all intersections of a given pair of curves, requires constant time.) Phase C sorts $O(n^2)$ intersection points on each of $O(n^2)$ critical curves, thus requiring $O(n^4 \log n)$ time. Phase D computes $\sigma(X)$ for one point X on each critical curve section of type I. Since there are only $O(n)$ type I curves, there are only $O(n^3)$ curve sections of type I. The computation of $\sigma(X)$ performed in the manner outlined above is easily seen to require $O(n \log n)$ time (cf. Lemma 3.1(c)). Hence the analysis of type I curve sections performed in this phase requires time at most $O(n^4 \log n)$. The analysis of curve sections of types II and III is simpler, but the construction of the map CROSSING in case (ii.2) is expensive, since it uses $O(n)$ time for each curve section which it handles, thus requiring in all $O(n^5)$ time. However, this can be improved by defining CROSSING for such curves so that it records the crossing rule only symbolically, using one entry rather than $O(n)$ entries per curve section. Assuming this improvement, treatment of curve sections of types II and III in phase D requires only $O(n^4)$ time. Phase E sorts sets of critical curve sections at each intersection point. Each curve section appears in two such sets. Hence the total size of all these sets is $O(n^4)$, and the total time required for this sorting operation is easily seen to be $O(n^4 \log n)$. Phase F builds the intersections of a single line segment L with $O(n^4)$ critical curve sections, and then finds the "first" and "last" intersection points on L, thus requiring $O(n^4)$ time. The second part of this phase is much cheaper, and uses only $O(n \log n)$ time. Phase G is easily seen to require $O(n^4)$ time. The combinatorial search performed in phase H is the most time consuming part of the algorithm. This is because the number of nodes that can be encountered during the search is $O(n^5)$, since there can be $O(n^4)$ "sides" of critical curve sections, and since each side can appear in $O(n)$ nodes, by Lemma 3.1(c). Thus we have

PROPOSITION 3.2.   *The running time of our algorithm is at most $O(n^5)$, where n is the number of wall edges. However, the parts of the algorithm that prepare for the combinatorial search require only $O(n^4 \log n)$ time.*

We note that these estimates are grossly pessimistic. Unless numerous walls are close to each other, the actual size of the connectivity graph is much less than $O(n^5)$. Thus in practice our algorithm will be much more efficient than the above analysis might indicate.

Remark. Several phases of the algorithm presented above could be made to run much more efficiently, using high-efficiency techniques drawn from computational geometry. However, in order to improve the overall complexity of the algorithm one would have to find a substantially different approach which avoids explicit construction and search of the connectivity graph, which is the step dominating the complexity of the algorithm.

4. AN EXAMPLE

We conclude this paper with a simple example that illustrates the use of our algorithm in planning a motion of a ladder in 2-space around obstacles. In this example we wish to move a long ladder inside a narrow corridor through a sharp corner, such that there is no room to turn the ladder around the corner, but there is

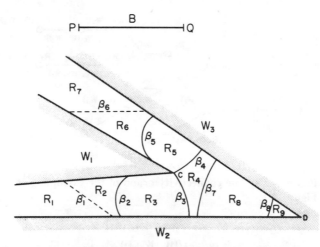

Figure 4.1. **The ladder B has to be moved from the region**
R_1 to the region R_7.

just enough room to push the ladder all the way inside the corner, then swing it and pull it back towards the other part of the corridor. The situation is shown in Figure 4.1. As shown in Figure 4.1, in this problem the space of free positions is decomposed into nine noncritical regions $R_1, \cdots, R_9$ bounded by critical curve sections $\beta_1, \cdots, \beta_8$. These curves are defined as follows:

β_1 is the type I curve determined by W_3,
β_2 is the type III curve determined by W_3 and C (a conchoid),
β_3 is the type III curve determined by W_3 and C (a conchoid),
β_4 is the type III curve determined by W_2 and C (a conchoid),
β_5 is the type III curve determined by W_2 and C (a conchoid),
β_6 is the type I curve determined by W_2,
β_7 is the type II curve determined by D (a circular arc,
β_8 is the type I curve determined by W_1 (a circular arc about C).

Applying the methods explained in section 1, we find easily that the noncritical regions $R_1, \cdots, R_9$ have the following characteristics:

$$\sigma(R_1)=\{[W_1,W_2],[W_2,W_1]\},$$

$$\sigma(R_2)=\{[W_1,W_2],[W_2,W_1]\},$$

$$\sigma(R_3)=\{[W_1,W_2],[W_2,W_3]\},$$

$$\sigma(R_4)=\{[W_1,W_2],[W_2,W_3],[W_3,W_1]\},$$

$$\sigma(R_5)=\{[W_2,W_3],[W_3,W_1]\},$$

$$\sigma(R_6)=\{[W_1,W_3],[W_3,W_1]\},$$

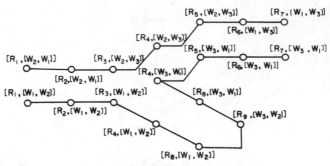

Figure 4.2.   The connectivity graph of the example given in Figure 4.1.

$$\sigma(R_7)=\{[W_1,W_3],[W_3,W_1]\},$$

$$\sigma(R_8)=\{[W_1,W_2],[W_3,W_1]\},$$

$$\sigma(R_9)=\{[W_3,W_2]\}.$$

Then, using the general crossing rule given at the end of subsection 1.1, the connectivity graph CG for this example can be constructed readily (see Figure 4.2). As shown there, CG has two connected components, and thus the manifold FP of free ladder positions/orientations has also two connected components. The first component of CG corresponds to positions/orientations of B in which the marked end P of B is farther away from the corner D than the other end Q; the second component of CG corresponds to positions/orientations of B in which P is nearer to D than Q. Since these two components are separate, it follows that there is no way to rotate B around the corner. However, it is possible to move B from one side of the corner to the other side. These motions can be reconstructed from the connectivity graph paths in an obvious manner.

CHAPTER 2

On the Piano Movers' Problem: II. General Techniques for Computing Topological Properties of Real Algebraic Manifolds

JACOB T. SCHWARTZ

Courant Institute of Mathematical Sciences
New York University

MICHA SHARIR

Department of Mathematical Sciences
Tel Aviv University

This paper continues the discussion, begun in the previous chapter, of the following problem, which arises in robotics: Given a collection of bodies B, which may be hinged, i.e., may allow internal motion around various joints, and given a region bounded by a collection of polyhedral or other simple walls, decide whether or not there exists a continuous motion connecting two given positions and orientations of the whole collection of bodies. We show that this problem can be handled by appropriate refinements of methods introduced by A. Tarski (1951) and G. Collins (1975), which lead to algorithms for this problem which are polynomial in the geometric complexity of the problem for each fixed number of degrees of freedom (but exponential in the number of degrees of freedom). Our method, which is also related to a technique outlined by J. Reif (1979), also gives a general (but not polynomial time) procedure for calculating all of the homology groups of an arbitrary real algebraic variety. Various algorithmic issues concerning computations with algebraic numbers, which are required in the algorithms presented in this paper, are also reviewed.

0. INTRODUCTION

The piano movers' problem (see Reif, 1979; Lozano-Perez & Wesley, 1979; Ignat'yev, Kulakov, & Pokrovskiy, 1973; Udupa, 1977; Schwartz & Sharir, 1983a) is that of finding a continuous motion which will take a given body or bodies B from a given initial position to a desired final position, but which is subject to certain geometric constraints during the motion. These constraints forbid the bodies to come in contact with certain obstacles or "walls," or to collide with each other. These walls can be curved, and the full collection of walls is not required to be

Work on this paper has been supported in part by the Office of Naval Research Contract N00014-75-C-0571; work by the second author has also been supported in part by the Bat-Sheva Fund of Israel.

The authors would like to thank Professor Dennis Arnon, Purdue University, for his generous help during the preparation of this paper.

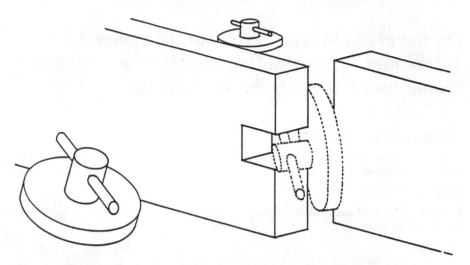

Figure 1. An instance of our case of the "piano movers" problem. The positions drawn in full are the initial and final positions of B; the intermediate dotted positions describe a possible motion of B between the initial and final positions.

connected (see Fig. 1). The problem that we set out to solve is: Given two configurations (i.e., positions and orientations of all subparts) of the bodies B in which none of these bodies touches any walls, and in which none of the bodies B collide, find a continuous wall- and collision-avoiding motion of all the B between these two configurations, or establish that no such motion exists.

This paper will present a general, though not very efficient, method for deciding on the existence of such a path (and for constructing such a path if it exists). Specifically, we will show that this problem can be handled by a variant of Tarski's famous algorithm (1951) for deciding statements in the quantified elementary theory of real numbers. Our approach is related to that outlined in an interesting paper of Reif (1979), and makes essential use of technical devices introduced by Collins (1975) and reviewed by Arnon (1981). As we shall see, these techniques also allow explicit, constructive calculation of the homology groups of an arbitrary real algebraic variety. In particular, the connectivity of such a variety can be calculated easily.

The paper is organized as follows. In Section 1 we begin to formulate the general mover's problem in which we are interested, as an abstract computational problem in algebraic topology. The algebraic machinery required to handle this problem is then developed in Section 2. The results obtained in this section can be used to calculate topological properties of algebraic varieties more general than those required for the solution of the mover's problem, and Section 2 also outlines some procedures for these calculations. Applications of the general theory of Section 2 to the mover's problem are then given in Section 3, yielding an algorithm which solves this problem in time polynomial in the number of geometric constraints on the motion of the body B, provided that the set of forbidden configurations for the body B is closed.

For the sake of completeness and clarity, we include a simple proof of the Collins construction on which our algorithms are based. This is given in Appendix A. Appendix B reviews various efficient techniques, some new, for exact calculations with algebraic numbers; such calculations appear repeatedly in our algorithms. Finally, Appendix C gives a technical details concerning the computations required to obtain the topological structure of the Collins decomposition. Note, however, that these, possibly very expensive, computations are not required for the simpler task of determining the connectivity of the space of free configurations of the body B.

1. AN ALGEBRAIC FORMULATION OF THE GENERAL MOVER'S PROBLEM

In this section we reformulate the general motion-planning problem in abstract algebraic terms, and reduce it to the problem of decomposing certain algebraic varieties into their connected components. A solution to this abstract problem is then developed in subsequent sections.

Like Reif, whose work is to be presented more fully in a forthcoming paper, we study the space of all collision-free positions of one or more hinged bodies B. We assume each body B to consist of a finite number of rigid compact subparts B_1, B_2, . . . , each bounded by various algebraic surfaces. These subparts can be connected to each other by various types of attachments, including the following:

(a) A point X on one part B_1 can be fastened to a point Y on another part B_2, in a manner which requires X and Y to be coincident but does not otherwise constrain the relative orientations of B_1 and B_2.

(b) The connection between X on B_1 and Y on B_2 can be a "hinge," i.e., can constrain B_2 to revolve around an axis V fixed in the frame of B_1.

(c) The connection between B_1 and B_2 can permit B_2 to slide, or to slide and rotate, along an axis V fixed in the frame of B_1.

Various other forms of affixment might be envisaged and can be treated in much the same way that we will treat the more common fastenings (a–c). As noted above, we are willing to consider any number of disjoint hinged bodies of this kind, which are to move in a coordinated fashion throughout an empty space bounded by a finite collection of walls, which can themselves be arbitrary algebraic surfaces. We can regard the walls as part of the given system of bodies, with the additional properties that (a) they need not be compact; and (b) they are constrained not to move at all.

Since we intend to proceed algebraically in what follows, our first task is to set up an appropriate algebraic parametrization of a superspace of the set of all allowed positions of the hinged body B. It is convenient to proceed as follows. The rotation group is a smooth 3-dimensional algebraic submanifold of the 9-dimensional Euclidean space of 3 by 3 real matrices. If $B = B_1$ is a single rigid body, we describe its position by giving a Euclidean motion T which takes B from some standard position to its given position. This transformation $Tx = Rx + X_0$ is defined by a pair $[X_0, R]$ consisting of a point X_0 in 3-dimensional Euclidean space E^3 and of a 3 by 3 rotation matrix R, and can therefore be regarded as a point in a smooth six-dimensional algebraic submanifold G of 12-dimensional Euclidean space E^{12}.

Next suppose that B is hinged, and that a second part B_2 of B is attached to B_1, say for the sake of definiteness in the manner (a). Then we can describe the overall position of the two parts B_1, B_2 of B as follows. As above, the position of B_1 is described by a Euclidean motion T which takes B_1 from a standard position to its actual position. By applying the inverse T^{-1} of T to both B_1 and B_2, we put B_1 into its standard position, and B_2 into a position which attaches a fixed one of its points X_2 to a point fixed on B_1. This position of B_2 is therefore defined by giving a Euclidean transformation T_2 such that $T_2 X_2 = X_2$. It is plain that the set of these transformations is in 1–1 correspondence with the set of rotation matrices R_2. Hence the overall position of B_1 and B_2 can be represented by a pair $[T, R_2]$, which once again varies over a smooth algebraic submanifold G_2 of a higher-dimensional Euclidean space E.

If instead B_2 is connected to B_1 in the manner (b), then much the same remarks apply, except that in this case the rotation matrix R_2 must satisfy $R_2 V = V$ for a certain 3-dimensional vector V. If B_2 can slide along an axis U fixed in B_1 but not rotate, its position is defined by a single real parameter u which defines the position of B_2 along this axis, etc. In all cases, the overall position of B_1 and B_2 is described by a pair $[T_1, T_2]$ of Euclidean motions, the first unconstrained, the second confined to some subgroup of the full Euclidean group. In all cases the allowed pairs form a smooth algebraic submanifold G_2 of some Euclidean space.

We can proceed similarly even if B consists of many parts hinged together in various ways. Suppose, for example, that B_2 is connected to B_1, and that a third part B_3 of B is connected to B_2. Then as above the overall position of B_1 and B_2 is defined by a pair $[T_1, T_2]$ of Euclidean transformations. T_1 maps B_1 from its standard to its actual position, and $T_1 T_2$ maps B_2 from its standard to its actual position. If we apply the inverse of $T_1 T_2$ to B_3, we put it into a position in which it is attached to a fixed point or axis of B_2 in one of the manners (a–c). Hence the actual position of B_3 is defined by a third Euclidean transformation T_3, belonging to a group of motions of one of the types we have already considered, and the mapping $T_1 T_2 T_3$ takes B_3 from its standard to its actual position.

These considerations make it clear that, irrespective of the manner in which the parts of a hinged body are connected together, the overall position of all its parts can always be defined by a point belonging to a smooth algebraic manifold G lying in a Euclidean space of some appropriate dimension.

Of course, the preceding considerations ignore all restrictions on the position of the parts of the bodies B imposed by the condition that none of these parts must collide. This point will be handled in Section 3 below, after the necessary algebraic machinery is introduced and developed in Section 2, which now follows.

2. TARSKI SENTENCES AND SETS; THE COLLINS DECOMPOSITION

By a *Tarski sentence* we mean a sentence, possibly containing free variables, which can be formulated in the decidable quantified language studied by Tarski (1951). In this language, variables designate real numbers and are quantified over the set of all reals. The operators allowed in the language are $+$, $-$, $*$, and $/$, designating the usual real arithmetic operators. The allowed comparators are $=$, $\neq$, $>$, $<$, $\geq$, $\leq$, all of which have their standard meanings. In addition quantifiers and Boolean connectives are allowed.

A Tarski sentence $Q(x_1, \ldots, x_n)$ containing exactly n free variables defines a subset of n-dimensional Euclidean space E^n, namely,

$$\Sigma_Q = \{[x_1, \ldots, x_n]: Q(x_1, \ldots, x_n)\}. \tag{1}$$

Sets of this form will be called *Tarski sets* (also known as *semi-algebraic sets*); the Q occurring in (1) is called the *defining formula* of Σ_Q. By a result given in the cited paper of Tarski, every Tarski set has a quantifier-free defining formula. A useful constructive proof of this result, involving a penetrating analysis of the geometric structure of Tarski sets, is given by Collins (1975) (see also Arnon (1981)), and we will base our analysis of the general mover's problem on Collins' results, which substantially improve Tarski's earlier work. We will also prove certain topological properties of the "cells" appearing in Collins' work, and will use these topological improvements to show that many standard topological properties (i.e., all the homology groups) of any algebraic variety are effectively calculable.

Using terminology slightly different from ours, Collins gives the following definitions and theorems:

DEFINITION 1. For any subset X of Euclidean space, a *decomposition* of X is a finite collection K of disjoint connected subsets Y of X whose union is X. Such a decomposition is a *Tarski decomposition* if each such subset Y is a Tarski set.

In what follows, E^r will denote the Euclidean space of r dimensions.

DEFINITION 2. A *cylindrical algebraic decomposition* of E^r is defined as follows. For $r = 1$ such a decomposition is just a partitioning of E^1 into a finite set of algebraic numbers and into the finite and infinite open intervals bounded by these numbers. For $r > 1$, a cylindrical algebraic decomposition of E^r is a decomposition K obtained recursively from some cylindrical algebraic decomposition K' of E^{r-1} as follows. Regard E^r as the Cartesian product of E^{r-1} and E^1, and accordingly represent each point p of E^r as a pair $[x, y]$ with $x \in E^{r-1}$ and $y \in E^1$. Then K must be defined in terms of K' and an auxiliary polynomial $P = P(x, y)$ with rational coefficients, in the following way.

(i) For each $c \in K'$, let $c \times E^1$ designate the *cylinder over c*, i.e., the set of all $[x, y]$ such that $x \in c$.

(ii) For each $c \in K'$ there must exist an integer n, such that for each $x \in c$ there are exactly n distinct real roots $f_1(x), \ldots, f_n(x)$ of $P(x, y)$ (regarded as a polynomial in y), and these roots must vary continuously with x. We suppose in what follows that these roots have been enumerated in ascending order. Then each one of the cells of K which intersects $c \times E^1$ must have one of the following forms:

(ii.a) $\{[x, y]: x \in c, y < f_1(x)\}$

(lower semi-infinite "segment" of $c \times E^1$).

(ii.b) $\{[x, f_i(x)]: x \in C\}$

("section" of $c \times E^1$).

(ii.c) $\{[x, y]: x \in c, f_i(x) < y < f_{i+1}(x)\}$

("segment" of $c \times E^1$).

(ii.d) $\{[x, y]: x \in c, f_n(x) < y\}$

(upper semi-infinite "segment" of $c \times E^1$).

All these cells are said to have c as their *base cell* in K'; K' is said to be the *base decomposition*, and P the *base polynomial*, of K. It is convenient to put $f_0(x) = -\infty$ and $f_{n+1}(x) = +\infty$, and then to designate the cells (ii.a), (ii.b), (ii.c), and (ii.d) as c_0^*, c_i, c_i^*, and c_n^*, respectively.

It obviously follows by induction that each of the sets constituting a cylindrical algebraic decomposition K of E^r is topologically equivalent to an open cell of some dimension $k \leqslant r$. We will therefore refer to the elements $c \in K$ as the (open) *Collins cells* of the decomposition K.

DEFINITION 3.   Let S be a set of functions of r variables, and K a cylindrical algebraic decomposition of E^r. Then K is said to be *S-invariant* if, for each c in K and each f in S, one of the following conditions holds uniformly for $x \in c$: either

(a) $f(x) = 0$ for all $x \in c$; or

(b) $f(x) < 0$ for all $x \in c$; or

(c) $f(x) > 0$ for all $x \in c$.

DEFINITION 4.   A point $p \in E^r$ is algebraic if each of its coordinates is a real algebraic number. A *defining polynomial* for p is a polynomial with rational coefficients whose set of roots includes all the coordinates of p.

THEOREM 1 (Collins).   *Given any finite set S of polynomials with rational coefficients in r variables, we can effectively construct an S-invariant cylindrical algebraic decomposition K of E^r into Tarski sets such that each $c \in K$ contains an algebraic point. Moreover, defining polynomials for all these algebraic points, and quantifier-free defining formulae for each of the sets $c \in K$, can also be constructed effectively.*

The proof of Collins' Theorem, which is not difficult, will be reviewed in Appendix A below.

In what follows we will find it useful to sharpen Collins' results in certain topological respects. For this, we have to impose an additional requirement on the decomposition; in certain unfavorable orientations of E^r, this condition can be false. However, as we shall show below, one can always restore this extra property by an appropriate rotation of the r-Euclidean space, and such a rotation can be easily calculated. This requirement is stated in the following:

DEFINITION 5. A Collins decomposition K is said to be *well-based* if the following condition holds. Let K' be the base decomposition and $P(b, x)$ the base polynomial of K. Then we require that $P(b, x)$ should not be identically zero for any $b \in E^{r-1}$. Moreover, we require that this same condition apply recursively to the base decomposition K'.

EXAMPLE. Consider the polynomial

$$P(x, y, z) = (x^2 + y^2)z + (x^2 - y^2)$$

in 3-dimensional space E^3. Then, since $P(0,0,z)$ vanishes identically, no P-invariant Collins decomposition of E^3 whose final step projects E^3 onto E^2 in the z-direction is well-based.

For the correctness of the topological assertions that we are about to make, it is essential that the decomposition K be well-based. Later we shall see how to rotate the given Euclidean space so as to make the decomposition well-based. For the moment we assume a well-based decomposition K, and show that it has certain useful topological properties.

LEMMA 1. *Let $c \in K$. Then the closure of c is a union of cells of K.*

Proof. To establish the lemma, we will use induction on the dimension r, and prove the following stronger

CLAIM. Let $r \geq 1$, and let K be a Collins decomposition of E^r. Then the closure of each cell in K is a union of cells of K. Moreover, for each cell c of K, each point z in the boundary of c, and each $\varepsilon > 0$, the open ε-ball about z contains a (relative) neighborhood U of z in $c \cup \{z\}$ such that $U - \{z\}$ is a connected subset of c.

Proof of claim. Our claim holds trivially for $r = 1$. Assume $r > 1$. With no loss of generality assume that c has either the form

$$c = c_j'^* = \{[b, x]: b \in c', f_j(b) < x < f_{j+1}(b)\}, \tag{1}$$

or

$$c = c_j' = \{[b, f_j(b)]: b \in c'\}, \tag{2}$$

for some $c' \in K'$ = the base decomposition of K. In either case the closure of c obviously contains c itself, and, in case (1), also the two "sections" c_j' and c_{j+1}' bounding c from above and from below, all of which are of course cells in K. Any other point in the closure of c must be of the form $[b, x]$, where b belongs to the boundary of c', and where $x_1(b) \leqslant x \leqslant x_2(b)$; here we write

$$x_1(b) = \liminf_{b' \in c', \, b' \to b} f_j(b') \quad \text{and} \quad x_2(b) = \limsup_{b' \in c', \, b' \to b} f_k(b'),$$

where $k = j + 1$ in case (1), $k = j$ in case (2). Note that

$$\liminf_{x \in s, \, x \to y} f(y)$$

designates the smallest limit of $f(y_i)$ for any sequence y_i of points of the set s for which $y_i \to y$, and similarly for lim sup. Conversely, any such point clearly belongs to the closure of c. Let e be the set of all such points. By our induction hypothesis, the boundary of c' is a union of cells in K'. Let $c'' \in K'$ be a cell of this boundary. Note that $x_1(b)$ and $x_2(b)$ either are roots of P or are $-\infty$ or $+\infty$. We will show that these functions are (equal if $k = j$, and) continuous in the entire closure of c, and hence are continuous for $b \in c''$ (here we give the *extended real axis* R^*, including the points $-\infty$, $+\infty$, its standard topology, which makes R^* homeomorphic to the compact unit interval $[-1, +1]$.)

The asserted continuity is an easy consequence of the following auxiliary lemma.

LEMMA 2. *Let* $c' \in K'$ *be as above, and let* $f_j(b)$ *denote the jth root of* $P(b, \cdot)$ *for* $b \in c'$. *Then there exists a unique continuous extension (in the sense of the extended topology of* R^**) of* f_j *to the entire closure of the cell* c'. *Moreover, this extension, which we will also denote by* f_j, *is either infinite or else a root of* P.

It will be shown below that Lemma 2 follows inductively from our other assertions. Assume for the moment that this has been proved. It then follows that the set

$$K(c'') = \{[b, x] : b \in c'', x_1(b) \leqslant x \leqslant x_2(b)\}$$

must be a union of cells in K. We can show this, and even provide a more explicit characterization of these cells, as follows. Let $c' \in K'$ be as above, and let $f_j(b)$ be the *j*th root (of m distinct real roots) of $P(b, \cdot)$ over c'. Let $c'' \in K'$ be any cell in the boundary of c'. Use the same symbol $f_j(b)$ to denote the continuous extension of f_j to $b \in c''$ which exists by Lemma 2. Then it follows from Lemma 2 that, for $b \in c''$, $f_j(b)$ is (either $-\infty$ or $+\infty$ or) one of the M roots of P over c''. In either case we can write $f_j(b) = F_J(b)$, where F_J is the *J*th root of P over c'' (by our convention, this also includes the extreme cases $J = 0$, $M + 1$); note that J is independent of b, because the roots F_J are isolated and f_j varies continuously over c''. Define a mapping

$$\rho(c', c'') : \{0..m + 1\} \to \{0..M + 1\}$$

by putting $\varrho(c', c'')(j) = J$ if $f_j(b) = F_J(b)$ for some, hence for all $b \in c''$. Then the assertion we need to prove is contained in the following somewhat more detailed

LEMMA 3. *Let $c' \in K'$ be as above. Then the cells of K which intersect the closure of the cell c'_j are c'_j itself, and all cells of the form c''_j, where c'' is contained in the closure of c' and $\varrho(c', c'')(j) = J$. The cells of K which intersect the closure of $c_j'^*$ are $c_j'^*$ itself, the two sections c'_j and c'_{j+1}, and all cells of the form c''_J, c''^*_J, $c''_{J+1}, c''^*_{J+1}, \ldots, c''_L$, where c'' is contained in the closure of c', $\varrho(c',c'')(j) = J$, and $\varrho(c', c'')(j + 1) = L$. Moreover, all of these cells are contained in the closure of c'_j (resp. $c_j'^*$).*

Proof. We prove only the second assertion, the first being even simpler. As noted in the paragraph immediately preceding the statement of Lemma 2, the closure of $c_j'^*$ consists of $c_j'^*$, c'_j, c'_{j+1}, and of the union of all the sets $K(c'')$ where c'' ranges over all the cells contained in the boundary of c'. By Lemma 2, and by the remarks preceding the present lemma, we have $x_1(b) = F_J(b)$, $x_2(b) = F_L(b)$, for $b \in c''$, where J, L are as in the statement of the present lemma. Thus $K(c'')$ is the union of all the cells $c''_J, \ldots, c''_L$. It is then clear that all these cells, together with c'_j and c'_{j+1}, are the only cells which can intersect (and hence be contained in) the closure of $c_j'^*$.   Q.E.D.

We can now complete the proof of Lemma 1. Indeed, the first part of the claim to be proved is now immediate from Lemma 3. To prove the second part, let c, c', $[b, x] \in c'$, and let ε be as in the claim. Let d, d', be the base cells of c, c' respectively. Then either $c = d_j$ for some j or $c = d_j^*$ for some j.

Assume first that $c = d_j$, for some j. Project the ε-ball U about $[b, x]$ onto a subset U' of E^{r-1}. By inductive hypothesis, U' contains a relative neighborhood V' of b in $d + \{x\}$ such that $V - \{x\}$ is connected. The continuity of $f_j(b)$ implies that for sufficiently small V' the set $\{[b, f_j(b)]: b \in V'\}$ is a connected neighborhood of $[b, x]$ in $c + \{[b, x]\}$ which is contained in U.

Next consider the case where $c = d_j^*$. Let J, L be as in Lemma 3, and let $0 \leqslant \alpha \leqslant 1$ be such that $x = \alpha F_J(b) + (1 - \alpha)F_L(b)$. Obtain U',V' as in the preceding paragraph, and consider the set

$$V = \left\{ \left[a, \beta f_j(a) + (1 - \beta)f_{j+1}(a)\right]: a \in V', 0 \leqslant \beta \leqslant 1, |\beta - \alpha| < \delta \right\}.$$

Note that V is relatively open in c. It also follows from the (uniform) continuity of f_j and f_{j+1} over V' that if V',δ are both sufficiently small, V will be a connected (relative) neighborhood of $[b, x]$ in $c + \{[b, x]\}$ which is contained in U.

Thus both assertions of our claim continue to hold in r dimensions, completing the inductive proof of Lemma 1.   Q.E.D.

Next we return to finish the proof of Lemma 2, which will complete our whole interlocking set of inductions.

Proof of Lemma 2. We give the extended real axis R^* the metric topology of the compact unit interval $[-1, +1]$ (which is homeomorphic to R^*) and write the distance function in R^* as $d(x, y)$. It is sufficient to prove that f_j is uniformly continuous on any bounded subset of c'. Suppose the contrary. Then there exist $\delta >$

0 and two sequences b_n, $b'_n \in c'$, such that both sequences converge to a point b_0 on the boundary of c', but for all n we have $d(f_j(b_n), f_j(b'_n)) \geq \delta$. By inductive hypothesis, for each integer m there exists a (relative) neighborhood N_m of b_0 in c' + $\{b_0\}$ which is contained in the $(1/m)$-ball about b_0 in E^{r-1}, and such that $N_m - \{b_0\}$ is connected. For all $m \geq 1$ define sets

$$R_m = \{[b, f_j(b)]: b \in N_m - \{b_0\}\},$$

and let R denote the intersection of the closures of these sets. R is obviously a compact set. We claim that it is connected. To see this we first observe that each of the sets R_m is connected, being the image of the connected set $N_m - \{b_0\}$ under the continuous function $b \rightarrow [b, f_j(b)]$. Next suppose that R is not connected. Then there exist two disjoint open sets U, V such that R is contained in the union of U and V and intersects both U, V. But this implies that for sufficiently large m, R_m is also contained in the union of U and V, for otherwise there would exist a sequence z_m, such that $z_m \in R_m - U - V$ for all m, which, by compactness, must converge to some $z \in R - U - V$, which is impossible. Since R_m is connected, we can assume with no loss of generality that R_m is contained in U for all sufficiently large m. But then R must also be contained in U, a contradiction which proves that R is connected. Since the projection of R into E^{r-1} consists of the single point b_0, R must be a vertical straight segment, all of whose points are obviously roots of $P(b_0, \cdot)$. However, since we have assumed that K is well-based, there are only finitely many such roots, and so R must consist of a single point $[b_0, x]$. Hence this point is the common limit of both sequences $\{[b_n, f_j(b_n)]\}$, $\{[b'_n, f_j(b'_n)]\}$, which is impossible. This shows that f_j is uniformly continuous over bounded subsets of c', and hence admits a unique continuous extension to the entire closure of c'.   Q.E.D.

  Remark. Lemma 2 is false if the decomposition is not well-based. Indeed consider the example given earlier. It is easily seen that $\{[0,0]\}$ is a base cell. Hence, if c is a 2-dimensional base cell c containing $[0,0]$ in its boundary, the single zero of $P(x, y, \cdot)$ for $[x, y] \in c$ can never admit a continuous extension to $[0,0]$.

  Lemma 2 has several consequences, which collectively show that a well-based Collins decomposition is free of local pathology.

  COROLLARY.   *Assume K is well-based. Let $c \in K'$, and let c' be a boundary cell of c. Let $g_k(b)$ denote the kth root of $P(b, \cdot)$ for $b \in c'$. For each $b \in c'$ let $J(b)$ denote the set of all indices j such that $g_k(b)$ is the limit of some sequence $f_j(b'_n)$, as $b'_n \in c$ approaches b. Then $J(b)$ is a constant depending only on c', but not on the particular point $b \in c'$.*

  LEMMA 4.   *Let $c \in K$, and let $c' \in K$ be a boundary cell of c. Then for each point $z \in c$ and each point $z' \in c'$ there exists a continuous path connecting z to z', which, except for its endpoint z', lies wholly in c.*

  Proof. Proceed by induction on r. Let d, d' be the base cells of c, c' respectively. Let $z = [b, x]$, $z' = [b', x]$, where $b \in d$, $b' \in d'$. If $d = d'$, then $c = d_j^*$ for some j, and c' is its top or bottom face. Assume c' to be the bottom face d_j of c; then the desired curve is constructed as follows. Let $q'(t)$ be a curve contained in d which connects b with b' (the recursive construction of Collins cells makes it quite easy to construct such a curve explicitly). Next let $0 < \alpha < 1$ be such that

$$x = \alpha f_j(b) + (1 - \alpha)f_{j+1}(b).$$

Then the desired curve $q(t)$ is simply $[q'(t), q^*(t)]$, where

$$q^*(t) = [\alpha(1 - t) + t]f_j(q'(t)) + [(1 - \alpha)(1 - t)]f_{j+1}(q'(t)), 0 \leqslant t \leqslant 1.$$

Otherwise, $d \neq d'$, and by inductive hypothesis there exists a continuous curve $q(t)$, $t \in [0, 1]$ such that $q(0) = b$, $q(1) = b'$, and $q(t) \in d$ for all $t < 1$. First suppose that c' is a section

$$\{[a, g_k(a)]: a \in d'\},$$

where $g_k(a)$ is the kth root of $P(a, \cdot)$ over d', and that c is a section

$$\{[a, f_j(a)]: a \in d\},$$

where f_j is the jth root of $P(a, \cdot)$ over d. Extending f_j continuously to the whole closure of d, so that $f_j(a) = g_k(a)$ for $a \in d'$, the curve we want is simply

$$p(t) = [q(t), f_j(q(t))], \qquad t \in [0, 1].$$

The other possible cases, i.e., those in which one or both of c and c' is a "segment" rather than a section, can be handled in essentially the same manner; we leave details to the reader.   Q.E.D.

Next we quote some standard definitions and results concerning finite cell complexes and their (singular) homology groups; see Cooke and Finney (1967).

DEFINITION 6.   A decomposition of a compact topological space S into finitely many disjoint sets $\{c_i\}$ is called a *cell complex* if

(a) Each c_i is homeomorphic to an open unit ball of some dimension d_i, which we call the dimension of c_i.

(b) For each integer dimension d, the union S_d of all the sets c_i of dimension $\leqslant d$ is closed.

(c) Each cell c_i of dimension d_i is open in the relative topology of S_{d_i}.

(d) For each cell c_i of dimension d_i, there exists a continuous mapping f_i of the unit closed ball B of dimension d_i onto the closure $\bar{c}_i$ of c_i, which maps the interior of B homeomorphically onto c_i.

DEFINITION 7.   The cell complex $\{c_i\}$ of the preceding definition is said to be *regular* if each of the mappings f_i is a homeomorphism of the closed ball B of dimension d_i onto the closure of the corresponding cell c_i.

THEOREM 2.   *The collection of compact cells of a (well-based) Collins decomposition of E^r forms a regular cell complex.*

We will, as usual, prove this theorem by induction on the dimension r, for which purpose the following easy lemma will be useful.

LEMMA 5. *Let B be the closed unit ball in E^r, and let f and g be two continuous real functions on B such that $g(b) < f(b)$ for each b in the interior of B. Put*

$$B^* = \{[b, x]: b \in B, g(b) \leqslant x \leqslant f(b)\}.$$

Then B^* is homeomorphic to the closed unit ball B_1 of E^{r+1}.

Proof. By shifting and contracting B^* along the x-axis we can assume that $g(b) = -f(b)$, and that $f(b) \leqslant \frac{1}{2}$ for all $b \in B$. For each $b \neq 0$ in B, let $\theta(b) = b/|b|$, so that θ is continuous for all $b \neq 0$. Put $r(b) = (1 - f^2(\theta(b)))^{1/2}$, and then let T map the point $[b, x]$ of B^* to

$$\left[br(b), x(1 - |b|^2 r^2(b))^{1/2}/f(b)\right].$$

(If $|b| = 1$ and $f(b) = 0$, then put $T(b, x) = [b, 0]$.) Since $br(b)$ is continuous for all b, and since $f(b) \neq 0$ for $|b| < 1$, this mapping is plainly continuous for all $[b, x] \in B^*$ such that $|b| < 1$. T is also continuous when $|b| = 1$. This is plain for each b such that $|b| = 1$ and $f(b) < 0$; on the other hand, if $f(b) = 0$, then since $|x' /f(b')| \leqslant 1$ for $[b', x'] \in B^*$, it follows that $T[b', x'] \to [b, 0]$ as $[b', x'] \to [b, 0]$ from within B^*. Moreover, T is $1-1$ on B^*. Indeed, if $T[b, x] = T[b', x']$ then plainly $\theta(b) = \theta(b')$, so that $r(b) = r(b')$, and hence $b = b', x = x'$. Since T is continuous and $1-1$ on the compact set B^*, it is a homeomorphism on B^*.

It is easily seen that the range $T(B^*)$ is the set B^+ of points $[a, y]$ in the closed unit ball $|a|^2 + y^2 \leqslant 1$ which satisfy the condition $|a| \leqslant r(a)$. The boundary of B^+ then consists of all points $[a, y]$ which either lie on the boundary of the closed unit ball B_1 of E^{r+1} and satisfy $|a| < r(a)$, or else are interior points of B_1 such that $|a| = r(a)$. It then follows that the boundary of B^+ has the property that each ray from the origin meets it in exactly one point. Hence B^+ is a star-shaped compact set, and it follows from Lemma 6 below that B^+, and hence also B^*, is homeomorphic to the closed unit ball B_1.   Q.E.D.

LEMMA 6. *Let A be a compact set in E^r which contains the origin 0 of E^r in its interior, and which is star-shaped relative to 0 (i.e., each ray from 0 intersects the boundary A' of A in exactly one point.) Then A is homeomorphic to the closed unit ball B of E^r.*

Proof. This is well known, but to prove it let θ designate an arbitrary point on the boundary of B, and let $r(\theta)$ be the length of the straight ray from 0 to A'. Then, since each such ray intersects A' in exactly one point, $r(\theta)$ obviously varies continuously with θ. The desired homeomorphism simply maps each nonzero $b \in B$ to $br(\theta(b))$, where $\theta(b) = b/|b|$.   Q.E.D.

Proof of Theorem 2. Suppose that Theorem 2 has been proven for a Collins decomposition K' of E^r, and let K be a Collins decomposition of E^{r+1} with base decomposition K'. Then any cell of K has the form of either c_j or c_j^* for some $c \in K'$, and for some integer j. In the case of a section cell $c_j = \{[b, f_j(b)]: b \in c\}$, we can simply note that, since f_j is continuous on the closure $\bar{c}$ of c, the projection of c_j to c extends to a homeomorphism of the closure of c_j with $\bar{c}$, which, by inductive assumption, is homeomorphic to a closed unit ball B of an appropriate dimension. Finally, using the same homeomorphism of $\bar{c}$ with B, the case of a segment cell c_j^* is covered by Lemma 5.   Q.E.D.

DEFINITION 8.   If c and c' are Collins cells belonging to a decomposition K, and if c' is contained in the closure of c, then we say that c' is a *face of c*.

We continue our analysis by quoting another standard definition from Cooke and Finney (1967).

DEFINITION 8. Given the finite regular cell complex K, an *incidence function* α on K is a function assigning one of the integers $\{-1, 0, +1\}$ to each pair, c, c' of cells of K, which satisfies the following conditions:

(i) $\alpha(c, c') \neq 0$ if c' belongs to the boundary of c and has dimension exactly one less than the dimension of c.

(ii) If c is of dimension 1, and the two 0-dimensional cells (i.e., discrete points) constituting the endpoints of c are c_1 and c_2, we have $\alpha(c, c_1) + \alpha(c, c_2) = 0$.

(iii) If c'' belongs to the boundary of a cell c of dimension d, and has dimension exactly $d - 2$, then

$$\sum_{c'} \alpha(c, c')\alpha(c', c'') = 0,$$

where the sum extends over all cells c' of dimension $d - 1$ whose closures are subsets of c^+ and supersets of the closure of c''. (It is well known (see, e.g., Cooke & Finney (1967) that for any c, c'' there are precisely two cells c' of this kind, if the cell complex is regular.)

Every regular cell complex A admits an incidence function, and by a standard result proved at length in the cited work of Cooke and Finney, any such incidence function can be used to compute the homology groups of A in a purely combinatorial manner. For the Collins cell decompositions which we consider, incidence functions are easy to define; we simply proceed inductively on the dimension, and use a variant of the standard "Cartesian product rule." More specifically, suppose that K and K' are as in the proof of Theorem 2 above, and that an incidence function α' has already been defined for the cells of K'. Extend this to the cells of K by putting

(a) $\alpha(c_j, c_j') = \alpha'(c, c')$, if c' belongs to the boundary of c and $\rho(c, c')(j) = J$;

(b) $\alpha(c_j^*, c_j) = -1$, $\alpha(c_j^*, c_{j+1}) = +1$;

(c) $\alpha(c_j^*, c_k'^*) = -\alpha'(c, c')$, if c' belongs to the boundary of c and $\rho(c, c')(j) \leq k < \rho(c, c')(j + 1)$;

and putting $\alpha(a, b) = 0$ in all other cases. It follows immediately from Lemma 3 that the function α defined this way satisfies condition (i) of Definition 8. It is trivial to verify that α also satisfies condition (ii) of that definition.

Next we verify that α satisfies condition (iii) of Definition 8. Let $c \in K$ be an s-dimensional cell, and let c'' be an $(s - 2)$-dimensional cell contained in the boundary of c. Let d, d'' be the base cells of c, c'', respectively. Several cases are possible:

(i) Suppose $c = d_j$ for some j. Then it follows from Lemma 3 that c'' must have the form d''_j. Furthermore any $(s - 1)$-dimensional face c' of c which contains c'' in its boundary must be of the form d'_L, for some $(s - 1)$-dimensional face $d' \in K'$ of d which contains d'' in its boundary and is such that $\varrho(d, d')(j) = L$ and $\varrho(d', d'')(L) = J$. Then, by (a) above and by induction hypothesis, we have

$$\sum_{c'} \alpha(c, c')\alpha(c', c'') = \sum_{d'} \alpha'(d, d')\alpha'(d', d'') = 0.$$

(ii) Next suppose that $c = d_j^*$. Then d must be $(s - 1)$-dimensional. Once more Lemma 3 implies that c'' either is an $(s - 2)$-dimensional face of d_j or d_{j+1} (both of which are $(s - 1)$-dimensional), or c'' is d''_k where d'' is $(s - 2)$-dimensional, or c'' is $d''_k{}^*$ where d'' is $(s - 3)$-dimensional. The first case is a special case of the second one, so we begin by assuming that the second case holds. Then it follows that d'' is a face of d, and that $\varrho(d, d'')(j) = J \leqslant k \leqslant L = \varrho(d, d'')(j + 1)$. Assume first that k lies strictly between J and L; then the only possible cells c' that 'come between' c and c'' are $d''_{k-1}{}^*$ and $d''_k{}^*$. Hence, in this case, clause (b) of our definition of the incidence function α implies

$$\sum_{c'} \alpha(c, c')\alpha(c', c'')$$

$$= \alpha(d_j^*, d''_{k-1}{}^*)\alpha(d''_{k-1}{}^*, d''_k) + \alpha(d_j^*, d''_k{}^*)\alpha(d''_k{}^*, d''_k)$$

$$= -\alpha'(d, d'') \cdot (-1 + 1) = 0.$$

Next suppose that $k = J < L$. Then the possible c' are the lower face d_j of c and the cell $d''_j{}^*$, and so we have

$$\sum_{c'} \alpha(c, c')\alpha(c', c'')$$

$$= \alpha(d_j^*, d_j)\alpha(d_j, d''_j) + \alpha(d_j^*, d''_j{}^*)\alpha(d''_j{}^*, d''_j)$$

$$= (-1) \cdot \alpha'(d, d'') + (-\alpha'(d, d'')) \cdot (-1) = 0.$$

A similar analysis covers the case $J < L = k$. Suppose finally that $k = J = L$. Then the possible c' are just the upper and lower faces d_j and d_{j+1} of c, and once again it is easy to verify *(iii)*.

Next consider the case in which d'' is an $(s - 3)$-dimensional cell, and $c'' = d''_k{}^*$. It is easily seen that in this case the only possible intermediate cells c' are of the form $d'_l{}^*$, where d' is a cell in K' intermediate between d and d'', and

$$\rho(d, d')(j) = J_1 \leqslant l < J_2 = \rho(d, d')(j + 1);$$
$$\rho(d', d'')(l) = L_1 \leqslant k < L_2 = \rho(d', d'')(l = 1).$$

However, it readily follows from Lemma 3 that for each intermediate cell d' there exists exactly one l satisfying these inequalities. Hence, by inductive hypothesis, we have

$$\sum_{c'} \alpha(c, c')\alpha(c', c'') = \sum_{d'} (-\alpha'(d, d'))(-\alpha'(d', d'')) = 0$$

Thus condition (iii) is established in all cases, so α is indeed an incidence function for K.

Since an incidence function for the cells of a Collins decomposition can be defined in this straightforward combinatorial manner, and since the homology groups of any finite regular cell complex can be computed combinatorially from such an incidence function, it follows that the homology groups can be computed by a purely finite procedure once we know what $(d - 1)$-cells are faces of each given d-cell. A technique for determining this, which is based on Lemma 3 and its corollary, will be described below. Assuming this, we have the following result:

THEOREM 3.   *For each j, the (singular) homology group $H_j(V)$ of the real algebraic variety V defined by any set Π of polynomial equations $P(x_1, \ldots, x_n) = 0$ with rational coefficients, can be computed in a purely rational manner from the coefficients of the polynomials P.*

In particular, since V is a regular cell complex the number of connected components of V can be formed simply by tracing sequences of cells which are faces of each other.

Collins gives estimates (which we reconstruct in Appendix A below) for the complexity of his cell decomposition procedure; in what follows we shall extend these to estimates of the work needed to determine whether one Collins cell c' is a face of another cell c, in the special case in which the dimension of c is the same as the dimension of the space E^r being decomposed. This will show that the connectivity analysis required to solve the general movers problem can be performed in time polynomial in the total degree of the set Π of polynomials (but exponential in the number r of variables appearing in these polynomials.) Appendix C discusses the more difficult adjacency analysis in case c has a lower dimension than r. At the present moment we do not know whether this more complex task can also be handled also in polynomial time.

To complete the foregoing arguments we still need to show how a well-based Collins decomposition can be defined for any algebraic variety. Let P be an r-variate polynomial, for which we wish to construct a P-invariant Collins decomposition of E^r. Following Hironaka (1975), we define a *good direction* to be a unit vector v (i.e., a point in the unit sphere S of E^r) such that P does not vanish identically on any line parallel to v. A simple technique for constructing a good direction, which is given by Lazard (1977), is as follows. Let $Q(x_1, \ldots, x_r)$ be the homogeneous part of P of highest degree $= n$. If v is a direction on which Q does not vanish (i.e., $Q(v) \neq 0$), then v is a good direction for P. Indeed, for any $x \in E^r$ and for sufficiently large real t, $P(x + tv)$ behaves asymptotically as $Q(tv) = t^n Q(v)$ $\neq 0$. By an easy lemma of Schwartz (1980), if $c > 1$, then the number of integer points in a cube of side $I \geq cn$ at which Q vanishes is at most $c^{-1}I^r$. Taking c substantially larger than 1, a good direction v can be found quite rapidly simply by picking a v at random from such a cube, testing the condition $Q(v) \neq 0$, and repeating the choice until a v satisfying this condition is found. Once having found such a v, we rotate E^r so that v becomes the rth axis, and we begin the Collins construction by projecting in the direction of v. As will be seen in Appendix A, this recursive construction generates a new polynomial Q in the remaining $(r - 1)$ variables from P and v; Q plays exactly the same role for the required base decomposition of E^{r-1} that P plays for the decomposition of E^r. This observation allows us to apply the above way of finding a good direction recursively, and in this way we build a well-based Collins decomposition.

Remark. The result of Hironaka (1975) which asserts the triangulability of real algebraic varieties, follows immediately from what has gone before. Note also that the fact that the cells of a Collins decomposition form a regular cell complex was anticipated by Professor P. Kahn of Cornell in a 1978 letter to Collins. See Kahn (1979).

To use the Collins cell decomposition associated with a set of polynomials to analyze the connectivity of a space defined by algebraic equalities and inequalities, we must tackle the problem of deciding when a Collins cell c' of dimension $d - 1$ forms part of the boundary of a d-dimensional Collins cell c. In analyzing this question, we shall first handle the relatively simple case in which c is of maximal dimension, i.e., is of the same dimension as the Euclidean space E^r which is being decomposed; this is all that is needed for the "movers problem" proper (see Section 3 for details). As usual, we proceed by induction on the dimension r; i.e., suppose that the decomposition K of E^r being considered has the base decomposition K', and that for each $(r - 2)$-dimensional cell c' of K' we know the two $(r - 1)$-dimensional cells c_1 and c_2 of K' of whose boundary c' forms part. We will also suppose that for each cell $c \in K'$ considered, an algebraic point $p(c)$ belonging to c is known, and that whenever c' forms part of the boundary of c, a vector v pointing from $p(c')$ into c, i.e., a vector v such that $p(c') + \delta v \in c$ for all sufficiently small δ, is known. We will shortly note that it is easy to carry the construction of such vectors forward inductively.

Any r-dimensional cell in E^r must have the form c_j^* for some $(r - 1)$-dimensional cell $c \in K'$, and then, by Lemma 3, the $(r - 1)$-dimensional cells lying in the boundary of c_j^* are its top and bottom cells c_j and c_{j+1}, together with all cells of the form $c_j'^*$, where c' belongs to the boundary of c and where $\varrho(c, c')(j) \leq J < \varrho(c, c')(j + 1)$. The vector $[0, \ldots, 0, -1]$ (resp. $[0, \ldots, 0, +1]$) points from c_{j+1} (resp. from c_j) into c_j^*. Moreover, if $v = [v_1, \ldots, v_{r-1}]$ points from an $x \in c'$

into c, then $[v_1, \ldots, v_{r-1}, 0]$ points from $c_j'^*$ into c_j^* if the former is part of the boundary of the latter. Hence it is trivial to carry the necessary vectors v forward, and the boundary determination problem presently under consideration reduces to that of calculating the map $\varrho(c, c')$ for an $(r-1)$-dimensional cell c and an $(r-2)$-dimensional cell c' of K'.

To make this calculation, we begin by observing that the corollary to Lemma 2 implies that to compute ϱ it suffices to compute, for each root $f_j(b)$ over c', the number k of roots of P into which $f_j(b)$ splits as b moves slightly into c. Moreover, we can make this calculation for an arbitrary $b \in c'$.

Given $b = p(c')$, and the vector v which points from b into c, the points $b' = b + tv$ will lie in c for sufficiently small positive values of t. Moreover, the bivariate polynomial

$$Q(x, y) = P\big(b + xv, f_j(b) + y\big)$$

has coefficients which are algebraic since they depend algebraically on the (algebraic) coordinates of b, and this polynomial vanishes at the origin $[0, 0]$. Let x vary in a sufficiently small interval $(0, x_0]$, and for each such x define the polynomial

$$R_x(y) = Q(x, y).$$

As is well known (see, e.g., Van der Waerden 1960, Ch. 1), for sufficiently small x all the roots y of $R_x(y)$ which lie near the origin are expressible by fractional power series of the form

$$y = C_1 x^{1/k} + C_2 x^{2/k} + \cdots$$

$$- C_1 x^{1/k} + o(x^{1/k}),$$

where $k \leq n = $ the degree of R_x as a polynomial in y. Thus, if we substitute x^{n+1} for x and if we let x be small enough, it follows that each root y of $R_{x^{n+1}}(\cdot)$ lying near the origin must belong to the interval $[-x, x]$, because the sum of the above fractional power series is $O(x^{(n+1/k)}) = o(x)$. It therefore suffices to determine the number N of zeroes of $R = R_{x^{n+1}}$ in the neighborhood $[-x, x]$ of 0. To find N, we can use the Sturm technique (see Appendix B for a review of Sturm sequences), that is, compute the Sturm sequence of R at x and at $-x$, and then N is $S(-x) - S(x)$, where $S(a)$ is the number of sign changes in that sequence evaluated at a. (The Sturm sequence is obtained as the remainder sequence of R and its y-derivative R'.) Since R and R' depend polynomially on the parameter x, extra care must be taken to ensure that, as this sequence is being calculated by repeated divisions, no leading coefficient of any of the polynomials in that sequence vanishes for the values of x that we consider (if such a term did vanish, subsequent divisions might involve completely different polynomials). However, since all these coefficients of R are polynomials in x, and since there are only finitely many such coefficients, none of their zeroes will belong to the open interval $(0, x_0]$, if x_0 is chosen to be small enough.

Therefore we can find the number N of roots in which we are interested by calculating the Sturm sequence

$$F_x^{(0)}(y), \ldots, F_x^{(k)}(y) \tag{1}$$

of $R_{x^{n+1}}$ as a sequence of polynomials in y with coefficients which are rational functions in x; and since we can multiply all those coefficients by a common denominator, we can even assume that these coefficients are polynomials in x. After calculating the polynomials (1), we must find their signs at the points $-x$ and x, for x small enough. If we substitute $y = x$ in the sequence (1), we obtain a sequence of polynomials $G_0(x), \ldots, G_k(x)$ in x, the signs of whose members are to be computed for x positive and sufficiently close to 0; and an exactly similar statement holds if we substitute $-x$ for y. Let $G(x)$ be any one of these polynomials; then the sign of $G(x)$ is either the sign of the coefficient in G of the nonzero term of the smallest degree, or, if all the terms of G vanish, is 0. This gives us N, i.e., the number of roots into which $f_j(b)$ splits as b moves into c. As already observed, by collecting these numbers for all the roots $f_j(b)$ of P for $b \in c'$, and by using Lemma 3 in a straightforward manner, we can reconstruct the map $\varrho(c, c')$.

The analysis required to determine when a Collins cell c' of dimension $k - 1$ is a face of a cell c of dimension $k < r$ is somewhat more difficult, for which reasons we prefer to present it in Appendix C below. Note that several two- and three-dimensional cases of the results proved in this section were established by Arnon (1981). Lemma 1 of this section is related in this way to Arnon's Corollary 3.3.25 and Theorem 3.4.11, Lemma 2 to his Theorem 3.6.16, and Lemma 3 to his Theorems 3.3.8 and 3.3.14.

3. A GENERAL ALGORITHM FOR THE PATH-PLANNING PROBLEM

As noted previously, the position of one of the hinged bodies B we consider is always described by a point T in a smooth algebraic manifold G, and the set of points occupied by a particular rigid part B_i of B is the range on B_i of a Euclidean transformation T_i whose coefficients can be expressed as polynomials in the components of T. We assume that each part B_i of each of our bodies is a compact Tarski set, and moreover that each face, edge, or other significant feature of such a part B_i is also a compact Tarski set (the walls, however, are not assumed to be compact but merely closed Tarski sets). The set F of forbidden configurations for the body B is then defined as follows. First we note that if B_i and B_j are two parts of B which are hinged together in any of the ways specified above, then there exist some face or other closed feature B_i' (resp. B_j') of B_i (resp. B_j) which always touch each other. Any other touch between B_i and B_j at a point outside B_i' or B_j' is then considered to be forbidden (in particular, if B_i and B_j are not directly hinged to each other, then it is forbidden for them to intersect at all). This defines a set F_0 of points of G which represent forbidden configurations of B, and we define F to be the closure of F_0. F

clearly constitutes a closed Tarski set, and hence a real closed semi-algebraic subset of G. (All this continues to apply in cases involving several disconnected, independently moving bodies, some of which can represent (immobile) walls.) Our path-planning problem is therefore that of deciding whether or not two points of $G - F$ belong to the same component of $G - F$, where F is a real closed semi-algebraic subset of G, and, if so, to construct a path that connects these points in $G - F$. (*Note:* If F is not required to be closed, the problem is still decidable, but the technique we are about to present may become less efficient; additional comment on this point is found below.)

To show how to decide this question, we can proceed as follows. Let Q be a quantifier-free defining formula for the set F of forbidden positions. Since the condition $a > b$ (resp. $a \geq b$) can be written as $a - b > 0$ (resp. $a - b > 0$ or $a - b = 0$), etc., we can suppose without loss of generality that Q is a boolean combination of clauses $P > 0$ and $P = 0$, P designating some arbitrary polynomial in the appropriate number r of variables. Let S designate the set of all polynomials appearing in Q or as defining equations of the manifold G, and use Collins' theorem to construct an S-invariant cylindrical algebraic decomposition K of E^r. Then it is clear that G, F, and $G - F$ are all unions of collections of cells $c \in K$.

It follows from the general topological results presented in the preceding section that two points p and q of $G - F$ can be connected by a continuous are in $G - F$ if and only if there exists a chain $c_1, \ldots, c_k$ of cells of $G - F$ such that $p \in c_1, q \in c_k$, and such that for each j, either c_{j+1} is a face of c_j, having one dimension less than that of c_j, or vice versa. (This follows from well-known properties of the homology group $H_0(G - F)$; see, e.g., Cooke & Finney, 1967.) Note that the recursive construction of the Collins cells gives us an effective way of connecting any two points in the same cell by a continuous arc lying wholly within that cell, while Lemma 4 gives us an equally explicit way of constructing an arc from any point p in a cell c' forming part of the boundary of another cell c to any point $q \in c$.

All in all, therefore, the results we have proved concerning the Collins decomposition give us a constructive way of determining whether the two given points p and q belong to the same (arcwise connected) component of $G - F$, and of finding a smooth arc connecting them when they do.

If F is not assumed to be closed, then to check for the existence of a chain of cells connecting the given points we may need to employ the more costly test for adjacency of Collins cells of arbitrary dimension described in Appendix C. At present we do not know whether this test can be carried out in time polynomial in the geometric complexity of the problem. By making the technical assumption that F is a closed subset of G we avoid this difficulty, since then connectivity in $G - F$ can be determined by following chains of cells of codimension at most 1. To see this, assume for the moment that G can be represented as a whole Euclidean space E^k for some k (we will explain shortly how such a representation can be constructed). Then, if F is closed it follows that $G - F$ is a smooth manifold of dimension k in E^k, so that no submanifold of $G - F$ of dimension $k - 2$ or less can disconnect any connected component of $G - F$ (for this well-known fact see, e.g., Lemma 1.9 of Schwartz & Sharir, 1983a). Hence in this case two points in $G - F$ are connected to one another if and only if the two cells containing them can be connected by a chain of cells all of which are either k-dimensional or $(k - 1)$-dimensional. This can be done using the relatively efficient "maximal dimension" adjacency testing method presented at the end of the preceding section.

To see that G can be represented in this manner, first note that by the general discussion of the algebraic representation of a collection of hinged bodies given in Section 1, G can be represented as the Cartesian product of a finite number of spaces, each of which is either a full Euclidean space E^k, or a rotation group of dimension 1, 2, or 3. These groups can in turn be represented as the circle S^1, the 2-dimensional sphere S^2, and the 3-dimensional sphere S^3, respectively, with the third representation being double-valued, i.e., each rotation in 3-space is represented by two antipodal points in S^3 (for more details concerning this representation, see, e.g., Hamilton, 1969). Moreover, with the exception of one point, S^2 and S^3 can be mapped algebraically onto E^2 and E^3 respectively by appropriate stereographic projections. Omission of the exceptional points of one or more such stereographic mappings will not affect the connectivity of the open manifold in which we are interested since the points omitted all lie on submanifolds of G of codimension at least 2. Therefore, with no loss of generality, we can assume that G is represented as a product of a Euclidean space E^k by a finite number of circles S^1. Each such circle can be mapped algebraically (with the exception of one point) onto a line by a stereographic projection, but here the omitted point can affect the connectivity of the resulting image of $G - F$. To overcome this small technical difficulty, assume first that only one circle is involved, i.e., $G = E^k \times S^1$. Map S^1 onto a line R_1 by projecting it from a point $X_1 \in S^1$, and also onto another line R_2 from another point X_2. We then obtain two distinct representations G_1 and G_2 of G, and can construct corresponding Collins decompositions K_1 and K_2 for each of them. Next we can analyze the connectivity of G_1 (resp. G_2) by constructing an appropriate connectivity graph CG_1 (resp. CG_2) in the simplified manner described above, and finally we can merge these graphs into one graph by adding edges which connect a cell $c_1 \in CG_1$, to a cell $c_2 \in CG_2$ whenever c_1 and c_2 have a common point of $G - F$ (this property of cells can be checked for easily if we take care to include the equation defining the subspace $E^k \times \{X_2\}$ (resp. $E^k \times \{X_1\}$) among the algebraic equations from which K_1 (resp. K_2) is generated). The connectivity of $G - F$ can then be determined by analyzing chains of edges in this merged graph. Cases in which G is the product of E^k by more than one circle can be handled in a similar manner.

Note. Another minor technical point to be noted is that the representation of the full 3-dimensional rotation group as S^3 is bivalent, so that a path between two specified rotations R_1 and R_2 of some subpart B_i of B can correspond either to a path between a point $\xi_1 \in S^3$ representing R_1 and a similar point ξ_2 representing R_2 or to a path between ξ_1 and $-\xi_2$. Thus in order to determine whether R_1 and R_2 can be connected we have to check for the existence of one of several paths.

Once a well-based Collins decomposition has been constructed, the connectivity analysis can proceed, as already noted, via a simple search through the *connectivity graph* whose nodes represent the Collins cells of highest dimension, and whose edges indicate cell adjacency. The computational cost of such an analysis is plainly linear in the size of the Collins decomposition. Collins has shown (see also Appendix A below) that the number of cells in a cell decomposition K is $O((2n)^{3r+1} \cdot m^{2r})$, where m is the number of polynomials defining the sets G and F, and where n is the maximum degree of any one such polynomial. Note that n is related to the degree of any single geometric constraint, and that r is related to the number of degrees of freedom of the bodies B. If we fix r, it follows that the number of cells in K, as well as the number of adjacent pairs of cells, is polynomial in m and n, i.e., in the geometric complexity of the problem, that is, in the number of different walls, faces, and other features of the system B of bodies, and in their algebraic degrees. Moreover, the time required to construct the Collins decomposition, and to test for adjacency of cells of maximal dimension, can also be shown to involve a number of operations on algebraic numbers which is also polynomial in m and n. As is well known (see Appendix B for details), each such operation can be accomplished in time polynomial in the degree of the polynomials defining these algebraic numbers. Taking all this into account, we obtain the following result:

THEOREM 4. *The mover's problem for (algebraic) bodies having a fixed number of degrees of freedom whose set of forbidden configurations is closed can be solved in time polynomial in the number of geometric constraints present in the problem.*

Remarks. (1) The computational cost of our solution of the movers problem is still exponential in the number of degrees of freedom of the bodies B. That this complexity growth is probably inherent is indicated by a theorem of Reif (1979), which asserts that the mover's problem for a robot B with many jointed arms (all free to rotate around a common axis) is PSPACE-complete.

(2) A comparison of Theorem 4 and the discussion preceding it with the more elaborate technique used in Schwartz and Sharir (1983a) to solve certain 2-dimensional cases of the motion-planning problem efficiently reveals a significant similarity between the two approaches. In both approaches the free space of configurations of B is partitioned into cells, and these cells are connected to each other whenever they are physically adjacent to each other. This imposes a combinatorial graph structure on these cells, whose connected components reflect the connected components of $G - F$. Moreover, in both cases these cells are constructed recursively by adding one dimension at a time. Also, the cells appearing in Schwartz and Sharir (1983a) can be shown to consist each of a finite union of Collins cells in the associated decomposition. We will not pursue these observations in this chapter, but they will reappear in a subsequent report on efficient algorithms for other special cases of the mover's problem.

APPENDIX A: THE COLLINS DECOMPOSITION—AUXILIARY REMARKS

In this appendix we review the construction which leads to the proof of Collins' Theorem 1, and add various auxiliary observations. We begin with the following remark. Let $P_b(z)$ be a polynomial of fixed degree n whose complex coefficients depend continuously on a parameter b which varies in some connected set S. Suppose that the number of distinct roots of $P_b(z)$ is independent of b. Then these roots vary continuously with b. This follows immediately from the fact that the unique root ϱ of P_b lying in any small circle C can be expressed by a quotient of Cauchy integrals over this circle. More specifically we have

$$\rho = \int_C \frac{\zeta P'(\zeta)\, d\zeta}{P(\zeta)} \Big/ \int_C \frac{P'(\zeta)\, d\zeta}{P(\zeta)}.$$

Next suppose that the polynomial P_b also has real coefficients for each value of b. Then the number m of real roots of P_b is also independent of b, and for each $j \leq m$ the jth largest real root of P_b depends continuously on b. To establish this, let S_k be the set of points b for which there exist exactly k real roots. Take a point b_0 in S_k, let $r_1 \ldots r_l$ be the distinct complex roots of P_{b_0}, with $r_1 \ldots r_k$ real and the remaining roots nonreal. Draw disjoint small circles $C_j, j = 1 \ldots l$, around these roots. Then for b sufficiently near b_0 each of those circles will contain exactly one of the roots of P_b, and each root of P_b will lie in one such circle. Since complex roots of P_b must occur in conjugate pairs, it follows that (if they are sufficiently small) the circles $C_1, \ldots, C_k$, and only these, contain real roots of P_b, which proves our assertion.

Next, suppose, in addition to the assumptions made above, that $R_b(x)$ is a second polynomial with real coefficients depending continuously on $b \in S$, and that for each $b \in S$ all the zeroes of R_b are contained in the set of zeroes of P_b. Then for each j, R_b is nonzero and of constant sign in the open interval $I_j(b)$ between the jth and the $(j + 1)$st largest real zeroes of P_b, and arguing by continuity and from the connectedness of S it is clear that the sign of R_b on the interval $I_j(b)$ is independent of b.

A simple variant of the Collins technique, sufficient for our purposes, but a bit less efficient than the one developed by Collins, can be described as follows. Assume that we are given a (finite) collection $\{P_i(b, x)\}$ of polynomials in $k + 1$ variables whose coefficients are all rational. (We continue to suppose that b designates a vector of k real variables, and that x designates the last of the $k + 1$ variables on which P_i depends; accordingly, we will treat the multivariate polynomials P_i as polynomials in x with coefficients belonging to the ring of rational polynomials in the k other variables b.) Let P denote the product of all these polynomials. We can then construct a family of polynomials $\{Q(b)\}$ in the k variables b with the property that for each (connected) k-dimensional set S over which each of the polynomials $Q(b)$ maintains a constant sign (zero, positive or negative), the number of distinct zeroes of $P(b, \cdot)$ is constant. Suppose for the moment that this has been done. Then the preceding remarks imply that the distinct real roots of $P(b, \cdot)$ over each such set S can be enumerated from smallest to largest so that for each j the jth root $f_j(b)$ varies continuously over S. On the other hand, once the family $\{Q(b)\}$ has been formed, we can partition E^k into connected sign-invariant subsets S by a recursive application of the Collins technique to the collection $\{Q(b)\}$. The complexity of the total procedure then depends on the number of polynomials Q needed to ensure the invariance of the number of distinct real roots of P over each connected set on which they maintain a constant sign, and on their maximal degree.

To construct the required polynomials Q, we have only to use the following well-known observation: Let P' denote the x-derivative of P, and let $R = \text{GCD}(P, P')$. Let n be the degree of P, and m be the degree of R. Then P has $n - m$ distinct roots. Hence it suffices to introduce enough polynomials $Q(b)$ to ensure that the degrees of P and R are constant on any connected set on which the Q's are sign-invariant.

To do this we first recall some facts concerning resultants and subresultants of polynomials; for which we see Brown and Traub (1971). Let $A(x)$ and $B(x)$ be two polynomials in x, having degrees a and b, respectively. Fix any $j \geq 0$, and consider the equation

$$A(x)U_j(x) = B(x)V_j(x) \tag{$*$}$$

in two polynomials U_j, V_j, having degrees $b - j - 1$ and $a - j - 1$, respectively. The unique factorization theorem for polynomials implies that $(*)$ has a nonzero solution if and only if A and B have $j + 1$ common roots. By expanding $(*)$ in terms of the coefficients of U_j and V_j, we obtain a system of $a + b - j$ linear equations in $a + b - 2j$ unknowns. We prefer to reduce this system to a square system, to which end we use the following observation. Suppose that we already know that $(*)$ admits a nonzero solution for all $i = 0, \ldots, j - 1$, so that A and B have at least j common roots. Replace $(*)$ by the weaker condition

$$A(x)U_j(x) - B(x)V_j(x) = C_j(x), \tag{$**$}$$

where $C_j(x)$ is an arbitrary polynomial whose degree is at most $j - 1$. This system involves exactly as many equations as unknowns, so that Eq. $(**)$ then has a nonzero solution if and only if $\psi_j(A, B) = 0$, where $\psi_j(A, B)$ is the determinant of the $(a + b - 2j) \times (a + b - 2j)$ matrix of the homogeneous system of linear equations representing the condition that highest $(a + b - 2j)$ powers of x in the left-hand side of $(**)$ have zero coefficients. (The determinant $\psi_j(A, B)$ is known as the jth *principal subresultant coefficient* of A and B; $\psi_0(A, B)$ is the *resultant* of these polynomials. See Brown and Traub (1971) for more details.) If $\psi_j(A, B) \neq 0$, then $(**)$, and hence also $(*)$, has only trivial solutions, so that A and B have exactly j roots in common. On the other hand, if $\psi_0(A, B) = 0$, then there exist U_j, V_j, and C_j satisfying $(**)$. However, we already know that A and B have at least j roots in common. Hence $C_j(x)$ must be divisible by their product. But since $C_j(x)$ is of degree at most $j - 1$, it must be identically 0, so that U_j and V_j also satisfy $(*)$, and so have at least $j + 1$ roots in common. Thus, given the two polynomials A and

B, we can determine exactly how many roots they have in common by computing $\psi_j(A, B)$ for increasing j until $\psi_j(A, B)$ becomes nonzero. This establishes the following:

LEMMA 1. *The number of common roots of two polynomials $A(x)$ and $B(x)$ is j, where j is the smallest integer such that $\psi_\theta(A, B) \neq 0$.*

In particular, the degree of the polynomial $R(b, x)$ introduced above is the least j such that $\psi_j(P, P') \neq 0$. Note also that the process just described depends on the knowledge of the degree of A and B (more precisely, on the maximal degree of A and B). Hence if A and B also depend on some parameter b (as does happen in the case in which we are interested), these degrees may vary if the leading coefficients of these polynomials become zero. All these considerations lead us to the following:

LEMMA 2. *Let $P(b, x)$ be of degree n in x. For each $j = 1, \ldots, n$ let $P_j(b, x)$ denote the sum of terms of P whose degree in x is $\leq j$, and let $Q_j(b)$ denote the leading coefficient of P_j. Also let $R_{jk}(b) = \psi_k(P_j, P'_j)$, for $k = 0, \ldots, j - 2$. Let M be the collection of all polynomials $Q_j(b)$ and $R_{jk}(b)$. Then on each connected set S on which all polynomials in M maintain a constant sign, the number of distinct real roots of $P(b, \cdot)$ is constant.*

To bound the computational cost of all this, let $Q^*(b)$ denote the product of all nonzero polynomials in M. Then the degree of Q^*, as a polynomial in any of the components y of b, is easily seen to be $O(dn^3)$, where d is the degree of P in y. Indeed this product involves $O(n^2)$ polynomials, which are determinants of matrices of size $2n \times 2n$ at most, each element of which is of degree d in y.

Using the preceding remarks, the Collins decomposition can be built up in the following recursive manner. Let $S = \{P_i(b, x)\}$ be any set of polynomials in $k + 1$ variables whose coefficients are all rational. Let P be the product of all the nonzero P_i, and let $Q(b)$ be the product of all nonzero polynomials appearing in Lemma 2. Applying Collins's construction recursively, let K be a Q-invariant cylindrical algebraic decomposition of the Euclidean space E^k. Let c be any one of the cells of K. Then Lemma 2 implies that the number of distinct real roots of $P(b, \cdot)$ remains constant as b varies in c. Hence, if $f_1(b), \ldots, f_m(b)$ designate the real roots of P over c in ascending order, then all the functions $f_j(b)$ are continuous in b for $b \in c$, and the collection of sets (ii.a)–(ii.d) of Section 2 partition the cylinder $c \times E^1$ in such a way so that the collection of all these sets over all base sets $c \in K$ defines an S-invariant $(k + 1)$-dimensional cylindrical algebraic decomposition.

Remark. Using the technique for selecting a good direction described in Section 2 we can reduce M somewhat. Let R denote the homogeneous portion of P of highest degree l. Rotating axes, we can ensure that R has a nonzero term involving x^l. In this case M need only include Q_l and the subresultants R_{lk}, so that Q^* will be of degree dl at most.

We omit the somewhat more refined argument, given by Collins, which shows how to find effectively quantifier-free defining formulae for the cells of the Collins decomposition.

As noted by Arnon (1981), it is easy to write a Tarski statement which asserts that y is the jth real root of P in ascending order. This is simply

$$P(y) = 0 \,\&\, (\exists y_1, \ldots, y_{j-1})|$$
$$\times (y_1 < y_2 \,\&\, y_2 < y_3 \,\&\, \cdots \,\&\, y_{j-2} < y_{j-1}$$
$$\&\, P(y_1) = 0 \,\&\, \cdots \,\&\, P(y_{j-1}) = 0 \,\&\, y_{j-1} < y$$
$$\&\, (\forall x)(P(x) < 0 \,\&\, x < y \Rightarrow)$$
$$\times ((x = y_1 \vee x = y_2 \vee \cdots \vee x = y_{j-1})). \tag{1}$$

Collins also notes that it is easy to find an algebraic point in each cell in the decomposition K, in the following recursive way. Let K' be the base decomposition of K. Proceeding recursively, obtain such a point for each $c' \in K'$. Let $b \in c \in K$ be such a point. Then the points

$$[c, y_1 - 1], [c, y_m + 1],$$
$$[c, y_j], \quad j = 1, \ldots, m,$$
$$\left[c, \frac{y_j + y_{j+1}}{2}\right], \quad j = 1, \ldots, m - 1,$$

are all algebraic and there is one such point in each cell intersecting $c \times E^1$.

EXAMPLE. We illustrate the technique described above by finding a P-invariant decomposition of the 2-dimensional plane, where

$$P(x, y) = x^3 + y^3 - 3xy$$

(This is one of the examples analyzed by Arnon (1981).) We begin by projecting E^2 onto E^1 in the y-direction. Since the leading coefficient of P (as a polynomial in y) is constant, in the first step of the Collins decomposition it is sufficient to construct the following polynomials in x (we delete the common factors of their coefficients):

$$\psi_0(P, P_y) \approx x^6 - 4x^3,$$
$$\psi_1(P, P_y) \approx x.$$

Moreover, since the second polynomial is a factor of the first, only $Q(x) = \psi_0$ need be retained. The real roots of Q are 0 and $4^{1/3}$, so that the base decomposition K' of the decomposition we seek has 5 cells, namely,

$$c_0^* = (-\infty, 0),$$
$$c_1 = \{0\},$$
$$c_1^* = (0, 4^{1/3}),$$
$$c_2 = \{4^{1/3}\},$$
$$c_2^* = (4^{1/3}, +\infty).$$

Next we determine how many distinct real roots $P(x, y)$ (as a polynomial in y) has over each of these cells. To do this we compute the Sturm sequence of P and P_y (see Appendix B below for a review of Sturm sequences), which is

$$f_0(y) = y^3 - 3xy + x^3,$$
$$f_1(y) = y^2 - x,$$
$$f_2(y) = 2xy - x^3,$$
$$f_3(y) = 4x - x^4.$$

From this sequence one easily finds out that $P(x, \cdot)$ has one root over c_0^*, one root over c_1, three roots over c_1^*, two roots over c_2, and one root over c_2^*.

Since the decomposition we have considered is well-based, it is a regular cell complex, and its topology will be completely determined once the ϱ maps on its base cells are found. The maps ϱ can be computed by using the technique described in Section 2. Omitting details, one finds that the map $\varrho(c_0^*, c_1)$ maps 1 to 1; the map $\varrho(c_1^*, c_1)$ maps all three roots 1,2,3 to 1; the map $\varrho(c_1^*, c_2)$ maps the two upper roots 2,3 to the upper root 2, and the lower root 1 to the lower root 1; and, finally, the map $\varrho(c_2^*, c_2)$ maps the single root 1 to 1.

APPENDIX B: ON EXACT SYMBOLIC COMPUTATIONS WITH ALGEBRAIC NUMBERS

This appendix addresses the problem of how to perform the exact calculations with algebraic numbers required for the algorithms described in this paper, for which numerical approximate solutions may not be acceptable, since such calculations may lead to incorrect conclusions, e.g., in comparing approximate quantities we may wind up putting them in an order which is different from the order of the original numbers, if these numbers are very close to each other. Of course, the algorithms to be described will never be able to give an "exact" value of an algebraic number. Nevertheless they can be used whenever an answer to some discrete query involving algebraic numbers is needed, as in the Collins decomposition related technique sketched in this paper.

This kind of problem, i.e., how to perform exact calculations involving algebraic numbers, has been studied by many authors (see Akritas, 1980; Heindel, 1971; Collins & Loos, 1976; Rump, 1976). In this appendix we will review the methods used to perform calculations of this kind, describe various improvements of techniques that have appeared in the literature, and present a few additional techniques.

In the following discussion, we ignore all those (possibly substantial) computational costs which can (and will) arise from the growth in size of the integers with which the algorithms to be described must deal; that is, we will measure cost by assigning each operation on integers (and hence each elementary operation on rational numbers) a nominal cost of 1. (Note, however, that much prior research has concentrated on obtaining more realistic cost estimates for such algorithms, taking into account the possible growth of coefficients during certain operations on polynomials, such as computation of the GCD of two polynomials, the Sturm sequence of a polynomial, the sequence of derivatives of a polynomial, etc. (see Brown & Traub, 1971; Heindel, 1971; Collins & Loos, 1976). These more refined estimates have shown that the extra cost incurred in such operations is still polynomial in the degree and the size of the coefficients of the polynomial(s) involved. Our significantly more optimistic cost measure is like the one used by Aho, Hopcroft, and Ullman, 1974.)

Some of the results presented below rely on the weak but useful lower bound on the smallest possible distance between two distinct real roots of a polynomial. This is the content of the result of Mahler (1964) (see also Mignotte, 1976) which the following definition and theorem summarize.

DEFINITION 1.   (a) Let P be a polynomial over the complex field. Then $|P|$ is defined to be the sum of the absolute values of all the coefficients of P.

(b) The *squarefree part* P^* of P is the quotient of P by the greatest common divisor GCD(P, P') of P and its derivative P'.

As already observed, P^* and P have exactly the same roots, but all the roots of P^* are simple. If $P = P^*$, i.e., if P has simple roots only, then P is said to be *squarefree*. With the significant reservation noted above the squarefree part of a polynomial P of degree n can be calculated in time $O(n \log^2 n)$ by using fast techniques for the required GCD computation and division steps; see Aho, Hopcroft, and Ullman (1974, Ch. 8).

THEOREM 1 (Mahler).   *The minimum distance between two distinct roots of a squarefree polynomial P of degree n with integer coefficients is bounded below by*

$$\Delta(n, |P|) = \frac{1}{n^{(n+2)/2}|P|^{n-1}}.$$

Theorem 1 is important in what follows, since it guarantees that sufficiently precise approximate calculations with algebraic numbers (of the type to be considered below) will yield entirely precise results. However, in most of the following algorithms we will not have to compute the roots of a polynomial P to such a high degree of accuracy, unless roots of P actually happen to be that close to one another.

DEFINITION 2.   (a) Let P be the squarefree polynomial of degree n with integer coefficients. Then a *P-isolating interval* for a real root r of P is an interval with rational endpoints, which contains r in its interior, but does not contain any other root of P.

(b) Let P be as in (a). A *P-separation* of the real line is a partition of R into a union of disjoint P-isolating intervals.

Theorem 1 yields a lower bound on the size of a maximal P-isolating interval for a root of P. However, assuming that the roots of P are randomly distributed, the size of P-isolating intervals can be expected to be much larger than $\Delta(n, |P|)$.

Following a convenient convention, we can represent an algebraic number x by a pair consisting of a squarefree polynomial with integer coefficients having x as a root and of a P-isolating interval for x. To proceed in this way, it is obviously important to be able to find isolating intervals for all the real roots of a squarefree P (i.e., to find a P-separation of the real line) rapidly. This *root isolation problem* is considered by Heindel (1971), Akritas (1980), and Collins and Loos (1976). Akritas describes an isolation technique based upon systematic binary searching using the Descartes rule of signs, which he indicates can solve this problem for a polynomial P of degree n in time $O(n^5)$. (However, the details of his efficiency estimate are not entirely clear.) The estimates of Heindel and of Collins and Loos give the bound $O(n^{10} + n^7\log^3|P|)$, which is also polynomial in n and $|P|$, although with a relatively high exponent. Below we will sketch a root-isolation algorithm, essentially an improved variant of the older technique suggested by Heindel (1971), which uses a Sturm sequence-based technique and can accomplish root isolation using $O(n^3\log n)$ arithmetic operations (in our cost measure) in the worst case, but on the average will require only $O(n^2\log^2 n)$ such operations.

We begin by reviewing the beautiful classical theory of Sturm sequences (see Marden, 1949, p. 130ff.), which gives a very useful way of handling several of the problems that concern us. Let P be a univariate polynomial, and let P' be its derivative. The *Sturm sequence* of P is a sequence $\{f_i\}$ of polynomials such that $f_0 = P, f_1 = P'$, and such that for each $i > 1$, $-f_i$ is the remainder obtained by dividing f_{i-2} by f_{i-1}. For this sequence there plainly exists a sequence of quotient polynomials Q_{i-1} (with rational coefficients) such that

$$f_{i-2} = Q_{i-1}f_{i-1} - f_i, \qquad i > 1,$$

where the degree of f_i is strictly smaller than that of f_{i-1}. Since P is assumed to be squarefree, this process must terminate with some constant function f_k. The Sturm sequence has the property that, for any interval $[a, b]$, the number of roots of P in this interval is $S(a) - S(b)$, where $S(x)$ is the number of sign changes in the sequence $[f_0(x), f_1(x), \ldots, f_k(x)]$.

Next we describe a fast procedure for the computation of the Sturm sequence of a given polynomial P of degree n. At a first glance it might seem that this task will require time at least $O(n^2)$, since that many coefficients appear in the polynomials constituting this sequence. However, by representing the sequence in a more economical way, we can reduce this time to $O(n \log^2 n)$. To do this, we note that the degree of Q_i is the difference of the degrees of f_{i-1} and f_i. Thus if m_i denotes the degree of the quotient Q_i, $i = 1, \ldots, k - 1$, the sum of all the m_i's is n. Hence we can represent the Sturm sequence by the sequence $[f_0(x), f_1(x), Q_1(x), \ldots, Q_{k-1}(x)]$, which involves only $O(n)$ coefficients. Once this representation is available, we can use it to evaluate the whole Sturm sequence at any given x, as well as the number of sign changes in the Sturm sequence, in time $O(n)$. To do this, we first compute $f_0(x), f_1(x)$, and $Q_i(x)$ for $i = 1, \ldots, k - 1$ (which requires total time $O(n)$). Then, using the "backward formulae"

$$f_i(x) = Q_i(x)f_{i-1}(x) - f_{i-2}(x), \qquad i = 2, \ldots, k,$$

the Sturm functions can be evaluated at x in $O(n)$ additional time.

For this more efficient evaluation, we simply need all the quotient polynomials Q_i. Up to a sign change, these are exactly the quotients obtained during calculation of the GCD of P and P'. To compute all these quotients efficiently, we can use the fast polynomial GCD procedure described in Aho, Hopcroft, and Ullman (1974, Ch. 8). This procedure works as follows. Let $a(x)$, $b(x)$ be two polynomials of degree $\leq n$. Let the remainder sequence $\{r_i(x)\}$ and the quotient sequence $\{q_i(x)\}$ of $a(x)$ and $b(x)$ be defined so that $r_0(x) = a(x)$, $r_1(x) = b(x)$, $r_{i-1}(x) = q_i(x)r_i(x) + r_{i+1}(x)$, $i \geq 1$, and $\deg(r_{i+1}(x)) < \deg(r_i(x))$. For each i there exists a polynomial 2×2 matrix M_i such that

$$[r_i(x), r_{i+1}(x)] = [a(x), b(x)]M_i.$$

Furthermore, each M_i is the product of matrices of the form

$$N_j = \begin{bmatrix} -q_j(x) & 1 \\ 1 & 0 \end{bmatrix}, \qquad j \leq i.$$

The fast GCD algorithm first computes the two middle elements $r_j(x)$ and $r_{j+1}(x)$ in the remainder sequence, and then calls itself recursively with $r_j(x$ and $r_{j+1}(x)$ to process their remainder sequence, which of course coincides with the rest of the remainder sequence of a and b. To find r_j and r_{j+1}, the algorithm uses another recursive procedure HGCD which computes the matrix M_j. This second procedure uses the fact (see Aho, Hopcroft, & Ullman, 1974, Lemmas 8.6 and 8.7) that all the quotients $q_i(x)$, $i \leq j$, depend only on the most significant half of the polynomials a and b. HGCD thus discards the least significant halves of a and b, thereby obtaining polynomials a' and b' of degree at most $n/2$; it then calls itself recursively to compute the two middle elements $r_{j'}$, and $r_{j'-1}$ among the first j elements in the remainder sequence of a' and b', during which process it also calculates the matrix $M_{j'}$. The quotient $q_{j'}(x)$ and the matrix $N_{j'}$ are then calculated, after which HGCD calls itself once more with $r_{j'}$ and $r_{j'+1}$ as parameters to compute the product L of the matrices $N_{j'+1}, \ldots, N_j$. The matrix M_j is then computed as $M_{j'} \cdot N_{j'} \cdot L$, and recursively returned. This description should make it plain that the algorithm sketched computes all quotients $q_i(x)$ appearing in the quotient sequence of $a(x)$ and $b(x)$. With minor modifications, it can therefore be used to obtain the desired efficient representation of the Sturm sequence of a given polynomial. Since the fast GCD algorithm runs in time $O(n \log^2 n)$, this is also the time required for the calculation of the Sturm sequence.

This gives us the following:

THEOREM 2.   *The number of distinct real zeroes of a polynomial with rational coefficients of degree n, lying in any given interval $[a, b]$, can be found on $O(n)$ arithmetic operations, after preprocessing which requires $O(n \log^2 n)$ arithmetic operations.*

We can now describe a root-isolation procedure as follows. We first compute the Sturm sequence of the given polynomial P, in the manner just described. Next we find an upper bound b and a lower bound a such that all real roots of P lie in the interval $[a, b]$. For example (see Marden, 1949), we can take

$$b = -a = \max\left\{1 + \frac{p_i}{p_n} : i = 0, \ldots, n - 1\right\}$$

where p_i is the coefficient of the ith power of x in P. Let $N = S(b) - S(a)$ denote the number of distinct real roots of P. We perform a binary search of the interval $I = [a, b]$ to find a point $c \in [a, b]$ which separates it into two subintervals each containing at least one root of P. It follows from Mahler's theorem that such a point will be found after at most $O(n \log n)$ bisections of I, at each of which we have to evaluate S, so that to find c will take $O(n^2 \log n)$ steps. (However, assuming random distribution of the roots of P, the expected number of required bisections will be $O(\log n)$, so that c will be found on the average after $O(n \log^2 n)$ steps.) We then apply the same process to each of the intervals $[a, c]$ and $[b, c]$, and continue in this manner until isolating intervals for all the roots of P have been found. Obviously only $N \le n$ intervals will have to be processed, so that the whole procedure requires $O(n^3 \log n)$ steps in the worst case, and $O(n^2 \log^2 n)$ steps on the average; note that these time bounds also dominate the time required for the initial computation of the Sturm sequence.

An alternative, and possibly more attractive, technique for root isolation has been described by Collins and Loos (1976), and is based on Newton's approximation technique. Their technique proceeds inductively by first obtaining root-separating intervals for the derivative of P, making sure that none of these intervals contains a root of the second derivative of P. On any interval in the complement of the union of these intervals P' has constant sign, so that it is trivial to check whether such an interval contains a root of P. On any of the P'-isolating intervals I, P is either convex or concave throughout I. Hence Newton's technique will converge (very rapidly) to a root of P in I if such a root exists. This enables one to partition I rapidly into P-isolating intervals which do not contain a zero of P', thus allowing iteration of the process. More details can be found in Collins and Loos (1976).

Once any root-isolation procedure is available, we can use it to perform various exact computations in algebraic numbers. Before describing efficient procedures for such computations, we first note a simple but useful generalization of Sturm's theorem. Specifically, let $A(x)$ and $B(x)$ be two given polynomials, where A is squarefree. Form the generalized Sturm sequence of A and B (which, up to sign changes, coincides with the remainder sequence of A and B) as follows. Put $f_0(x) = A(x)$, $f_1(x) = B(x)$, and, for each $i \ge 1$,

$$f_{i-1}(x) = q_i(x)f_i(x) - f_{i+1}(x),$$

so that the last element in this sequence will be the GCD of A and B. Let $S(x)$ denote the number of sign changes in the sequence $[f_0(x), \ldots, f_k(x)]$. As in the case of standard Sturm sequences, it is easy to see that each time we cross a zero of some function $f_i(x)$, $i \geq 1$, $S(x)$ remains unchanged. However, each time we cross a zero x_0 of $f_0 = A$ from left to right, $S(x)$ decreases by sign $(A'(x_0)B(x_0))$. (It is easy to check that this statement remains true even if x_0 is a zero of $B(x)$.) Hence, in any given interval $[a, b]$,

$$S(a) - S(b) = \sum \text{sign}(A'(x)B(x)),$$

where the summation extends over all distinct roots of A lying in (a, b). (In the case of a standard Sturm sequence, $B(x) = A'(x)$, so that the above sum is equal to the number of distinct real roots of A in (a, b).)

In particular, if the interval (a, b) is known to contain just one root r of $A(x)$, then $S(a) - S(b)$ is 0, 1, or -1, depending on the signs of $A'(r)$ and $B(r)$. Since A has been assumed to be squarefree, the sign of $A'(r)$ is easily calculable from the signs of $A(a)$ and $A(b)$. Hence the sign of $B(r)$ can also be calculated. That is, given an algebraic number r, represented as the ith root of a squarefree polynomial A, for which an A-separation of the real line is available, and another polynomial B, we can find the sign of $B(r)$. (Note that (an efficient representation of) the generalized Sturm sequence can be computed and evaluated using precisely the same techniques prescribed for standard Sturm sequences.) This technique improves that described by Rump (1976).

This generalized Sturm technique is applicable to a variety of other problems. For example, we can use it to determine the multiplicity of the real roots of a given polynomial P. To do this, we first compute the squarefree part P^* of P, and then isolate the real roots of P^* (i.e., the distinct real roots of P). For each root r of P, we use the above procedure to determine sign $P'(r)$. If this is nonzero, then r is a simple root. Otherwise r has multiplicity 2 at least; we then repeat our procedure to find the sign of $P''(r)$, and so on, until a nonzero sign is obtained, from which the multiplicity of r is immediately calculable. Using this technique, the multiplicities of all real roots of a polynomial of degree n can be found in time $O(n^2 \log^2 n)$ (assuming that a P-separation is available). Indeed, the total number of roots is n, and to determine each multiplicity, an $O(n \log^2 n)$ procedure is applied.

A very similar procedure can be used to compare two real algebraic numbers a and b, given as roots of the squarefree polynomials P and Q, respectively. To do this, obtain a P-separation and a Q-separation of the real axis, and merge them into one partitioning. Let $I = [c, d]$ be a P-isolating interval for a, and let $J = [e, f]$ be a Q-isolating interval for b. If I and J are disjoint, then the manner in which a and b compare is immediately obvious. Suppose then that I and J intersect. Several cases can arise; we will treat only the case in which $c < e < d < f$, since the other cases can be handled in a similar manner. Since P is squarefree, and since (c, d) contains only one root of P, $P(c)$ and $P(d)$ have different signs. By evaluating the sign of $P(e)$ we can determine whether a lies in the subinterval (c, e) or in the subinterval (e, d) of I. If a lies in (c, e) then we must have $a < b$. Similarly, if b lies in (d, f) we also have $a < b$. In the remaining case, both a and b (but not other root of P or Q)

must lie in the interval (e, d). Using the generalized Sturm technique explained above, we can compute the sign of $P(b)$, and this shows at once how a and b compare. As before, this procedure takes $O(n\log^2 n)$ time, where n is the maximal degree of P and Q, provided that P- and Q-separations of the real axis are available.

Similar procedures can be used to perform various other exact computations with algebraic numbers. Suppose, for example, that we need to compare the sum (or product) of two algebraic numbers a and b to a third such number. One can of course compute a polynomial $R(x)$ having $a + b$ (or ab) as a root. However, it may be undesirable to do this since this can generate a polynomial whose degree is $\deg(A)\cdot\deg(B)$ (where $A(x)$ (resp. $B(x)$) is a polynomial having a (resp. b) as a root), and repeated computations of this sort may result in polynomials of extremely large degrees. To avoid this problem, a recursive symbolic representation of algebraic numbers might be more advantageous: We can specify an algebraic number r either as a polynomial $A(r_1, \ldots, r_k)$ in k other algebraic numbers, or as a root of a polynomial $B(x)$, whose coefficients are algebraic numbers $r_1, \ldots, r_k$; where each of these numbers is in turn represented in this same fashion, until numbers explicitly representable by polynomials with rational coefficients are finally reached. Such semi-symbolic representations can be used to perform computations of the kind discussed above. Consider the typical problem of determining the sign of an algebraic number r specified in this recursive fashion. Suppose to be specific that r is specified as a polynomial $A(r_1, \ldots, r_k)$. Let $B(r_k) = 0$ be an equation for r_k, and isolate r_k as a root of B. Then use the technique explained above to determine the sign of $A(r_1, \ldots, r_k)$, which we regard as a polynomial in r_k with coefficients which are algebraic. This will require that we determine the sign of various polynomials in $r_1, \ldots, r_{k-1}$, which we can do by using the same technique recursively.

Fast Algorithms for the Computation of Principal Subresultant Coefficients
Schwartz (1980; see also Moenck, 1973) describes a variant of fast GCD algorithm described above which computes the resultant of two polynomials having degrees $\leq n$ in $O(n\log^2 n)$ time. Since the Collins decomposition technique involves numerous calculations of resultants and subresultants, it is of interest to note that the algorithm described in Schwartz (1980) can be generalized in a straightforward manner to yield a rather similar fast procedure for the computation of subresultants as well. For the convenience of the reader we will describe the necessary modifications in full detail.

First we recall the definition of a principal subresultant coefficient. Let

$$A(x) = a_m x^m + a_{m-1} x^{m-1} + \cdots + a_0,$$

and

$$B(x) = b_n x^n + b_{n-1} x^{n-1} + \cdots + b_0$$

be two polynomials of degrees m and n, respectively. For $0 \leqslant j \leqslant \min(m, n)$, the jth principal subresultant coefficient $\psi_j(A, B)$ of A and B is the $(m + n - 2j) \times (m + n - 2j)$ determinant

$$
\begin{vmatrix}
a_m & a_{m-1} & \cdots & & a_{m-n+j+1} & & \cdots & & a_{2j-n+1} \\
0 & a_m & a_{m-1} & & & & \cdots & & a_{2j-n+2} \\
& & & \cdots & & & & & \\
0 & 0 & \cdots & & a_m & & \cdots & & a_j \\
b_n & b_{n-1} & & \cdots & & b_{n-m+j+1} & & \cdots & b_{2j-m+1} \\
0 & b_n & b_{n-1} & & & & & \cdots & b_{2j-m+2} \\
& & & \cdots & & & & & \\
0 & 0 & & \cdots & b_n & & & \cdots & b_j
\end{vmatrix}
$$

the first $n - j$ rows of which involve coefficients of A, and the last $m - j$ rows of which involve coefficients of B. (Here we use the convention that $a_i = 0$ if $i < 0$.) As noted earlier, this is precisely the determinant of the linear transformation

$$T: [U(x), V(x)] \rightarrow (A(x)U(x) + B(x)V(x))/x^j,$$

where U is a polynomial of degree $n - j - 1$, V is a polynomial of degree $m - j - 1$, and where, as usual, the remainder after the indicated division is discarded. If $j = 0$, $\psi_j(A, B)$ is just the resultant of A and B.

Let Q be the quotient obtained by dividing B by A, and let $R = B \bmod A$ be the corresponding remainder. Suppose that $n \geqslant m$, and let $k \geqslant n - m$. By subtracting an appropriate upper row from each lower row in $(*)$ we see that $\psi_j(A, B) = \psi_j(A, B - x^k A)$. Using this last formula repeatedly, it follows that, if both sides of the following equation are considered as $(M + n - 2j) \times (m + n - 2j)$ determinants, we have

$$\psi_j(A, B) = \psi_j(A, B \bmod A).$$

Moreover, expanding the second determinant by minors of the first $n - m + 1$ rows, we obtain

$$\psi_j(A, B) = L(A)^{n-m+1} \psi_j(A, B \bmod A),$$

where $L(A)$ denotes the leading coefficient of A, and where the determinant on the right-hand side is now an $(2m - 1 - 2j) \times (2m - 1 - 2j)$ determinant. However, since the remainder $B \bmod A$ can be of any degree k lower than m, a more appropriate reduction is

$$\psi_j(A, B) = L(A)^{n-k} \psi_j(A, B \bmod A) \tag{1}$$

where this time $\psi_j(A, B \bmod A)$ is an $(m + k - 2j) \times (m + k - 2j)$ determinant. To rewrite this in the symmetric form which covers the case $m \geqslant n$, first note that

$$\psi_j(A, B) = (-1)^{(m-j)(n-j)} \psi_j(B, A). \tag{2}$$

Hence, if $m \geqslant n$, and if k is the degree of $A \bmod B$, we have

$$\psi_j(A, B) = (-1)^{(n-j)(m-k)} L(B)^{m-k} \psi_j(A \bmod B, B). \tag{3}$$

Finally, if $j = m \leqslant n$, then

$$\psi_j(A, B) = L(A)^{n-m}. \tag{4}$$

It is also appropriate to put

$$\psi_j(A, B) = 0. \tag{5}$$

if $\min(m, n) < j$. Note that the last equality is consistent with the reduction formulae (1) given above, in the sense that $\psi_j(A, B) = 0$ if either $\deg(A \bmod B) < j$ or $\deg(B \bmod A) < j$.

The technique for resultant calculation given in Schwartz (1980) depends only on the identities (1)–(5) and hence can be adapted to the calculation of principle subresultant coefficients. Fleshing out this summary remark, we shall now present an efficient technique for the simultaneous calculation of all subresultants of a given pair of polynomials P and Q. To this end, we make the following definition.

DEFINITION 3. Let a pair of polynomials $w = [P, Q]$ of degree d, d' with coefficients in a field F be given, and let $d = \max(d, d')$. Write $P \bmod Q$ for the remainder of P upon division by Q. Then the *RQ-sequence* $RQ(w)$ of w is the sequence t_i, $i = d, d - 1, \ldots, 0$, of quadruples

$$t_i = [[P_i, Q_i], a_i, b_i, M_i],$$

defined as follows:

(1) P_i, Q_i are polynomials, a_i is a quantity of F, b_i is always $+1$ or -1, and M_i is a 2×2 matrix of polynomials with coefficients in F.

(2) $t_d = [[P, Q], 1, 1, I]$, where I is the 2×2 identity matrix.

(3) $\max(\deg(P_i), \deg(Q_i)) \geqslant i \geqslant \min(\deg(P_i), \deg(Q_i))$ for $i > 0$.

(4a) If $\min(\deg(P_i), \deg(Q_i)) < i$, then $t_{i-1} = t_i$;

(4b) Otherwise, if $\deg(P_i) = i$, then (dropping remainders in all polynomial divisions) we have

$$t_{i-1} = [[P_i, Q_i \bmod P_i], a_i e_i, b_i f_i, N_i M_i],$$

where $n = \deg(Q_i)$, $k = \deg(Q_i \bmod P_i)$, $f_i = 1$, $e_i = L(P_i)^{n-k}$, and

$$N_i = \begin{bmatrix} 1 & 0 \\ -Q_i/P_i & 1 \end{bmatrix};$$

(4c) Otherwise $\deg(Q_i) = i$, and then

$$t_{i-1} = [[P_i \bmod Q_i, Q_i], a_i e_i, b_i f_i, N_i M_i],$$

where $m = \deg(P_i)$, $k = \deg(P_i \bmod Q_i)$,

$$e_i = ((-1)^i L(Q_i))^{m-k},$$

$$f_i = (-1)^{m-k},$$

and

$$N_i = \begin{bmatrix} 1 & -P_i/Q_i \\ 0 & 1 \end{bmatrix}.$$

The following lemma generalizes Lemma 3 of [12].

LEMMA 3. *Let* $w = [P, Q]$, t_i, *etc., be as in the preceding definition. Then the sequence* $RQ(w)$ *has the following properties:*
(i) $[P_i, Q_i] = M_i[P, Q]$.
(ii) $\deg(M_k) \leqslant d - \max(\deg(P_i), \deg(Q_i)) \leqslant d - i$.
(iii) *For each* $i \leqslant d$ *and each* $0 \leqslant j \leqslant \min(\deg(P_i), \deg(Q_i))$, $\psi_\theta(P, Q) = a_i\psi_j(P_i, Q_i)$ *if* j *is even, and* $\psi_j(P, Q) = a_i b_i \psi_j(P_i, Q_i)$ *if* j *is odd.*

Proof. All this is clear for $i = d$. A step from t_i to t_{i-1} via (4a) of the preceding definition clearly preserves the validity of (i)–(iii). Now suppose that rule (4b) applies to the step from t_i to t_{i-1}. Property (i) is clearly preserved. Moreover,

$$\deg(M_{i-1}) \leqslant \deg(Q_i) - \deg(P_i) + \deg(M_i)$$
$$\leqslant \deg(Q_i) - \deg(P_i) + d - \deg(Q_i)$$
$$= d - \deg(P_i) = d - \max(\deg(P_{i-1}), \deg(Q_{i-1})).$$

Concerning (iii), it follows by (1) and (2) that

$$\psi_j(P, Q) = a_i\psi_j(P_i, Q_i)$$
$$= a_i(L(P_i))^{(\deg(Q_i)-\deg(Q_i \bmod P_i))} \cdot \psi_j(P_i, Q_i \bmod P_i)$$
$$= a_{i-1}\psi_j(P_{i-1}, Q_{i-1}),$$

if j is even, and similarly if j is odd.

Suppose finally that (4c) is used to obtain t_i from t_{i-1}. Again, property (i) is clearly preserved, and property (ii) follows by an argument symmetric to the one used above, in which P_i and Q_i are interchanged. Concerning (iii), using (1) and (3) we obtain, assuming j is even,

$$\psi_j(P, Q) = a_i\psi_j(P_i, Q_i)$$
$$= ((-1)^i L(Q_i))^{(\deg(P_i)-\deg(P_i \bmod Q_i))} \cdot \psi(P_i \bmod Q_i, Q_i)$$

$$= a_{i-1}\psi_j(P_{i-1}, Q_{i-1}),$$

and similarly if j is odd. This proves the lemma.   Q.E.D.

LEMMA 4.   *Let* $w = [P, Q]$, $w^* = [P^*, Q^*]$ *be two pairs of polynomials. Suppose that* $\max(\deg(P), \deg(Q), \deg(P^*), \deg(Q^*)) = d$, *and suppose that the terms of order not less than* $d - 2i$ *in* P, Q *agree with the corresponding terms in* P^*, Q^*. *Then the first* $i + 1$ *terms of the sequence* $RQ(w) = [t_d, t_{d-1}, \ldots]$ *have precisely the same components* a_i, b_i, M_i *as the corresponding terms* a_i^*, b_i^*, M_i^* *of the sequence* $RQ(w^*) = [t_d^*, t_{d-1}^*, \ldots]$.

Proof. Except for the equality of b_i and b_i^*, the proof is completely identical to that of Lemma 4 of Schwartz (1980). The equality of b_i and b_i^* is also an easy consequence of that same proof.   Q.E.D.

Lemmas 3 and 4 justify the following principal subresultant calculation algorithm, whose underlying idea is to compute the terms a_i, b_i, and M_i by stepping through the sequence $[t_i]$ by steps of increasing length, each of length double to that of the preceding step. Lemma 4 implies that each of these steps can use polynomials of substantially lower degree than that of the original polynomials. However, as soon as we have calculated a_i, b_i, and M_i for some $i \leq d$ for which $j = \deg(P_i) \leq \deg(Q_i)$, we can get $\psi_j(P, Q)$ directly. Indeed, using (4) and Lemma 3(iii), we have

$$\psi_j(P, Q) = a_i \psi_j(P_i, Q_i)$$
$$= a_i \cdot L(P_i)^{(\deg(Q_i)-j)},$$

if j is even, and similarly if j is odd (and of course analogous formulae are available if $j = \deg(Q_i) \leq \deg(P_i)$.) Note that if $t_i \neq t_{i'}$ then $\min(\deg(P_i), \deg(Q_i)) \neq \min(\deg(P_{i'}), \deg(Q_{i'}))$, so that for each j either there exists no i for which $j = \min(\deg(P_i, Q_i))$, or the i's for which this equality holds all have the same value t_i, and hence the above formula defines ψ_j unambiguously. Observe finally that, by (5), all ψ_j which do not appear in the above formula for any value of i are 0.

In view of these comments, the algorithm proceeds as follows. Initialize four static lists, AL, BL, CL, and DL to the null list each.

(1) Call an auxiliary routine HSBRSL(P, Q, 0), with final parameter $l = 0$. This routine will build up the first half of the RQ-sequence $[t_i]$, for $d \geq i \geq d/2 = d'$, and will return the quantities $e_{d'}$, $f_{d'}$, and $M_{d'}$.

(2) Compute $[P_{d'}, Q_{d'}] = M_{d'}[P_d, Q_d]$.

(3) Let $m = \deg(P_{d'})$, $n = \deg(Q_{d'})$. If $m \leq n$ then append the quantities $e_{d'}, f_{d'}$, $L(P_{d'})^{n-m}$, m respectively to the end of the four lists AL, BL, CL, DL. Similarly, if $m \geq n$ then append the quantities $e_{d'}, f_{d'}, L(Q_{d'})^{m-n}, n$, to the end of these four respective lists.

(4) Call the whole procedure recursively, passing $P_{d'}, Q_{d'}$ to it as inputs to complete the construction of the whole sequence.

The subprocedure HSBRSL(P, Q, l) required is rather similar to the auxiliary procedure HGCD used in the last polynomial GCD algorithm in Aho, Hopcroft, and Ullman (1974). It consists of the following steps.

(1) Discard the least significant half of the coefficients of P and Q; that is, let $P = P*x^k + P'$, $Q = Q*x^k + Q'$, where $k = \max(\deg(P), \deg(Q))/2$, and where $\deg(P')$, $\deg(Q') < k$.

(2) Apply HSBRSL to $P*$, $Q*$, and $l + k$, to obtain $e_{k'}$, $f_{k'}$, and $M_{k'}$ for $k' = 3k/2$ (by Lemma 4, these are the same as the corresponding quantities associated with the original P and Q).

(3) Compute $[P_{k'}, Q_{k'}] = M_{k'}[P, Q]$.

(4) As in step (3) of the main procedure, let $m = \deg(P_{k'})$, $n = \deg(Q_{k'})$. If $m \leq n$ then append the quantities $e_{k'}$, $f_{k'}$, $L(P_{k'})^{n-m}$, $m + l$, to the end of the four lists AL, BL, CL, DL respectively. Similarly, if $m \geq n$ then append the quantities $e_{k'}$, $f_{k'}$, $L(Q_{k'})^{m-n}$, $n + l$, to the end of these four respective lists.

(5) Let $P_{k'} = P_k*x^{k''} + P''$, $Q_{k'} = Q_k x^{k''} + Q''$, where $k'' = k/2$, and where $\deg(P'')$, $\deg(Q'') < k''$.

(6) Call HSBRSL with $P_{k'}^*$, $Q_{k'}^*$, and $l + k''$, to obtain e_k^*, f_k^*, and M_k^*.

(7) Return e_k^*, f_k^*, and $M_k^* \cdot M_{k'}$.

When the algorithm just sketched has terminated, we can perform one final scan through the lists AL, BL, CL, and DL, to accumulate the scalar quantities a_i by b_i by repeated multiplication of the e_is and the f_is, respectively. Using the following technique we can also calculate all the principle subresultant coefficients ψ_j. For each iteration step k, let γ_k, m_k be the kth components of CL and DL, respectively, and let a_{k-1} and b_{k-1} be the product of all preceding components in the lists AL and BL, respectively. If m_k is even, put

$$\psi_{m_k} = a_{k-1}\gamma_k,$$

and if m_k is odd, put

$$\psi_{m_k} = a_{k-1}b_{k-1}\gamma_k.$$

At the end of this final iteration, all remaining undefined subresultants are set to 0.

The computational cost of the procedure just sketched is evidently the same as the original procedure given in Schwartz (1980). To estimate this cost we note that multiplication and division of polynomials of order m can be accomplished in time $O(m \log m)$. During each call to the subprocedure HSBRSL(P, Q, l) that we have just described polynomials of order at most $m = \max(\deg(P), \deg(Q))$ need to be multiplied and divided (at steps (3) and (7)). The total time $T(m)$ required to apply HSBRSL to two polynomials of degree m therefore satisfies $T(2m) = 2T(m) + O(m \log m)$ and hence has the bound $T(m) = O(m \log^2 m)$. Similar considerations show that the time used by the main procedure to build up the RQ-sequence has the same bound, so that this estimate also bounds the time required to calculate all the subresultants of two polynomials of maximal degree m.

APPENDIX C: ADJACENCY OF COLLINS CELLS OF GENERAL DIMENSION

In this appendix we complete our calculation of Collins cell adjacency by considering cells whose dimension is less than that of the whole space E^r being decomposed. The problem here is to determine when a Collins cell c' of dimension $k - 1$ is a face of a cell c of dimension $k < r$. Our approach will again be based upon consideration of the base decomposition K' of K. However, the present case is somewhat more complex than the case $k = r$ considered previously, and this makes it necessary to use a bit more (largely classical) machinery drawn from the theory of algebraic curves. Accordingly, we recall the following definitions and lemmas, for which see Keller (1974, Ch. 5).

DEFINITION 1. (a) A *fractional (Laurent) series* is a formal series of the form

$$y(x) = \sum_{i=m}^{\infty} a_i x^{i/D} \tag{1}$$

whose coefficients are complex numbers, and in which D is an integer. Such a series is said to be *convergent* if it converges in the neighborhood of $x = 0$. Note that we allow m to be negative. The series (1) is said to be *truncated* if it contains only finitely many terms.

(b) If $a_m \neq 0$, then m/D is called the *leading exponent* of the series y.

(c) If the leading exponent e of the series (1) satisfies $e \geq k$ (resp. $e > k$), we will write $y = O(x^k)$ (resp. $y = o(x^k)$).

It is easily seen that the standard definitions of addition, multiplication, etc., for power series make the collection of all fractional series into a field. This statement, like many others made in the next few paragraphs, is true irrespective of whether we consider convergent fractional series only, or allow arbitrary, nonconvergent series, and treat operations on them in a purely formal manner. Moreover, this statement remains true even if the coefficients of the series (1) are required to lie in some subfield of the complex numbers.

In what follows we will designate the field of fractional power series by Fr.

LEMMA 1 (see Keller, 1974, Ch. 5). *The field of fractional series with complex (or with algebraic) coefficients is algebraically closed.*

Thus if $P(y) = a_n y^n + \cdots + a_o$ is a polynomial in y with coefficients in the field Fr and if $a_n \neq 0$, P can be factored as

$$P(y) = a_n(y - r_1) \ldots (y - r_n), \tag{2}$$

where $r_1, \ldots, r_n$ are themselves fractional series, namely, the roots of $P(y) = 0$.

In what follows we will need to work in purely finite manner with fractional series representing the roots of various polynomials. Our ability to do so without ambiguity will rest upon various extensions of the following simple lemmas, which are also noted by Kung and Traub (1978).

LEMMA 2. *Let $P(x, y)$ be a polynomial with complex coefficients in two variables, of total degree $d > 0$. Let its degree n in y be nonzero, and suppose that, when regarded as a polynomial $P_x(y)$ in y, P has no factor in common with its y-derivative $P'_x(y)$ and has a leading coefficient $a_n(x)$ which is not zero for $x = 0$. Regard $P_x(y)$ as a polynomial in y with coefficients in Fr, and let $r_1(x), \ldots, r_n(x)$ be its roots (which are elements of Fr). Then*
  (a) If $r_i(x) - r_j(x) = O(x^{d(2n-1)})$, then $i = j$.
  (b) We cannot have $P'_x(r_i(x)) = o(x^{d(2n-1)})$ for any root $r_i(x)$.

Proof. Consider the discriminant $D(x)$ of the polynomial $P_x(y)$, i.e., $D = \mathrm{Res}(P_x, P'_x)$ is the resultant of P_x and P'_x. This is a polynomial of degree $d(2n - 1)$ in x, and since P_x and P'_x have no factor in common it is nonzero. Hence $D(x) = o(x^{d(2n-1)})$ is impossible. By well-known identities (see Van der Waerden, 1939, Secs. 30, 31),

$$D(x) = \prod_{i \neq j} \big(r_i(x) - r_j(x)\big) \tag{3}$$

and also

$$D(x) = \prod_{i \neq j} P'_x(r_i(x)). \tag{4}$$

Since the leading coefficient of y in $P_x(y)$ does not vanish at $x = 0$, it is easy to see (by substitution) that no fractional series $r(x)$ with negative leading exponent can satisfy $P_x(r(x)) = 0$. Hence $r_i(x) = O(1)$ for all i, and therefore if assertion (a) of our lemma were violated, (3) would imply that $D(x) = o(x^{d(2n-1)})$, which is impossible. If we use (4) instead of (3) in this argument, (b) follows in the same way. Q.E.D.

LEMMA 3. *Let $P_x(y) = a_n(x)y^n + a_{n-1}(x)y^{n-1} + \cdots + a_0(x)$ be a polynomial in y, of degree n, with coefficients in Fr, and suppose that $a_n(x)$ does not vanish at $x = 0$. Let $k \geq 1$, and let $y_0 = \Sigma a_i x^{i/D}$ be an element of Fr such that $P_x(y_0(x)) = O(x^{nk})$. Then there exists a root $r_i(x)$ of $P_x(y) = 0$ such that $y_0(x) - r_i(x) = O(x^k)$.*

Proof. Factor $P_x(y)$ in the manner (2). Then it is plain that if $y_0 - r_i$ has leading exponent less than k for all i, the leading exponent of $P_x(y_0(x))$ must be less than nk, contradicting our assumption. Q.E.D.

Lemma 3 asserts that any power series y_0 which comes close enough to making $P_x(y_0)$ equal to zero must lie quite close to a root of P_x. It is also worth noting that if $P_x(y)$ has no repeated roots, then, once $P_x(y_0)$ has been made small enough, any desired number of coefficients of a root of $P_x(y) = 0$ can be calculated rapidly from the coefficients of y_0 by purely rational operations. A technique for calculating fractional power series which approximate each of the roots of $P_x(y)$ near $x = 0$ to an arbitrary degree of precision is described by Kung and Traub (1978). Their technique first uses the Newton's polygon method to obtain an initial (truncated) fractional series approximation for each of the roots of P_x, and then uses a variant of Newton iteration (similar to that described by Lipson, 1976) to extend each of these series to an arbitrary degree of precision. Kung and Traub give a complexity bound of $O(n\,N \log N)$ for their procedure, where n is the y-degree of the polynomial P_x, and where N is the number of terms sought in each fractional series. This bound holds asymptotically if n is held fixed and N increases, but it ignores the cost of applying the Newton polygon procedure (which indeed is independent of N) in order to obtain the initial collection of terms required to ensure convergence of Newton's approximation method. At present our best estimates of this cost are still exponential, leading us to pose the following:

Open problem. Given a polynomial $P(x, y)$ with complex coefficients, of y-degree n and total degree d, does there exist a procedure for calculating truncated fractional series approximations for each of the roots of $P(x, y)$ (regarded as a polynomial in y) near $x = 0$ up to terms of order $O(x^{d(2n-1)})$, in time polynomial in n and d?

As will become clear from the subsequent discussion, an affirmative answer to this problem would imply that adjacency of Collins cells of general dimension, and hence also the homology groups of any algebraic variety, can be calculated in time polynomial in the number and the maximal degree of polynomials defining a Collins decomposition (or an algebraic variety).

At any rate, once these fractional power series are available, we can use them to determine the adjacency of Collins cells. It is worth explaining what is involved by commenting briefly on the special case $r = 3$. Let K' be the 2-dimensional base decomposition of a well-based Collins decomposition K of E^3 whose base polynomial is $P(x, y, z)$. Let c be a 1-dimensional curved cell in K', and let $c' = [x_0, y_0] \in K$ be an endpoint of c. Let $Q(x, y)$ be the base polynomial of K'. It follows from the preceding discussion that in the vicinity of c' the curve c can be specified uniquely by a truncated fractional series $y(x)$ which approximates the corresponding root of $Q(x, y)$ near (x_0, y_0). If enough leading terms of this fractional series are known, then, to determine the number of roots of $P(x, y, \cdot)$ into which a given root z_0 of $P(x_0, y_0, z) = 0$ splits as we move from (x_0, y_0) into c, we can substitute the curve $y(x)$ for y in P, thus obtaining a polynomial $R_x(z) = P(x, y(x), z)$ in z with coefficients in Fr. Then, using a Sturm-based technique of the sort described in Appendix B, we can determine the number of roots of R_x into which z_0 splits for sufficiently small positive x. We can also compute truncated fractional series approximating each of these roots, and this enables us to carry the same form of analysis inductively to cases involving more than three variables.

However, several technical difficulties must be overcome in following this conceptual approach. First of all, Lemma 2 is not immediately applicable to R_x, since the coefficients of R_x are not simply polynomial but lie in Fr; moreover R_x need not be squarefree. This makes it a little harder to state how many terms of a fractional series suffice to characterize a root of $R_x(z)$ uniquely. Also, the truncated fractional series $y(x)$ gives only an approximate representation of the root of Q which traverses the curve c, and at first glance it is not clear how many terms need to be included in $y(x)$ to guarantee that the analysis we have outlined will carry forward inductively.

To overcome these purely technical difficulties, it is useful to consider systems of polynomial equations in several variables having the form specified in the following definition.

DEFINITION 2.   (a) A *triangular monic system* of polynomial equations is a system

$$P_2(x_1, x_2) = 0$$
$$P_3(x_1, x_2, x_3) = 0$$
$$\vdots \tag{5}$$
$$P_n(x_1, x_2, \ldots, x_n) = 0$$

of polynomial equations of total degrees $d_2, \ldots, d_n$ with complex coefficients, whose jth equation involves the variables $x_1, \ldots, x_j$ only, and which, regarded as a polynomial in x_j with coefficients polynomial in the remaining variables, has a constant leading term. The sequence $d_2, \ldots, d_n$ is the *degree sequence* of the triangular system (5).

(b) Given the triangular monic system (5), its *resultant system* is the sequence

$$R_2(x_1, x_2), \qquad R_3(x_1, x_3), \qquad \ldots, \qquad R_n(x_1, x_n) \tag{6}$$

of bivariate polynomials defined as follows. R_2 is simply P_2. R_3 is obtained from P_3 and P_2 by regarding them both as polynomials in the variable x_2 and forming their resultant. More generally, to form $R_j(x_1, x_j)$ proceed as follows: Regard $P_j(x_1, \ldots, x_j)$ and $P_{j-1}(x_1, \ldots, x_{j-1})$ as polynomials in x_{j-1}, and form their resultant, thus obtaining a polynomial $Q_1(x_1, \ldots, x_{j-2}, x_j)$ from which the variable x_{j-1} has been eliminated. Then form the resultant Q_2 of Q_1 and P_{j-2} (as polynomials in x_{j-2}), obtaining a polynomial from which both variables x_{j-2} and x_{j-1} have been eliminated. Continue repeatedly in this way, thus finally obtaining the desired polynomial $R_j = Q_{j-2}$, from which all variables but x_1 and x_j will have been eliminated.

(c) The sequence $\delta_2, \ldots, \delta_n$ of degrees of the resultant polynomials (6) is called the *resultant degree sequence* of the triangular system (5).

DEFINITION 3.   (a) A *solution curve* of the triangular monic system (5) is a continuous vector-valued function $f(t) = [f_2(t), \ldots, f_n(t)]$ with complex components defined for all sufficiently small nonnegative values of the parameter t, such that for all $j = 2, \ldots, n$ we have $P_j(t, f_2(t), \ldots, f_j(t)) = 0$.

(b) Let P_j and f be as in (a), let $d = [d_2, \ldots, d_n]$ be a sequence of rational numbers, and let $\alpha = [\alpha_2, \ldots, \alpha_n]$ be a sequence of truncated fractional series in the variable t, such that α_j contains terms of order $O(t^{d_j})$ at most. Then α is said to be an *order d descriptor* of the solution curve f if $f_j - \alpha_j = O(t^{d_j})$ for all $j = 2, \ldots, n$.

Our aim is to generalize Lemmas 2 and 3 to general triangular monic systems of more than two variables. For this, we need the following straightforward technical lemma, whose main purpose is to enable us to bound the order of finite descriptions of solution curves of (5) needed to characterize such a solution uniquely.

LEMMA 4. *As in Definition 3, let $P_2, \ldots, P_n$ be a triangular monic system and let $R_2, \ldots, R_n$ be its resultant sequence. Then each R_j can be factored as*

$$R_j(x_1, x_j) = A(x_1) R_j^0(x_1, x_j) R_j^1(x_1, x_j), \tag{7}$$

where A, R_j^0, and R_j^1 are all polynomials in their respective variables, and where R_j^0 is the squarefree part of R_j (regarded as a polynomial in x_j) represented by a polynomial whose coefficients (which are polynomials in x_1) have no common factor. Moreover, we can arrange this factorization so that the leading coefficient $C(x_1)$ of R_j^0 does not vanish at $x_1 = 0$.

Proof. As in Definition 2, regard $P_j(x_1, \ldots, x_j)$ and $P_{j-1}(x_1, \ldots, x_{j-1})$ as polynomials in x_{j-1} with polynomial coefficients, and form their resultant Q_1. If we use the fact that P_{j-1} is monic in x_{j-1}, it follows from the fundamental theorem of resultant theory that a tuple $[x_1, \ldots, x_{j-2}, x_j]$ is a solution of the system $P_1 = P_2 = \cdots = P_{j-2} = Q_1 = 0$ of equations if and only if there exists an x_{j-1} such that $[x_1, \ldots, x_{j-2}, x_{j-1}, x_j]$ is a solution of the system $P_i = 0$, $i = 2, \ldots, j$. Arguing repeatedly in this way, we see that $[x_1, x_j]$ is a solution of the equation $R_j(x_1, x_j) = 0$ if and only if there exist $x_2, \ldots, x_{j-1}$ such that $[x_1, x_2, \ldots, x_j]$ is a zero of P_i for all $i = 2, \ldots, j$. On the other hand, it follows since each P_i is monic that all the solutions of the system $P_i = 0$, $i = 2, \ldots, j$ which lie over a small neighborhood of the point $x_1 = 0$ remain bounded. Thus all the solutions of $R_j(x_1, x_j) = 0$ over such a neighborhood have the same property.

Next regard R_j is a polynomial in x_j, take its (x_j-) derivative R_j', form $T^1 = \text{GCD}(R_j, R_j')$, and use it to factor R_j as $R_j = T^0 T^1$. These are polynomials in x_j with coefficients rational in x_1 but then by multiplying through by an appropriate polynomial $C(x_1)$ we can write $C R_j = S^0 S^1$, where now S^0 and S^1 have coefficients which are polynomials in x_1. Since the ring of polynomials in two variables is a unique factorization domain, every prime factor of C divides one of the polynomials on the right, which is to say, divides all its coefficients. This remark allows us to divide through by all these prime factors, and if we then collect all common factors of the coefficients of R^0 and R^1 in $A(x_1)$, we arrive at the factorization (7).

It is now clear that all the solutions $[x_1, x_j]$ of $R_j^0 = 0$ which lie over a sufficiently small neighborhood of $x_1 = 0$ have second components x_j which remain bounded. By Lemma 1, these solutions $r_1(x_1), \ldots, r_m(x_1)$ can be written as fractional power series $\sum_{i=m} a_i x_1^{i/D}$, and since they remain bounded near $x_1 = 0$ none of these fractional series contain any term with negative exponents. Suppose that the leading coefficient $a_m(x_1)$ of R_j^0 vanishes at $x_1 = 0$. Then since R_j^0 can be factored as

$$R_j^0(x_1, x_j) = a_m(x_1)\Pi(x_j - r_i(x_1)),$$

it would follow that all the coefficients of R_j^0 vanish at $x_1 = 0$. But then x_1 would be a common factor of all coefficients of R_j^0, contrary to the way in which R_j^0 has been defined.   Q.E.D.

COROLLARY. *Let the polynomials P_j, $j = 2, \ldots, n$ form a triangular monic system, and let the resultant degree sequence of this triangular system be $[\delta_2, \ldots, \delta_n]$. Put*

$$d = [\delta_2(2\delta_2 - 1), \ldots, \delta_n(2\delta_n - 1)].$$

Then any two solution curves $f(t)$, $g(t)$ of this triangular system which have identical order d descriptors are identically equal for all sufficiently small nonnegative values of t.

Proof. Let R_j be the resultant system of the triangular system P_j. Arguing as in the proof of Lemma 4, we can conclude that the components f_j of f satisfy $R_j(x, f_j(x)) = 0$, and similarly for the components of g. Our assertion is therefore an immediate consequence of Lemmas 2 and 4.   Q.E.D.

DEFINITION 4. Let P_j and δ_j be as in the preceding corollary, let $d = [d_2, \ldots, d_n]$ be the degree sequence of the P_j, and define $d^* = \max_{j=2\cdots n} d_2 d_3 \cdots d_j \delta_j(2\delta_j - 1)$. Put $d_j^* = d^*/(d_2 \cdots d_j)$ for $j = 2, \ldots, n$ (so that d_j^* is a decreasing sequence of positive rational numbers). Suppose that for $j = 2, \ldots, n$ $a_j(x)$ is a truncated fractional series such that there exist auxiliary truncated fractional series $a_j^*(x)$ (all of whose exponents can be assumed to be $\leq d_j^*$) such that $P_j(x, a_2^*(x), \ldots, a_j^*(x)) = O(x^{d_{j-1}^*})$ and $a_j(x) - a_j^*(x) = O(x^{d_j^*})$. Then $\alpha = [a_2, \ldots, a_n]$ is called an *adequate descriptor for a solution of the triangular system* $P_2 = \cdots = P_n = 0$ *of equations.*

LEMMA 5. *Let d, d^*, and δ be as in the preceding definition. Given any adequate descriptor $\alpha = [a_2, \ldots, a_n]$ for a solution of the triangular system(5), then for all sufficiently small t there exists a solution $f(t) = [f_2(t), \ldots, f_n(t)]$ of the system $P_j(t, f_2(t), \ldots f_j(t)) = 0$, $j = 2, \ldots, n$ such that*

$$f_j(t) - \alpha_j(t) = O(t^{\delta_j(2\delta_j - 1)}). \tag{8}$$

Proof. Use the notations of Definition 4. We proceed by induction on j to establish the existence of a solution curve $f = [f_2, \ldots, f_n]$ which satisfies the formula

$$f_j(t) - \alpha_j(t) = O(t^{d_j^*}), \tag{8'}$$

for $j = 2, \ldots, n$. By definition of the d_j^* it is plain that (8') implies (8). The base case $j = 2$ for our induction is immediate from Lemma 3. Suppose that we have already shown the existence of $f_2(t), \ldots, f_{j-1}(t)$ such that $P_i(t, f_2(t), \ldots, f_i(t))$ $= 0$ for $i = 2, \ldots, j - 1$, and such that (8') holds for all such i. As in Definition 4, there exist auxiliary fractional series $\alpha_i^*(t)$ such that $P_j(t, \alpha_2(t), \ldots, \alpha_j(t)) = O(t^{d_j^*}-1)$ and $\alpha_i^*(t) - \alpha_i(t) = O(t^{d_i^*}) \leqslant O(t^{d_j^*}-1)$ (since the d_i^* are decreasing). Thus by (8') we have $\alpha_i^*(t) - f_i(t) = O(t^{d_j^*}-1)$, from which we see immediately that

$$P_j\big(t, f_2(t), \ldots, f_{j-1}(t), \alpha_i^*(t)\big) = O(t^{d_j^*-1}).$$

Hence by Lemma 3 there exists a root $f_j(t)$ of this same equation such that $f_j(t) - \alpha_j^*(t) = O(t^{(d_j^*-1/d_j)}) = O(t^{d_j^*})$ for small t, from which (8') is plainly seen to hold for j too.   Q.E.D.

COROLLARY. *Let $P_j, f = [f_2, \ldots, f_n]$, and $\alpha = [\alpha_2, \ldots, \alpha_n]$ be as in Lemma 5. Suppose that all the polynomials P_j have real coefficients. Then f_j is real if and only if all the nonzero coefficients appearing in α_j are real.*

Proof. Since all P_j are real, the whole situation being considered is invariant under complex conjugation. Thus the curve $\bar{f} = [\bar{f}_2, \ldots, \bar{f}_n]$ is also a solution curve of the system P_j, and $\bar{\alpha} = [\bar{\alpha}_2, \ldots, \bar{\alpha}_n]$ is an adequate descriptor for $\bar{f}$. It is plain from Definition 3 and from the corollary to Lemma 4 that $f = \bar{f}$ if and only if $\alpha = \bar{\alpha}$, from which the present corollary is immediate.   Q.E.D.

Our preparation is now sufficient for computations of the topological relationships which interest us to be feasible. The following definition takes the next step in this direction.

DEFINITION 5. Let K be a Collins decomposition of the Euclidean space E^r, let c be a d-dimensional cell in K, and let $c' \in K$ be a $(d - 1)$-dimensional face of c. Then an *incidence curve* for the pair c, c' is a triple consisting of

(i) a triangular monic system $P_2, \ldots, P_m$ of m polynomials with real algebraic coefficients,

(ii) a continuous solution curve $f(t) = [f_1(t), f_2(t), \ldots, f_r(t)]$, all of whose components are real, such that $f(t) \in c$ for all sufficiently small nonnegative t, and such that $f(0) \in c'$ [all constant components of the vector f must be algebraic numbers, and its nonconstant components $f_{i_1}(t), \ldots, f_{i_m}(t)$, which for convenience we shall write in left-to-right order as $g_1(t), \ldots, g_m(t)$, must satisfy $P_j(g_1(t), \ldots, g_j(t)) = 0$ for $j = 2, \ldots, m$; moreover, $g_1(t)$ must have the form $g_1(t) = t + \zeta$, where ζ is an algebraic number];

(iii) an adequate descriptor α for the solution $g_2, \ldots, g_m$ of the triangular monic system $P_2 = \cdots = P_m = 0$. [This (finite) descriptor will be used to represent the nonconstant components of the curve f in finite terms in a unique manner].

Our aim is to carry such a family of incidence curves through the whole inductive construction of the Collins decomposition, and at each stage to use them to determine which $(d - 1)$-dimensional cells c' are faces of a given d-dimensional cell c. To this end, let K be a well-based decomposition of E^r, and let K' be the base decomposition of K. Suppose inductively that for each cell b of K' and each of its faces b' of one less dimension an incidence curve for the pair b, b' is available. Let c be a d-dimensional cell of K, and let c' be one of its faces. Then, by Lemma 3 of Section 2, either $c = b_j^*$ for some $b \in K'$ and $c' = b_j$ or $c' = b_{j+1}$, or $c = b_j^*$ and $c' = b_j'^*$ where b' is a $(d-2)$-dimensional face of (the $(d-1)$-dimensional) cell b, or $c = b_j$ and $c' = b_j'$ where b' is a $(d-1)$-dimensional face of (the d-dimensional) cell b.

Let $P(x_1, \ldots, x_r)$ be the base polynomial of K; since K is well-based, the leading coefficient of P (regarded as a polynomial in x_r) is constant. If $c = b_j^*$ and $c' = b_j$, then we can take $[\zeta_1, \ldots, \zeta_{r-1}]$ to be any algebraic point in b, and then if ζ_r is the jth root of $P(\zeta_1, \ldots, \zeta_{r-1}, x_r)$ we can put $f(t) = [\zeta_1, \ldots, \zeta_{r-1}, \zeta_r + t]$, thereby defining an incidence curve. Of course, an entirely similar construction yields an incidence curve for the pair b_j^*, b_{j+1}.

Next suppose that $c = b_j^*$ and $c' = b_j'^*$, where b' is a $(d-2)$-dimensional face of b. Let $f' = [f_1(t), \ldots, f_{r-1}(t)]$ be an incidence curve for the pair b, b', let $g_1(t) = \zeta + t$, $g_2(t), \ldots, g_m(t)$ be the nonconstant components of this curve, enumerated in left-to-right order, and let $P_2, \ldots, P_m$ be the associated triangular monic system of polynomials. Put $\zeta_j = f_j(0)$ for $j = 1, \ldots, r-1$, so that the point $Z = [\zeta_1, \ldots, \zeta_{\varrho-1}]$ is an algebraic point in b'. Let P be as in the preceding paragraph, let η be the Jth real root of $P(\zeta_1, \ldots, \zeta_{r-1}, y) = 0$ (or a sufficiently small rational number if $J = 0$), and let η' be the $(J+1)$st root of this same equation (or a sufficiently large rational number if this equation has only J real roots). Put $\zeta_r = \frac{1}{2}(\eta + \eta')$, and $f(t) = [f_1(t), \ldots, f_{r-1}(t), \zeta_r]$. This is plainly an incidence curve for the pair c, c', whose nonconstant components are plainly the same as those of f'. We have an adequate descriptor for the nonconstant components of f', and plainly this can also serve as an adequate descriptor for f.

Finally, suppose that $c = b_j$ and $c' = b_j'$, where b' is a $(d-1)$-dimensional face of b. Again, let $f'(t) = [f_1(t), \ldots, f_{r-1}(t)]$, $[P_2, \ldots, P_{m-1}]$, and $\alpha' = [\alpha_2, \ldots, \alpha_{m-1}]$ be an incidence curve for the pair b, b'. As in Definition 5, let $g_1, \ldots, g_{m-1}$ be the nonconstant components of f', so that $g_1(t) = t + \zeta$ where ζ is algebraic, and put $\alpha_1(t) = t + \zeta$ for convenience. Extend the truncated fractional series α_j, $j = 2, \ldots, m-1$ to series α_j^* such that $g_j - \alpha_j^* = O(t^D)$, where D will be chosen below. Let P be the same base polynomial as in the two preceding paragraphs, let d be its degree, and let P^* be the polynomial obtained from P by replacing each variable x_j in it for which $f_j(t) = \zeta_j$ is an algebraic constant by ζ_j. Clearly, the degree of P^* is at most d. Having P^*, we can then find all distinct real roots $y_1, \ldots, y_l$ of the equation $P^*(\alpha_1(0), \ldots, \alpha_{m-1}(0), y) = 0$ and arrange these solutions in increasing order. By Lemma 2 of Section 2, these roots are in 1-1 correspondence with the sequence of "section" cells b'_k of K whose base cell is b'. Next, for each such y_k, we can find all the fractional series solutions of the monic equation $P^*(\alpha_1^*(t), \ldots, \alpha_{m-1}^*(t), y) = 0$ whose constant term is y_k; all terms of these series up to those of order t^D must be constructed. Let $\alpha(t)$ designate any one of the truncated fractional series constructed in this way. Since $g_j - \alpha_\theta^* = O(t^D)$ for $j = 1, \ldots, m-1$, we have $P^*(g_1(t), \ldots, g_{m-1}(t), \alpha(t)) = O(t^D)$ and thus by Lemma 3 there exists a solution g of $P^*(g_1(t), \ldots, g_{m-1}(t), g(t)) = 0$ such that

$\alpha(t) - g(t) = O(t^{D/d})$. Append the polynomial P^* to the triangular monic system P_2, $\ldots, P_{m-1}$, thus obtaining a larger system $P_2, \ldots, P_m$, and let $R_2, \ldots, R_m$ and $\delta_2, \ldots, \delta_m$ be respectively the resultant system and the resultant degree sequence of this extended system. Arguing as in the first paragraph of the proof of Lemma 4, we see that $R_m(g_1(t), g(t)) = R_m(t + \zeta, g(t)) = 0$. Hence if we choose $D = d\delta_m(2\delta_m - 1)$ it follows by Lemma 2 that this equation, together with the truncated fractional series α and the relationship $g - \alpha = O(t^{D/d})$, determine g uniquely, and in particular (arguing as in the proof of the corollary to Lemma 5) that g is real for $t \geq 0$ if and only if all the coefficients of the fractional series α are real.

Dropping all those fractional series α which involve any nonreal coefficient, we can then go on to append any of the remaining α to $\alpha_1, \ldots, \alpha_{m-1}$, thereby obtaining an adequate descriptor for the nonconstant components of the curve $f = [f_1, \ldots, f_{r-1}, g]$, where g is the unique (real) solution of $R_m(t + \zeta, g(t))$ represented by α. Let us agree to compare any two fractional series $\beta(t)$, $\gamma(t)$ with real coefficients by comparing coefficients of like terms lexicographically; this is equivalent to agreeing that $\beta \leq \gamma$ if $\beta(t) \leq \gamma(t)$ for all sufficiently small nonnegative t. Then it is plain that for each y_k the real solutions of $P^*(g_1(t), \ldots, g_{m-1}(t), g(t)) = 0$ are in 1-1 ordered correspondence with the truncated fractional series α which we retain and which satisfy $\alpha(0) = y_k$. The number of such series is therefore the number of distinct roots into which the root y_k of $P(\zeta_1, \ldots, \zeta_{r-1}, y) = 0$ splits as $f'(t)$ moves from the point $f'(0) = [\zeta_1, \ldots, \zeta_{r-1}]$ of b' to immediately neighboring points inside b. The sequences $P_2, \ldots, P_{m-1}, P^*, g_1, \ldots, g_{m-1}, g$, and $\alpha_1, \ldots, \alpha_{m-1}$, α clearly define incidence curves for all pairs of cells having the form b_j, b'_j, such that the latter is part of the boundary of the former.

This finishes our inductive construction of incidence curves, and also shows how the information on root splitting need to determine the mappings $\varrho(b, b')$, defined in the paragraph preceding Lemma 3 of Section 2, can be calculated; so that description of a technique for testing the adjacency of Collins cells of general dimension is now complete.

Remark. Given two Collins cells and Tarski sentences defining each of them, we can easily write a quantified Tarski sentence which is true if and only if they are adjacent. Thus in principle we can obtain an adjacency-testing algorithm using standard decision procedures for Tarski sentences. However, since the number of quantified variables required to define adjacency of Collins cells in this way can be large (compare, e.g., the explicit Tarski sentence defining a Collins cell given at the end of Appendix A), this technique will probably be much less efficient than that described in the present appendix.

CHAPTER 3

On the Piano Movers' Problem: III. Coordinating the Motion of Several Independent Bodies: The Special Case of Circular Bodies Moving Amidst Polygonal Barriers

JACOB T. SCHWARTZ

Courant Institute of Mathematical Sciences
New York University

MICHA SHARIR

Department of Mathematical Sciences
Tel Aviv University

We present an algorithm that solves the following motion-planning problem which arises in robotics: Given several 2-dimensional circular bodies B_1, B_2, . . . , and a region bounded by a collection of 'walls,' either find a continuous motion connecting two given configurations of these bodies during which they avoid collision with the walls and with each other, or else establish that no such motion exists. This paper continues other studies by the authors on motion-planning algorithms for other kinds of moving objects. The algorithms presented are polynomial in the number of walls for each fixed number of moving circles (for two moving circles the algorithm is shown to run in time $O(n^3)$ if n is the number of walls), but with exponents increasing with the number of moving circles.

0. INTRODUCTION

The piano movers' problem (see Reif, 1979; Lozano-Perez & Wesley, 1979; Ignat'yev, Kulakov, & Pokrovskiy, 1973; Udupa, 1977; Schwartz & Sharir, 1983a,b; Hopcroft, Joseph & Whitesides, 1982) is that of finding a continuous motion that will take a given body (or a group of bodies) from a given initial configuration to a desired final configuration, but which is subject to certain geometric constraints during the motion. These constraints forbid the body to come in contact with certain obstacles or 'walls,' and, in the case of a coordinated motion of more than one body, also forbid individual bodies to come into contact with each other. In a preceding paper (Schwartz & Sharir, 1983a), we have analyzed a simplified two

Work on this paper has been supported by ONR Grants N00014-75-C-0571 and N00014-82-K-0381, and by a grant from the U.S.-Israeli Binational Science Foundation.

dimensional version of this problem involving a polygonal body moving amidst polygonal walls. A subsequent paper (Schwartz & Sharir, 1983b) studies the case of an arbitrary number of moving bodies, some of which may be jointed, and shows that this general problem can be solved in time polynomial in the number of smooth surfaces of the walls and the bodies, and in the maximal degree of the equations defining them, but exponential in the number of degrees of freedom of the system of bodies. However, even for a fixed number of degrees of freedom, the algorithm presented in Schwartz and Sharir (1983b) although polynomial, is of complexity $O(n^e)$, where the exponent e can be quite high. Accordingly, this general algorithm is entirely impractical except possibly for the simplest cases. It therefore remains important to develop more efficient specialized algorithms for specific systems of bodies.

In this paper we consider such a special case, namely that in which the coordinated motion of several disjoint, independent circular bodies B_1, B_2, . . . , B_k moving in two dimensions must be planned. (cf. Figure 0.1 for an instance of this problem involving two circular bodies).

This problem can be viewed as a simplified prototype of other more realistic problems involving the coordinated motion of several bodies. In fact, a simple but powerful heuristic for planning motions of more complex rigid bodies is to enclose each of them within a circle and plan a motion of these circles (as suggested, e.g., by Moravec, 1980). Only if no collision-free motion of the enclosing circles exists will more precise algorithms taking into account the exact geometry of the moving bodies be brought to bear. Thus the solution presented in this paper to the coordinated motion-planning problem for circles will facilitate approximate solutions to more complex motion-planning problems for general rigid manipulators.

We will begin our analysis by considering the special case of two circular bodies, then go on to attack the somewhat more complicated case involving three circular bodies, and finally comment on attacking the general case of k bodies using recursive methods.

One might expect the motion-planning problem for two circles to be harder than the corresponding problem for one polygonal body considered in Schwartz and Sharir (1983a). Indeed, the problem considered in Schwartz and Sharir (1983a) involves only three degrees of freedom, whereas the motion of two circular bodies involves four degrees of freedom. It turns out however that we can give a relatively simple solution for the problem of two moving circles, by an algorithm of complexity lower than that of the algorithm presented in Schwartz and Sharir (1983a).

As in Schwartz and Sharir (1983a), our approach is based on a study of the space FP of all collision-free configurations of B_1 and B_2. The problem is to decompose FP into its connected components. In order to construct such a decomposition, we separate the original 4-dimensional problem into two 2-dimensional subproblems by projecting FP onto a two-dimensional space, either point of which corresponds to a fixed position X_1 of the center of B_1 in V. This leaves B_2 free to move over a subspace A of V that can be decomposed into connected components which can be assigned

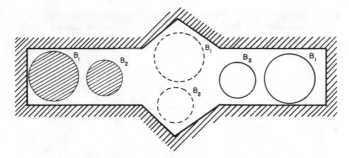

Figure 0.1. An instance of our case of the piano movers' problem. The shaded circles describe the initial configuration of B_1 and B_2, and the unshaded circles describe the desired final configuration; the intermediate dotted positions describe a possible motion of B_1 and B_2 between the initial and final configurations.

standard labels. For most positions of the center of B_1 the set of connected components of A changes only slightly and quantitatively if B_1 is moved slightly, but for certain 'critical' positions X of B_1 this set of components changes qualitatively if B_1 moves in the neighborhood of X. These critical positions lie along curves, called 'critical curves,' that can be characterized as follows: A critical curve is a locus of points X such that if B_1 is placed in V with its center C_1 at X, then a 'critical' contact between B_1, B_2, and the walls will occur at some position of B_2. More specifically, at such a critical contact, either B_2 touches both B_1 and some wall at diametrically opposite points of B_2, or B_2 touches B_1 and two other walls, etc.

The critical curves introduced in this way divide V into 'noncritical' subregions R (see Figure 0.2 for an example of critical curves and noncritical regions). It is then easy to state necessary and sufficient conditions for two configurations, for each of which C_1 lies in the same noncritical subregion R, to be reachable from each other via a continuous coordinated collision-avoiding motion during which C_1 remains within R (see Lemma 1.5). After stating these conditions we go on to study crossings of C_1 from one noncritical subregion R to another R', and show that to analyze these completely we have only to concern ourselves with finitely many possible types of crossings (see Lemmas 1.8 and 1.9). These observations enable us to reduce our original motion-planning problem to a finite combinatorial problem described by a finite 'connectivity graph' CG which characterizes all possible inter-region crossings, and to show that two given configurations of B_1 and B_2 are reachable from one another by a continuous collision-avoiding motion if and only if two associated vertices in the connectivity graph CG are reachable from one another in CG.

For the three-circle case, the manifold FP of free configurations of the system is 6-dimensional, but we can project FP onto the two-dimensional space of positions

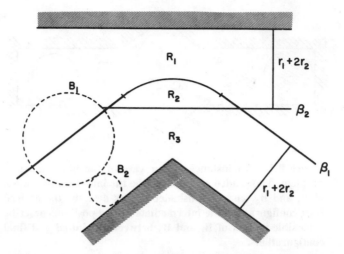

Figure 0.2. Two critical curves β_1 and β_2 dividing the admissible space into three noncritical regions R_1, R_2, R_3.

of the center C_1 of B_1. For each such fixed position X_1, the two remaining circles can move in a space $A(X_1)$ bounded by the original polygonal walls and by the fixed circle B_1. The connected components of the corresponding 4-dimensional space $P(X_1)$ of configurations of B_2 and B_3 can be found by the technique used in the case of two circular bodies, and using our analysis of the two-circle case we can give these components standard labels. Accordingly, we can classify points X_1 as being 'critical' if this labeling of the components of $P(X_1)$ changes discontinuously in the neighborhood of X_1, but as 'noncritical' otherwise. This approach to the three-circle case is recursive, and can be generalized to the case of arbitrarily many independent circular bodies. We sketch this general recursive approach briefly, but do not give details, as the complexity of these details would increase rapidly with the number of moving circles involved.

The present paper is organized as follows. Section 1 analyzes the two-circle problem and reduces it to combinatorial terms. Section 2 contains additional geometric details pertaining to the solution of this problem, and also sketching and analyzing the algorithm corresponding to this solution. Section 3 handles the three-circle problem, and Section 4 discusses the case of arbitrarily many circles.

1. ANALYSIS OF THE TWO-CIRCLE PROBLEM: TOPOLOGICAL AND GEOMETRIC RELATIONSHIPS; THE CONNECTIVITY GRAPH

Let B_1 and B_2 be two circular bodies with centers C_1, C_2, and radii r_1, r_2, respectively. We suppose that $r_1 \geq r_2$, and assume that the region V in which B_1 and B_2

are free to move is a two-dimensional open region with compact closure, bounded by finitely many polygonal walls which can be partitioned into a disjoint collection of simple polygonal closed curves. We label the various straight line segments (walls) constituting the boundary of V as W, W', etc. To avoid various minor technicalities, we assume that the complement V' of V is a two-dimensional region (called the wall region) having the same boundary as V. This excludes cases in which the complement of V contains one-dimensional 'slits' or isolated points. Furthermore, the assumption that the walls separating V' from V fall into a disjoint union of closed polygonal curves excludes cases in which a boundary point of V is an inner point of two distinct boundary curves. Assume that B_2 is placed in V with C_2 at some point X. X is called an *admissible point* if, when C_2 lies at this position, B_2 does not touch or penetrate any wall. Clearly, a point is admissible if its distance from the nearest wall is at least r_2. (A point satisfying the somewhat more stringent condition that it lies at distance at least r_1 from any wall is said to be *admissible for* C_1.) It is plain that the set of admissible points is a finite union of closed connected regions bounded by straight-line segments at distance r_2 from the walls bounding V, and by circular arcs at distance r_2 from convex corners formed by these walls. We call a segment at distance r_2 from a wall a *displaced wall*, and call a circular arc at distance r_2 from a convex wall corner a *displaced corner*. We will use A to designate the set of all admissible points, and A_1 to designate the set of all points admissible for the center C_1 of the larger circle B_1.

Since $r_1 \geq r_2$, the region in which the center of B_1 is free to move (i.e., the set A_1) is a subset of A (having very similar boundaries.) If the center C_1 of B_1 is put at a point X, then the set of positions Y available to C_2 is the set-theoretic difference of A and the circular domain bounded by the circle $\rho(X)$ whose center is X and whose radius is $r_1 + r_2$. Such a pair $[X, Y]$ of points, i.e., a position X of C_1 and a position Y of C_2 in which B_1 and B_2 neither meet each other nor any wall, is called a *free configuration* of the bodies B_1 and B_2. Similarly, a pair $[X, Y]$ consisting of a position X of C_1 and a position Y of C_2 is called a *semi-free configuration* if it is a free configuration, or if, at these positions, either B_1 and B_2 touch (but do not cross) each other, or one of these bodies touches, but does not penetrate into, some wall. The set FP of all free configurations of B_1 and B_2 is plainly an open 4-dimensional manifold, and the set SFP of all semi-free configurations is closed. To simplify our analysis we shall, without seriously restricting the problem, assume that the walls are not arranged in such a manner that either of the circles B_1 and B_2 can touch three points on the walls simultaneously, nor can either circle simultaneously touch two points on the walls at diametrically opposed points, nor can either circle simultaneously touch the walls at two points, one of which is a common endpoint of two walls. These last assumptions imply that no two displaced walls or displaced corners are ever tangent. Hence both A_1 and A are, like V, bounded by a finite collection of simple closed curves. Figure 1.0 exemplifies the concepts just defined.

To analyze the irregularly shaped 4-dimensional manifold SFP, it is convenient to project SFP into a more easily graspable space of fewer dimensions. A natural

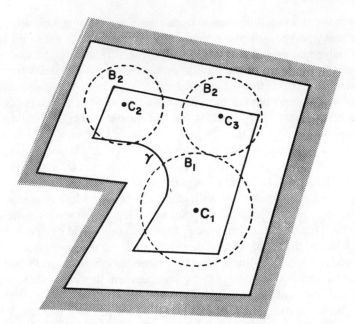

Figure 1.0.   The curve γ consists of a sequence of displaced walls and corners, and encloses the space A of admissible positions. The configuration $[C_1, C_2]$ is free whereas the configuration $[C_1, C_3]$ is semi-free.

choice in this case is to project *SFP* onto the two-dimensional region of admissible positions X for C_1, and then consider the set of positions available to B_2 for each such fixed position of C_1. This leads us to the following initial definition and lemma.

DEFINITION 1.1:   (a) For each wall edge W, let $\gamma(W)$ denote the displaced wall W (i.e., the locus of all points at distance r_2 from W.)

(b) For each convex corner E between two wall edges W_1 and W_2, let $\gamma(W_1W_2)$ denote the displaced corner E (i.e., the circular arc at distance r_2 from E which connects $\gamma(W_1)$ and $\gamma(W_2)$.)

(c) For each fixed position X of C_1, $P(X)$ will designate the set of all positions Y available to C_2. That is, $P(X)$ is $A - \rho^*(X)$, where $\rho^*(X)$ is the solid disc of radius $r_1 + r_2$ centered at X.

The following elementary geometric lemma is obvious from this definition, and from the fact that the boundary of A is a set of simple closed curves, each made up of straight and circular arcs.

LEMMA 1.1:   (a) For each X, the set $P(X)$ is the disjoint union of a finite collection of open connected planar regions, each of which is bounded by a finite number of straight edges and circular arcs. For each point Y on the boundary of such a region, the configuration $[X, Y]$ is semi-free.

(b) For each pair $K \neq K'$ of connected components of $P(X)$, the boundary of K can meet the boundary of K' only at a common corner, which is necessarily a point at which $\rho(X)$ is tangent to some displaced wall or corner.

(c) Every straight edge or circular arc in the boundary of a component K of $P(X)$ is either contained in one of the displaced walls or corners or is an arc of the circle $\rho(X)$. Every corner of K is convex, i.e., the interior angle of K at that corner is less than 180 degrees.

If the exterior boundary E of a component of $P(X)$ does not intersect $\varrho(X)$, then E consists of a sequence of displaced walls and corners, and can be obtained by starting at any segment S_1 on E, following S_1 until its intersection with a subsequent displaced wall or corner S_2, and so forth until we have traced out a simple closed curve. If E intersects $\varrho(X)$, then its intersection with $\varrho(X)$ is a set of arcs of $\varrho(X)$ (some of which may degenerate into points) and E consists of these arcs and of various connected displaced wall/corner portions which do not lie interior to $\varrho(X)$. These displaced wall and corner portions are also straight line segments and circular arcs, and the boundary E can still be obtained by tracing along these segments and arcs, and along arcs of $\varrho(X)$, in the manner just outlined. Note that several separate arcs belonging to a single displaced wall or corner, or to $\rho(X)$, can appear in E. We label the exterior boundary E (and also the component K of $P(X)$ which it bounds) with the circular sequence of displaced walls, corners, and arcs of $\varrho(X)$ to which each boundary segment of E belongs, and arrange this sequence in the order in

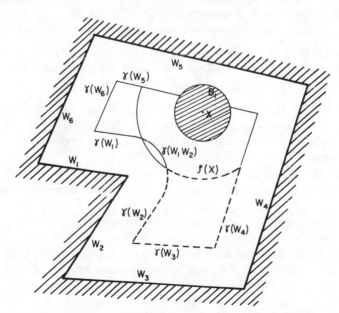

Figure 1.1.   Displaced walls and the circle ρ. The dashed connected component of P(X) has the labeling $[\gamma(W_1 W_2)$, $\gamma(W_2)$, $\gamma(W_3)$, $\gamma(W_4)$, $\rho]$.

which these boundary segments appear on E as E is traversed with K to its right. The appearance in E of an arc of ϱ is indicated simply by including the symbol ϱ in the labeling sequence. An example of this is shown in Figure 1.1.

DEFINITION 1.2: The circular sequence of displaced walls and corners and $\varrho(X)$ containing the portions of the exterior boundary of a component K of $P(X)$ arranged in the order of traversal outlined above is called the *labeling* of K, and is written $\lambda(K)$.

Remark: A connected component K of $P(X)$ need not be simply-connected. In multiply-connected cases it will be convenient in what follows to ignore the interior boundaries of K and to derive the labeling of K from its exterior boundary only.

Suppose that $\varrho(X)$ does not pass through any point at which two displaced walls or corners meet, and moreover that $P(X)$ is not tangent to any displaced wall or corner $\gamma(W)$ (so that, if $\varrho(X)$ intersects a curve $C = \gamma(W)$ or $C = \gamma(W_1W_2)$, then $\varrho(X)$ and C are transversal at their point(s) of intersection.) Then it is clear that if X undergoes a sufficiently small displacement, the displaced walls and corners which $\varrho(X)$ intersects remain exactly the same, also that all points of intersection between $\varrho(X)$ and these displaced walls/corners move only slightly, and finally that the arcs into which these points divide $\varrho(X)$ and these displaced walls/corners change only slightly. (That is, all these geometric objects depend continuously on X in a sufficiently small neighborhood of such a point.) Hence the number of components of $P(X)$, and also the labeling of these components, remains unchanged when such a point X undergoes a small displacement.

The following definition and lemma capture these basic facts and a few others.

DEFINITION 1.3: A point X is called *critical* if $\varrho(X)$ either passes through some point at which two displaced walls or corners meet, or is tangent to a displaced wall or corner at some point. If X is not critical, it is called *noncritical*.

LEMMA 1.2: (a) The set of critical points is closed, and consists of the union of a finite collection of curves of the three following kinds:
(i) Straight-line segments which lie at distance r_1+2r_2 from a wall W (i.e., segments at distance r_1+r_2 from the displaced wall $\gamma(W)$);
(ii) Circular arcs which lie at distance r_1+2r_2 from a convex corner at which two walls W_1,W_2 meet (i.e., arcs at distance r_1+r_2 from the displaced wall corner $\gamma(W_1W_2)$);
(iii) Circular arcs which lie at distance r_1+r_2 from a convex corner at which two displaced walls or corners meet.
(b) Removal of the closed set of critical points decomposes the set A_1 of admissible locations of the center C_1 of B_1 into a finite number of disjoint connected open regions $R_1,R_2, \ldots$, (which we will call the *noncritical regions* of the present case of our movers' problem). As X varies in such a region R, both the number of components of $P(X)$, and the labeling of each of these components, remain invariant, and these components remain at positive distance from each other.

Proof: Part (a) is obvious from the definition of critical points. Part (b) is also trivial, and follows from part (a) and from the preceding discussion. Q.E.D.

The following lemma shows that each connected component of $P(X)$ is defined uniquely by the labeling which we have assigned to it (and even by a small portion of this labeling).

LEMMA 1.3: Let X be a noncritical point admissible for C_1. Let K,K' be two connected components of $P(X)$, and suppose that there exist two boundary segments δ_1,δ_2, both of which are portions of a displaced wall, a displaced corner, or of $\varrho(X)$, and that δ_1,δ_2 appear as consecutive components in both circular sequences $\lambda(K)$, $\lambda(K')$. Then $K = K'$. In particular, if $\lambda(K) = \lambda(K')$ (up to a circular shift) then $K = K'$.

Proof: By definition, the two curves identified by δ_1 and δ_2 must meet at a corner D of K and at a corner D' of K'. By the final statement of Lemma 1.2(b) we must have $D \neq D'$, since K and K' are at positive distance from each other. Since δ_1 and δ_2 are both either straight or circular arcs, they can have at most two points of intersection, which must therefore be D and D' respectively. It follows from the definition of λ that the curve δ_2 follows the curve δ_1 as we trace the boundary of K (resp. K') in the vicinity of D (resp. D') with K (resp. K') remaining to the right. However, since all the corners of both K, K' are convex (cf. Lemma 1.1(c)), this can happen at one of the points D,D' but not at both. This contradiction implies that $K = K'$. Our second assertion follows immediately, since both sequences $\lambda(K)$, $\lambda(K')$ label the exterior boundary of a region bounded by straight lines and circles, which cannot consist of a single circle since $P(X)$ always lies on the convex (outer) side of any circular portion of its boundary. Thus $\lambda(K)$ and $\lambda(K')$ must be sequences of length at least 2. Q.E.D.

DEFINITION 1.4: Let X be a noncritical point admissible for C_1. Define $\sigma(X)$ to be the set

$$\{\lambda(K) : K \text{ a connected component of } P(X)\}$$

Let $T \in \sigma(X)$, and let S be a contiguous subsequence of T containing at least two curve labels. (Two such subsequences will be called *equivalent* at X if they are both subsequences of the same circular label sequence $\lambda(K)$.) We let $\psi(X,S)$ denote the unique connected component K (cf. Lemma 1.3) of $P(X)$ for which $\lambda(K)$ contains S.

The following lemma is an immediate consequence of Lemma 1.3, Definition 1.4, and of the observations made in the paragraph preceding Lemma 1.2.

LEMMA 1.4: Let R be a connected open noncritical region of points admissible for C_1. For all $X \in R$ the sets $\sigma(X)$ are identical, and for each T belonging to such a set $\sigma(X)$, $X \in R$, the function $\psi(X, T)$ is continuous (in the Hausdorff topology of sets) for $X \in R$.

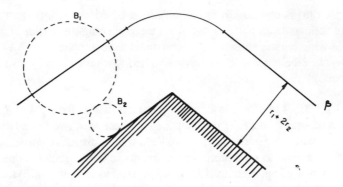

Figure 1.2.   Type I critical curves.

DEFINITION 1.5:   For each noncritical region R we put $\sigma(R) = \sigma(X)$, where X is a point chosen arbitrarily from R.

As already noted (cf. Lemma 1.2), the critical curves of our problem fall into the two following categories.

Type I: For each wall edge W (resp. for each pair W_1, W_2 of adjacent wall edges) the locus of all points at distance $r_1 + 2r_2$ from W (resp. from the corner at which W_1 and W_2 meet) is a type I critical curve (Figure 1.2).

Type II: Let δ_1, δ_2 be a pair of displaced walls or corners which intersect at a point D, and suppose that D is an admissible point (for C_2). Then the circle at radius $r_1 + r_2$ about D is a type II critical curve.

Note that this category of critical curve also includes the case in which δ_1 and δ_2 meet at a common endpoint at 180 degrees (i.e., when δ_1 is a displaced wall and δ_2 is a displaced endpoint of that wall, or vice versa.) The curve β' in Figure 1.3 is an example of such a curve. These two types of critical curves will be analyzed in

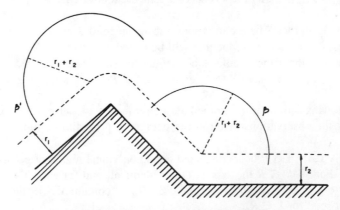

Figure 1.3.   Type II critical curves.

greater detail in the next section, where a way of treating various degenerate cases will also be described.

LEMMA 1.5:   Let R be a connected open noncritical subregion of the set A_1 of all admissible points for C_1. Suppose that X and X' are both points in R. Then one can move continuously through FP from a given free configuration $[X,Z]$ to another such configuration $[X',Z']$, via a motion during which C_1 remains in R, if and only if the connected component $\varkappa(Z,X)$ of $P(X)$ to which Z belongs has a labeling λ $(\varkappa(Z,X))$ equal to the labeling $\lambda(\varkappa(Z',X'))$ of the connected component $\varkappa(Z',X')$ of $P(X')$ to which Z' belongs.

Proof: It is clear from Lemma 1.4 that $\varkappa(U,Y)$ changes continuously as $[Y,U]$ moves continuously through FP with Y remaining in R, so that the label $\lambda(\varkappa(U,Y))$ cannot change during such a motion. This proves the 'only if' part of the present lemma.

For the converse, put $\varLambda = \lambda(\varkappa(Z,X)) = \lambda(\varkappa(Z',X'))$. Take a curve $c(t)$, $0 \leq t \leq 1$, that connects X to X' in R, and consider the mapping

$$f(t) = \psi(c(t),\varLambda), \ 0 \leq t \leq 1,$$

which, by Lemma 1.4, is continuous. Note that the connected component $f(t)$ varies with t only due to the motion of $\varrho(c(t))$. Choose some fixed t_0 in the interval $0 \leq t_0 \leq 1$. Then the boundary of the connected component $K = f(t_0)$ of $P(c(t_0))$ must contain a point U at which two displaced walls or corners meet. Indeed, as observed previously, the boundary of K consists entirely of arcs of $\varrho(c(t))$ and of such displaced walls and corners, all of which are straight line segments and circular arcs, and K lies on the convex (outer) side of any circular arc of its boundary. Thus the boundary of K must include at least two arcs other than $\varrho(c(t_0))$, and hence must have a corner U outside $\varrho(c(t_0))$. By Lemma 1.2, any value $c(t)$ for which $\varrho(c(t))$ passed through U would lie on a critical curve, contradicting the hypotheses of the present lemma. It follows that this can never happen, i.e., U must be a corner of the boundary of all the connected components $f(t)$, $0 \leq t \leq 1$, and during the motion $c(t)$ of C_1 the circle $\varrho(c(t))$ never passes through U. Hence there exists a free position V near U inside $f(t)$ for all $0 \leq t \leq 1$. But then the required motion of the two circles B_1 and B_2 can be constructed as follows:

(a)   Move B_2 inside $\varkappa(Z,X) = f(0)$ from Z to V;
(b)   Move B_1 along the curve $c(t)$ from X to X';
(c)   Move B_2 from V to Z' along a path inside $\varkappa(Z',X') = f(1)$. Q.E.D.

DEFINITION 1.6:   Let R be a connected open noncritical region. Then (a) $C(R)$ is the set of all free configurations $[X,Z]$ such that $X \in R$. (b) For each pair $\xi = [R,\varLambda]$, where $\varLambda \in \sigma(R)$, we define $C(\xi)$ to be the set of all $[X,Z] \in C(R)$ such that Z belongs to the connected component $\psi(X,\varLambda)$.

It is obvious from Lemma 1.5 that the connected components of $C(R)$ are the sets $C(\xi)$ of the form defined in (b).

Next we consider what happens when the center C_1 of B_1 crosses between noncritical regions R_1, R_2 separated by a critical curve. The following simple lemma, taken from Schwartz and Sharir (1983a), rules out extreme cases that would otherwise be troublesome.

LEMMA 1.6: Let $p(t) = [x(t), z(t)]$ be a continuous curve in the open four-dimensional manifold FP of free configurations of B_1 and B_2. Suppose that the end-points $[X, Z]$, $[X', Z']$ of p are specified. Let $\{X_1. .X_n\}$ be any finite collection of points in the 2-dimensional space V not containing either X or X'. Then by moving p slightly we can assume that, during the motion described by p, C_1 never passes through any of the points $X_1. . X_n$.

Proof: The subset of FP for which C_1 lies at one of the points $X_1 . . . X_n$ is a finite union of submanifolds of dimension 2, and these can never disconnect the four-dimensional manifold, FP, even locally. (See Schwartz, 1968.) Q.E.D.

Remark: A similar argument, based on Sard's lemma (see Schwartz, 1968) shows that, by modifying any given free motion very slightly, we can always ensure that the curve $x(t)$ traced out by C_1 during the motion $p(t)$ has a nonvanishing tangent everywhere along its length, and that, given any finite set $\beta_1, \ldots, \beta_n$ of smooth curves in two-dimensional space, we can assume that the tangent to $x(t)$ lies transversal to β_j at any point in which $x(t)$ intersects β_j (see Schwartz, 1968). Moreover, we can assume that the position $z(t)$ of C_2 is constant and $x(t)$ is linear in t for all points along p lying in a sufficiently small neighborhood of each such intersection.

These observations imply that, in order to characterize the connected components of the four dimensional manifold FP, it is sufficient to analyse what happens as $x(t)$ crosses between regions R_1, R_2 along a line L transversal to a critical curve β separating these two regions, such that L does not pass through any point common to two critical curves. Moreover, we can suppose that B_2 maintains a constant position in the neighborhood of each such crossing.

LEMMA 1.7: Suppose that (a portion of) the critical curve β forms part of the boundary of a noncritical region R, and that S is a (subsequence consisting of at least two successive components of an) element of $\sigma(R)$. Put $\xi = [R, S]$. Let $X \in \beta$, and let $Y_n \in R$ and $Y_n \to X$. Then the (region-valued) sequence $\psi(Y_n, S)$ converges (in the Hausdorff topology of sets) to a unique closed set, which we denote by $\phi(X, \xi)$, whose interior is contained in $P(X)$ and is a union of connected components of $P(X)$. If $T \in \sigma(R)$, $T \neq S$, and $\eta = [R, T]$, then int($\phi(X, \xi)$) and $\phi(X, \eta)$ are disjoint for each $X \in \beta$. For each $Z \in V$ the set $\{X \in \beta : Z \in \text{int}(\phi(X, \xi))\}$ is open in β.

Proof: We omit this proof, which is given in full in Schwartz and Sharir (1983c).

LEMMA 1.8: Suppose that (a portion of) a smooth critical curve β separates two connected noncritical regions R_1 and R_2 and that $R_1 \cup R_2 \cup \beta$ is open. Let S_1 (resp. S_2) be a subsequence (containing at least two components) of an element of $\sigma(R_1)$ (resp. $\sigma(R_2)$). Put $\xi_1 = [R_1, S_1]$ and $\xi_2 = [R_2, S_2]$, and let $C_1 = C(\xi_1)$, $C_2 = C(\xi_2)$. Then the following conditions are equivalent:

Condition A: There exists a point $X \varepsilon \beta$ such that the open sets $\mathrm{int}(\phi(X,\xi_1))$, $\mathrm{int}(\phi(X, \xi_2))$ have a non-null intersection.

Condition B: There exists a smooth path $c(t) = [x(t),z(t)] \in FP$ which has the following properties:

(i) $c(0) \in C_1$, $c(1) \in C_2$;

(ii) $x(t) \in R_1+R_2+\beta$ for all $0 \leqslant t \leqslant 1$;

(iii) $x(t)$ crosses β just once, transversally, when $t=t_0$, $0<t_0<1$, and $z(t)$ is constant for t in the vicinity of t_0.

Proof: Suppose first that there exists a path $[x(t),Z]$ in the open 4-dimensional manifold FP of free configurations of B_1 and B_2 satisfying (i) – (iii) of Condition B. (By Lemma 1.6 we can assume without loss of generality that Z is constant throughout this whole path.) Let $K(t)$ denote the open connected component of $P(x(t))$ containing Z. Since $c(0) \in C_1$, it follows from Lemma 1.5 that for $t<t_0$ we have $K(t) = \psi(x(t),S_1)$. Similarly, for $t>t_0$ we have $K(t) = \psi(x(t),S_2)$. Moreover since for $t<t_0$ we have $Z \in \psi(x(t),S_1)$, it follows from Lemma 1.7 that $Z \in \phi(X,\xi_1)$. However, since Z is a free position for C_2, and since the boundary of $\phi(X,\xi_1)$ consists of positions which are semi-free but not free, we must have $Z \in \mathrm{int}(\phi(X, \xi_1))$. Similar reasoning applied to $t>t_0$ shows that Z also lies in $\mathrm{int}(\phi(X,\xi_2))$. Hence these interiors have a non-null intersection, thus establishing Condition A.

Next, suppose that Condition A holds. Let $Z \in P(X)$ be a point in the intersection of the sets $\mathrm{int}(\phi(X,\xi_1))$ and $\mathrm{int}(\phi(X,\xi_2))$. Since $[X,Z] \in FP$ and β is a smooth curve with a nonvanishing tangent (as will be shown below), we can draw a short curve $x(t)$ crossing β at X from R_1 to R_2, such that $x(t)$ satisfies (ii) and (iii) of Condition B, and such that for all t the condition $[x(t),Z] \in FP$ is satisfied. It follows from Definition 1.6 and the remark following it that there exist Λ_1,Λ_2 belonging to $\sigma(R_1),\sigma(R_2)$ respectively such that $[x(t),Z] \in C([R_1,\Lambda_1])$ for $t<t_0$, and $[x(t),Z] \in C([R_2,\Lambda_2])$ for $t>t_0$. By Lemma 1.7 we have $Z \in \phi(X,[R_i,\Lambda_i])$ for $i=1,2$. However, since by Lemma 1.7 the interior of $\phi(X,[R_i,\xi_i])$ *is disjoint from* $\phi(X,[R_i,\xi_i])$ if $\Lambda_i \neq \xi_i$, we must have $\Lambda_i = \xi_i$ for $i = 1,2$. Therefore $c(t) \in C(\xi_1)$ for $t<t_0$ and $c(t) \in C(\xi_2)$ for $t>t_0$, showing that Condition B holds. Q.E.D.

Next we show that if condition A of the preceding Lemma holds for one point lying on a portion β' of β not intersected by any other critical curve, this same condition holds for all points of β'. This fact, closely related to similar assertions derived in Schwartz and Sharir (1983a,b), allows us to derive crossing rules for β' without having to be concerned with the particular point at which we cross.

LEMMA 1.9: Let the smooth critical curve β separate the two noncritical regions R_1 and R_2. Let β' be a connected open segment of β not intersecting any other critical curve, and suppose that $\beta' \cup R_1 \cup R_2$ is open. Let S_1,S_2, ξ_1,ξ_2 be defined as in Lemma 1.8. Then the set of $X \in \beta'$ for which the open sets $\mathrm{int}(\phi(X, \xi_1))$ and $\mathrm{int}(\phi(X,\xi_2))$ have a non-null intersection is either all of β' or is empty.

Proof: Let M be the set of all $Y \in B'$ for which the sets $\mathrm{int}(\phi(Y,\xi_1))$ and $\mathrm{int}(\phi(Y, \xi_2))$ have a point in common. Since M is the union of all sets of the form

$$\{Y \in \beta : Z \in \text{int}(\phi(Y,\xi_1))\} \cap \{Y \in \beta : Z \in \text{int}(\phi(Y,\xi_2))\},$$

for $Z \in V$, and since by Lemma 1.7 each of these sets is open, it follows that M is open. Hence we have only to show that M is also closed. Suppose the contrary; then there exists an $X \in \beta'$ such that $\text{int}(\phi(X,\xi_1))$ and $\text{int}(\phi(X,\xi_2))$ are disjoint but for which there also exists a sequence Y_n of points on β' converging to X such that for all n the sets $\text{int}(\phi(Y_n,\xi_1))$ and $\text{int}(\phi(Y_n,\xi_2))$ intersect each other.

By Lemma 1.7, for each $n \geq 1$ the sets $\text{int}(\phi(Y_n,\xi_j))$, $j=1,2$, are unions of connected components of $P(Y_n)$. Thus, passing to a subsequence if necessary, we may assume that for each $n \geq 1$ both sets $\text{int}(\phi(Y_n,\xi_j))$, $j=1,2$, contain a connected component K_n of $P(Y_n)$ for which $\lambda(K_n)$ is constant. We will prove that these sets K_n converge in the Hausdorff topology of sets to some set $\{D\}$, where D is the intersection of two displaced walls or corners γ_1,γ_2. For this, note that the boundary of all the K_n must contain some fixed corner D at which a certain fixed pair γ_1,γ_2 of displaced walls meet. If the circles $\varrho(Y_n)$ intersect both these displaced walls at a sequence of points converging to D, then it is clear that K_n converges to $\{D\}$ in the Hausdorff metric. Suppose therefore that U is a small circular neighborhood of D of radius δ, not intersecting any fixed displaced wall other than γ_1,γ_2, such that either

(a) $\varrho(Y_n)$ does not intersect either γ_1 or γ_2 within U, or
(b) $\varrho(Y_n)$ intersects one of γ_1,γ_2 (for definiteness, say γ_1) at a sequence of points converging to D, but does not intersect γ_2 within U.

Let p_1 and p_2 be points on γ_1,γ_2 respectively at distance $\delta/2$ from D. In case (a) it is clear that the whole interior of the region bounded by γ_1,γ_2, and by the circle of radius $\delta/2$ about D, lies outside all the circles $\varrho(Y_n)$ and hence belongs to both $\phi(X, \xi_j)$, $j=1,2$. Similarly, in case (b) the whole part of U lying between γ_2 and a circle of the same radius as $\varrho(Y_n)$ tangent to γ_2 at D and containing γ_2 in its exterior, belongs to both $\phi(X,\xi_j)$, $j=1,2$. Hence in both cases we have

$$\text{int}(\phi(X,\xi_1)) \cap \text{int}(\phi(X,\xi_2)) \neq \emptyset,$$

contrary to assumption.

This proves that $K_n \to \{D\}$, and implies that for large n, the circle $\varrho(Y_n)$ must intersect both of the arcs γ_1,γ_2, at a sequence of points converging to D, but not identical to D. But then X must lie on the circular type II critical curve β_0 having D as center and radius r_1+r_2. Since $\varrho(Y_n)$ does not pass through D, Y_n does not lie on β_0. Hence $\beta \neq \beta_0$, so that X lies on two distinct critical curves, contrary to assumption. This proves that M is closed, and then as noted the lemma follows immediately. Q.E.D.

COROLLARY: Let R_1,R_2,β be as in Lemma 1.9, and let $T_i \in \sigma(R_i)$, $i=1,2$. Let $X_1 \in \beta$. Then $\text{int}(\phi(X_1,[R_1,T_1]))$ and $\text{int}(\phi(X_1,[R_2,T_2]))$ have a non-null intersection if and only if there exists an intersecting pair $S = W_1W_2$ of r_2-displaced walls or corners, through which intersection $\varrho_{12}(X_1)$ does not pass, which bound an open angle $\alpha < \pi$ (not containing any other r_2-displaced wall) and having the following property: For any (and, equivalently, for every) $X_1' \in R_j$, $j=1,2$, there exists $\in > 0$ such that all points interior to α and lying within distance $> \varepsilon$ from its apex belong to a component of $P(X_1')$ whose external boundary has the label T_j.

Proof: First suppose that such a pair S exists, and let X_2 lie in α and be near enough to the apex of α to be disjoint from $\varrho_{12}(X_1')$ for all X_1' in a small neighborhood U of X_1. Then it is clear that any sufficiently small neighborhood V of X_2 will be included in $\psi(X_j',[R_j,T_j])$ if $X_j' \in U$ and $X_j' \in R_j$, $j=1,2$. Hence V is a subset of both $\text{int}(\phi(X_1,[R_1,T_1]))$ and $\text{int}(\phi(X_1,[R_2,T_2]))$, proving that these sets have a nonempty intersection.

Conversely, suppose that these sets have a nonempty intersection. Then, by Lemma 1.7, they have a connected component K of $P(X_1)$ in common. Since K lies exterior to each of its bounding circles, some corner of K must lie off $\varrho_{12}(X_1)$. At this corner, two displaced walls or corners W_1 and W_2 must meet and bound an angle $\alpha < \pi$; all points X_2 interior to this angle and close enough to its apex belong to K. Take such a point X_2; then it is clear that for X_1' sufficiently close to X_1, $X_2 \in \psi(X_1',[R_j,T_j])$ if $X_1' \in R_j$, $j=1,2$; hence X_2 remains in $\psi(X_1'',[R_j,T_j])$ as long as X_1'' can be connected to X_1' by a path in R_j for which the circle $\varrho_{12}(X_1'')$ does not pass through X_2. If X_2 lies near enough to the apex of α, which is a point disjoint from $\varrho_{12}(Y_1)$ for all noncritical Y_1, the X_1'' having this property will approximate the whole of R_j. Q.E.D.

As in Schwartz and Sharir (1983a), we can now define a finite graph, called the *connectivity graph* for the case of two independent circular bodies, whose edges describe the way in which the components of the sets $C(R)$ connect as we cross between adjacent noncritical regions R.

DEFINITION 1.7: The connectivity graph CG of an instance of our case of the movers' problem is an undirected graph whose nodes are all pairs of the form $[R,T]$ where R is some connected noncritical region (bounded by critical curves) and where T is an equivalence class of labels which are all subsequences (having length at least 2) of the same element of $\sigma(R)$ (i.e., they all label the same connected component of $P(X)$, for each $X \in R$). The graph CG contains an edge connecting $[R_1,T_1]$ and $[R_2,T_2]$ if and only if the following conditions hold:
(1) R_1 and R_2 are adjacent and meet along a critical curve β.
(2) For some one of the open connected portions β' of β contained in the common boundary of R_1 and R_2 and not intersecting any other critical curve, and for some (and hence every) point $X \in \beta'$ there exist $S_1 \in T_1$, $S_2 \in T_2$ such that the sets $\text{int}(\phi(X,[R_1,S_1]))$ and $\text{int}(\phi(X,[R_2,S_2]))$ have a non-null intersection.

To illustrate this concept, consider the example given in Figure 0.1. Figure 1.4 shows the partitioning of the space A_1 into noncritical regions. The corresponding connectivity graph is shown in Figure 1.5, where each node in CG is labeled by the

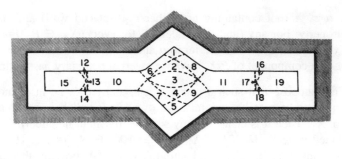

Figure 1.4.   The noncritical regions of the example shown in
Fig. 0.1.

number identifying the corresponding noncritical region R, and by one of the
symbols $L,R,U,$ or D designating respectively a connected component of $P(X)$ (for
$X \in R$) lying on the left, right, upper or lower side of A.

We can now state the main result of this section.

THEOREM 1.1:   There exists a continuous motion c of B_1 and B_2 through the
space FP of free configurations from an initial configuration $[X_1,Y_1]$ to a final
configuration $[X_2,Y_2]$ if and only if the vertices $[R_1,T_1]$ and $[R_2,T_2]$ of the connectivity graph CG introduced above can be connected by a path in CG, where R_1, R_2
are the noncritical regions containing X_1, X_2, respectively, and where T_1 (resp. T_2)
is the marking of the connected component in $P(X_1)$ (resp. $P(X_2)$) containing Y_1
(resp. Y_2).

Remark: We assume here that neither X_1 nor X_2 lies on a critical curve. If either
X_1 or X_2 lies on such a curve, we first move X_1 (or X_2) slightly into a noncritical
region, and then apply the above theorem.

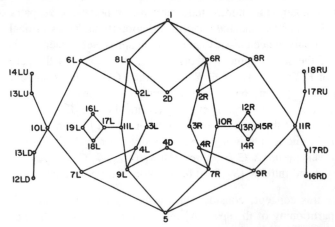

Figure 1.5.   The connectivity graph of the example shown in
Fig. 0.1.

Proof: Suppose that there exists a path connecting $[R_1,T_1]$ to $[R_2,T_2]$ in CG. Let $[R,T]$, $[R',T']$ be two adjacent nodes along that path. Then Lemma 1.5 implies that $R \neq R'$, and Lemma 1.8 implies that there exists a short path in FP connecting points in $C([R,T])$ to points in $C([R',T'])$. Since by Lemma 1.5 any two points in $C([R,T])$ can be connected to each other by a path in FP, one can construct a path in FP connecting $[X_1,Y_1]$ to $[X_2,Y_2]$ by an appropriate concatenation of 'crossing paths' between two domains $C(\xi_1)$, $C(\xi_2)$, and of internal paths within such domains.

Conversely, if there exists a path $p(t) = [x(t),z(t)]$ in FP connecting the two configurations $[X_1,Y_1]$ and $[X_2,Y_2]$, then we can assume, using Lemma 1.6 and its corollary, that this path is such that $x(t)$ crosses critical curves only finitely many times, transversally, avoiding intersections between critical curves, and that $z(t)$ is constant near each such crossing. Lemma 1.8 and the definition of CG then imply that by tracing the domains $C(\xi)$ through which p passes, one obtains a path in CG connecting $[R_1,T_1]$ and $[R_2,T_2]$. Q.E.D.

2. ADDITIONAL GEOMETRIC AND ALGORITHMIC DETAILS

In this section, we will study the critical curves and their associated crossing rules in more detail. As will be shown below, the crossing patterns that can arise are quite similar to those described in Schwartz and Sharir (1983a) for a single polygonal body. Specifically, assuming that no two critical curves coincide, exactly one of the three following crossing patterns can arise as we cross a critical curve β at a point X not lying on any other critical curve (in what follows R_1 and R_2 are the regions lying on the two sides of β near X, and for specificity we assume that $P(X)$ for $X \in R_1$ contains at least as many components as $P(X)$ for $X \in R_2$):

(i)   One component of $P(X)$ may shrink to a point, and then disappear, in which case $\sigma(R_1)$ consists of all labels in $\sigma(R_2)$ plus an extra label marking this component.

(ii) Two connected components of $P(X)$ may join each other at a point as X approaches β, and then merge with each other as β is crossed. In this case, $\sigma(R_1)$ consists of all the labels in $\sigma(R_2)$, plus the labels of the two components that merge as we cross into R_2, less the label of the component into which these two components merge.

(iii) The labeling T_1 of one component of $P(X)$ may change to another labeling T_2 as β is crossed. In this case, $\sigma(R_1)$ and $\sigma(R_2)$ differ by just two components, one of which appears in $\sigma(R_1)$, the other in $\sigma(R_2)$.

The crossing rules then assume the following simple and general form: Connect $[R_1,T]$ to $[R_2,T]$ for each $T \in \sigma(R_1) \cap \sigma(R_2)$, and connect each $[R_1,T_1]$, $T_1 \in \sigma(R_1) - \sigma(R_2)$, to each $[R_2,T_2]$, $T_2 \in \sigma(R_2) - \sigma(R_1)$.

Next we give additional details concerning the structure of the various critical curves described in the section 1, state 'crossing rules' for each type of curve as

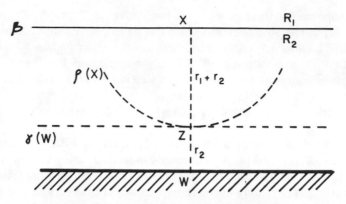

Figure 2.1. Crossing a type I critical curve.

special cases of the general crossing rule just given, and explain how to deal with degenerate cases in which several critical curves become coincident.

As noted earlier, the critical curves fall into the following two categories.

Type I curves: These consist of concatenated sequences of line segments and circular arcs at distance $d = r_1 + 2r_2$ from a convex wall section W. Ignoring the exceptional case (treated below) in which there exist two parallel walls exactly $2d$ apart, we can easily see what happens as we cross such a curve β at a point X not lying on any other critical curve. Specifically (see Figure 2.1), when we cross at X from R_1 to R_2, there appears exactly one new position Z at which the stationary boundary curve $\gamma(W)$ and the moving curve $\varrho(X)$ touch each other. If B_1 is placed with its center at a point $X' \in R_1$ near X, these two curves will not meet, whereas if B_1 is placed with its center at X'' in R_2 near X, these two curves will intersect each other. This implies that there exists a neighborhood N of Z such that $N \cap P(X')$ is connected, whereas $N \cap P(X'')$ is not. Hence when we cross β at X from R_1 to R_2, either a connected component of $P(X')$ splits into two separate components, or an interior boundary curve of some connected component K of $P(X')$ comes to touch another boundary curve (interior or exterior) of the same K. To ease the statement of the crossing rule that applies in this case, we find it convenient to represent the node $[R,T]$ of CG corresponding to a connected component K of $P(X)$ for any $X \in R$ as follows: T is represented as a collection of circular lists, each having the form $[t_1, \ldots, t_m]$, where each t_i is either a wall section or is B_1. Each such list describes a connected (interior or exterior) portion γ^* of the boundary of K, specifically by listing the wall sections which contain the boundary curve segments comprising γ^*, which we arrange in the order in which they appear as γ^* is traversed with K to the right.

Keeping this convention in mind, let $[R_1,T_1]$ be the node designating the component K, where first we suppose that X lies on the R_1-side of β. Then let X cross β from the R_1-side, and suppose that the point Z of contact between two boundaries that appears during the crossing lies on the boundary curve segments CS_1 (which is a portion of $\gamma(W)$) and CS_2 (which is a portion of $\varrho(X)$). Two cases can occur:

(i)   The labels of CS_1 and CS_2 may appear as two components of the same circular list in T_1, say as t_1 and t_i respectively in the list $L = [t_1, \ldots, t_m]$. It is easily checked that in this case K always splits into two subcomponents, and to represent this topological fact combinatorially we split L into two (circular) sublists $L_1 = [t_1, \ldots, t_i]$ and $L_2 = [t_i, \ldots, t_m, t_1]$. If L labeled the exterior boundary of K, then L_1 and L_2 are labelings for the exterior boundaries of the two new components. On the other hand, if L labeled an interior portion of the boundary of K, then one of the new lists, say L_1, labels the exterior boundary of a new component, whereas the exterior boundary of the second component is the same as the exterior boundary of K itself. (Note: some additional geometric analysis, whose details we leave to the reader, will be needed to assign the remaining interior boundary portions to one or another of the two new components.) Overall, we obtain two collections of circular lists T_2, T_2', defining two nodes $[R_2, T_2]$ and $[R_2, T_2']$ belonging to CG, and we connect $[R_1, T_1]$ to both these nodes. As usual, we also connect $[R_1, T]$ to $[R_2, T]$ for all other T appearing in connectivity graph nodes $[R_1, T]$.

(ii)   The labels of CS_1 and CS_2 may appear as components of two different circular lists L_1 and L_2 in T_1. In this case it is easily seen that K is not split across β, and that only its labeling changes due to the merging of two portions of its boundary into one. In this case we simply merge the two lists L_1, L_2 into one circular list L by re-linking CS_1 in L_1 to CS_2 in L_2 and vice versa. Replacing the two circular lists L_1 and L_2 in T_1 by the single list L gives us a new collection T_2 and $[R_2, T_2] \in CG$. We then link $[R_1, T_1]$ to $[R_2, T_2]$, and also connect all other nodes $[R_1, T] \in CG$ to $[R_2, T] \in CG$.

Type II Curves: These require a somewhat different treatment than curves of type I. Recall that a type II curve is a circular arc β of radius $r_1 + r_2$ centered at a corner point Z at which two boundary curves $\gamma(W_1)$, $\gamma(W_2)$ intersect. Here we distinguish between the two following subcases:

(a)   The interior angle at Z (between $\gamma(W_1)$ and $\gamma(W_2)$) is less than $180°$ but greater than $90°$.

The type I critical curves γ_1, γ_2 that touch β partition it into three segments β_1, β_2, β_3 (see Figure 2.2(a)). If we cross the portion β_3 of β at some point X, then the point at which $\varrho(X)$ intersects $\gamma(W_2)$ approaches Z, coincides with Z on β_3, and in R_3' $\varrho(X)$ intersects $\gamma(W_1)$ rather than $\gamma(W_2)$. In this case, the corresponding component of $P(X)$ simply changes its labeling. More specifically, let $[R_3, T] \in CG$ be the node designating K. Then one of the lists in T contains three consecutive labels B_1, W_2 and W_1. By removing W_2 from this list we obtain a new T' such that $[R_3', T']$ $\in CG$ describes K on the R_3'-side of β_3. We thus connect $[R_3, T]$ to $[R_3', T']$ and also connect every other $[R_3, T''] \in CG$ to $[R_3', T''] \in CG$. A completely symmetric situation arises as we cross β_1, with the triple W_2, W_1, B_1 of consecutive list entries on the R_1-side of β_1 replaced by the pair W_2, B_1 on the R_1'-side of β_1.

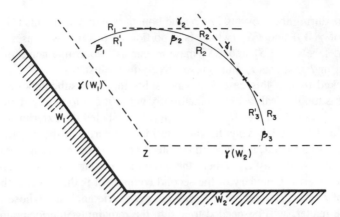

Figure 2.2(a). Crossing a type II critical curve (obtuse case).

However, if we cross the curve portion β_2, the situation is quite different: As the center X of B_1 crosses β_2 from R_2 to R_2', a small connected component K of $P(X)$ about Z shrinks to the single point Z, and then disappears. In this case we simply do not connect the node $[R_2,T]$ in CG to any of the nodes of R_2', but connect every other node $[R_2,T']$ to the corresponding node $[R_2',T']$.

(b)   The interior angle at Z (between $\gamma(W_1)$ and $\gamma(W_2)$) is less than or equal to 90 degrees. This case is really a special case of case (a) just considered. Here only the curve section β_2 appears, and the appropriate crossing rule already seen in (a), i.e., that which applies when a small component about Z shrinks to the point Z and then disappears, applies in this case too.

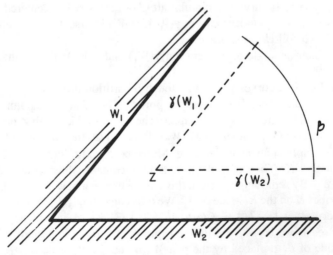

Figure 2.2(b). Crossing a type II critical curve (acute case).

Finally, we need to consider various extreme cases ignored in the preceding discussion. Since all our critical curves are straight segments and circular arcs, they can have nonisolated intersection only if they coincide. Two circles can only coincide if they have the same center and radius, so that no two critical circular arcs can coincide. Hence the only coincidence of critical curves than can arise is that of two type I straight segments overlapping each other. Plainly, this can happen only if there exist two parallel walls at distance $2(r_1+2r_2)$ apart.

The crossing rules applicable in this case are easily derived by imagining such a pair of walls to be shifted by some randomly chosen infinitesimal amount. This splits the corresponding coincident critical line into two lines separated by an infinitesimal distance; and then any crossing of the original line amounts to crossing both these infinitely close split lines. An infinitely thin strip region will then appear between the two lines split in this way from a single critical line. After this splitting, all coincidences of critical curves will have been removed, and then the crossing rules stated above will apply. Then the crossing rule applying in the case of coincident critical curves derive from those of the infinitely displaced case by treating connections across the coincident curves as successive connections across the two infinitely separated critical curves introduced by this 'splitting' procedure.

Schwartz and Sharir (1983c) summarize the techniques described so far in this paper by sketching an algorithm which solves the motion-planning problem for two circular bodies.

If carefully implemented, the complexity of this algorithm is $O(n^3)$. This follows from the fact that the total number of displaced walls and corners and of critical curves is $O(n)$. Since all of these curves are straight or circular arcs, it follows that they can intersect in at most $O(n^2)$ points; consequently there are at most $O(n^2)$ possible noncritical regions, and for each of these regions R the set $\sigma(R)$ can contain at most n labels. The size of CG is therefore $O(n^3)$, and careful implementation of the steps outlined above allows one to construct and search through this graph in total time $O(n^3)$.

An Example:

We conclude our description of the two-circle movers' problem by an example which involves two circles moving through the inside V of a regular pentagon. The sizes of the circles are chosen so that when B_1 is placed nearly touching the midpoint of one edge AB of the pentagon, B_2 can barely fit in the space between B_1 and the two edges CD and DE of the pentagon (see Figure 2.3(a)). The instance of the problem that we wish to solve is to move the circles from the positions shown in Figure 2.3(a) to those shown in Figure 2.3(b).

As is demonstrated by the preceding discussion, this task can be accomplished by the following sequence of motions: (i) Move B_1 from its initial position parallel to AB until it almost touches both AB and BC; (ii) Move B_2 parallel to DE until it almost touches both DE and AE; (iii) Move B_1 parallel to BC until it almost touches both BC and CD; (iv) Move B_2 parallel to AE until it almost touches both AE and AB; (v) Move B_1 parallel to CD until it reaches its target position; (vi) Move B_2

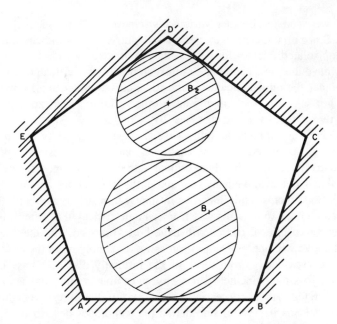

Figure 2.3(a).   An Example of the 2-circle Movers' Problem: Initial Positions.

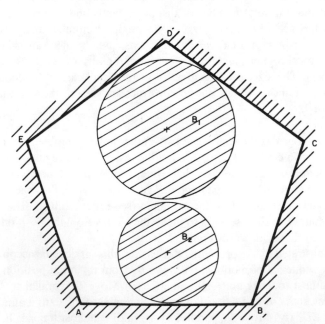

Figure 2.3(b).   Final Positions of the Circles in the Example.

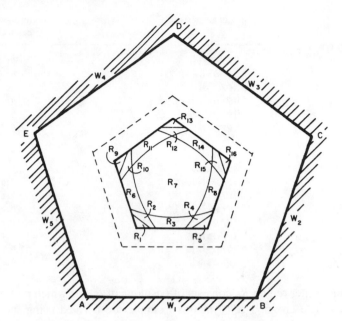

Figure 2.4.   The critical curves and noncritical regions of our example.

parallel to *AB* until it reaches its target position. Furthermore, again as shown by our algorithm, there is no shorter sequence of motions of B_1 and B_2 which accomplishes the required motion.

Remark: This example can be easily generalized to yield examples in which two circles move inside a regular $(2k+1)$-gon. To rotate them around the polygon (never moving two circles simultaneously) will require a sequence of $O(k)$ alternating motions of each of the circles.

To apply our technique to this instance of the two-circle problem, we first displace each side of the pentagon into *V* by the amount r_2, thereby obtaining the boundary arcs of the region *A* of B_2-admissible positions, shown as dashed lines in Figure 2.4. Let us denote the r_2-displacement of the side W_j of the pentagon by γ_j, $j=1, \ldots, 5$. Next we displace the sides of our pentagon into *V* by the amount r_1, to obtain the boundary of the region A_1 of B_1-admissible positions. Within this region we draw the type I critical curves, i.e., the sides of the pentagon displaced by the amount r_1+2r_2 into *V*, and the type II critical curves, i.e., the circular arcs of radius r_1+r_2 about the corners of *A*. These critical curves partition A_1 into 16 noncritical regions. The table below lists the characteristics of each of these noncritical regions.

noncritical region	characteristic	corresponding node(s) in connectivity graph
R_1	$\{[\varrho\gamma_4\gamma_3\gamma_2]\}$	n_1
R_2	$\{[\varrho\gamma_4\gamma_3], [\varrho\gamma_3\gamma_2]\}$	n_2^a, n_2^b
R_3	$\{[\varrho\gamma_4\gamma_3]\}$	n_3
R_4	$\{[\varrho\gamma_5\gamma_4], [\varrho\gamma_4\gamma_3]\}$	n_4^a, n_4^b
R_5	$\{[\varrho\gamma_5\gamma_4\gamma_3]\}$	n_5
R_6	$\{[\varrho\gamma_3\gamma_2]\}$	n_6
R_7	$\{\}$	
R_8	$\{[\varrho\gamma_5\gamma_4]\}$	n_8
R_9	$\{[\varrho\gamma_3\gamma_2\gamma_1]\}$	n_9
R_{10}	$\{[\varrho\gamma_3\gamma_2], [\varrho\gamma_2\gamma_1]\}$	n_{10}^a, n_{10}^b
R_{11}	$\{[\varrho\gamma_2\gamma_1]\}$	n_{11}
R_{12}	$\{[\varrho\gamma_2\gamma_1], [\varrho\gamma_1\gamma_5]\}$	n_{12}^a, n_{12}^b
R_{13}	$\{[\varrho\gamma_2\gamma_1\gamma_5]\}$	n_{13}
R_{14}	$\{[\varrho\gamma_1\gamma_5]\}$	n_{14}
R_{15}	$\{[\varrho\gamma_1\gamma_5], [\varrho\gamma_5\gamma_4]\}$	n_{15}^a, n_{15}^b
R_{16}	$\{[\varrho\gamma_1\gamma_5\gamma_4]\}$	n_{16}

Note that the center region R_7 consists of positions of B_1 for which no free position of B_2 exists. The crossing rules in this example are derived using the principles outlined above, and are depicted in the connectivity graph CG shown in Figure 2.5. It is also easy to check that the initial (resp. final) configuration of the two circles (shown in Figure 2.3) belongs to the cell of FP represented by the node n_3 (resp. n_{13}) of CG. Since CG is connected, there exists a continuous motion of the two

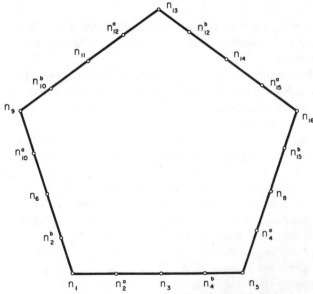

Figure 2.5.   The Connectivity Graph of our Example.

circles between the initial and the final configurations. In fact, CG contains just two paths between n_3 and n_{13}, corresponding to motions of the circles in which the line connecting their centers rotates in a clockwise or counterclockwise direction. The actual motion of the two circles can be reconstructed from these paths in the manner described in Theorem 1.1. One of these motions is easily seen to coincide with the motion described above, and the other is simply a 'mirror image' of this motion.

As another example of the technique presented in this paper, consider the problem instance shown in Figure 0.1. The noncritical regions and the connectivity graph CG of this example have been shown in Figures 1.4 and 1.5, respectively. Since CG is connected in this case, there exists a collision-free motion between any pair of free configurations of the circles. The initial and final configurations shown in Figure 0.1 belong respectively to the cells $15R$ and $19L$ of CG. One possible path connecting these two nodes in CG is

$$15R - 12R - 13R - 10R - 6R - 2D - 8L - 11L - 17L - 16L - 19L$$

We leave it to the reader to transform this path to a continuous wall-avoiding motion of B_1 and B_2 from their initial to their final configuration.

3. COORDINATING THE MOTION OF 3 INDEPENDENT CIRCULAR BODIES

Having now treated the case of 2 independent circles, we go on to study motion-planning algorithms for 3 independent circles. The approach adopted will illustrate a more general recursive technique for successive elimination of degrees of freedom, which can be applied to other motion planning problems.

Denote the three circular bodies with whose motion we are concerned as B_1, B_2, B_3, and suppose that their respective radii are $r_1 \geqslant r_2 \geqslant r_3$. These bodies are constrained to move in a 2-dimensional region V bounded by polygonal walls and obstacles, as described in the preceding sections, and must avoid collision with the walls and with each other. For $i=1,2,3$, let A_i denote the open subset of V consisting of all *admissible positions* for the center C_i of B_i (that is, all positions X whose distance from any of the walls is greater than r_i). We will refer to positions in A_i as *B_i-admissible positions*. We make simplifying assumptions concerning V that are similar to those made for the two-circle case, namely we assume that the boundary of each set A_i (which consists of walls and corners displaced into V by distance r_i) consists of a collection of disjoint simple closed curves.

A triple $[X_1, X_2, X_3]$ of points $X_j \in A_j, j=1, \ldots, 3$, is called a *free configuration* if when each circle B_j is placed with its center C_j at X_j, no circle touches any wall or any other circle. Similarly, a *semi-free configuration* $[X_1, X_2, X_3]$ is a configuration at which zero or more contacts (but no penetrations) between circles and walls or between two circles occur. As before, FP denote the space of all free configurations and SFP the space of all semi-free configurations. Plainly FP is an open 6-dimensional manifold, whereas *SFP* is a 6-dimensional manifold with boundary. As usual, our aim is to partition *FP* into its connected components.

As in the two-circle case, we attack this problem by projecting *FP* into a subspace of fewer dimensions. Specifically, let $P(X_1)$ denote the 4-dimensional subspace consisting of all free positions of the centers of B_2 and B_3, when B_1 is fixed with its center at X_1. Then each such projected set $P(X_1)$ needs to be partitioned into its connected components, and the dependence of these components on X_1 must be studied. We can use the preceding results concerning the coordinated motion of two circles to obtain the desired partitioning of $P(X_1)$. For this, note that each position of B_1 with its center C_1 at X_1 can be viewed as defining an additional barrier for the collision-free motion of B_2 and B_3. Although this new barrier is not polygonal, our analysis of the motion of two circles applies even when some of the barriers are displaced polygonal, rather than simple polygonal, curves. Since B_1 can be viewed as the single point X_1 displaced by distance r_1, this remark applies to the case at hand. Hence, for each fixed value of X_1, decomposition of $P(X_1)$ into connected components can proceed as follows.

(i)   Displace all walls (including B_1) by distance r_3. The subset of *V* lying outside these displaced walls is the set $A_3'(X_1)$ consisting of all B_3-admissible positions in the presence of B_1 with its center at X_1.

(ii)   Displace all walls (including B_1) by distance r_2, to obtain a set $A_2'(X_1)$ having analogous meaning.

(iii)   Draw all B_2-critical curves. These are either walls (or the circular periphery of B_1) displaced by distance r_2+2r_3, or are corners of $A_3'(X_1)$ displaced by distance r_2+r_3. These curves partition $A_2'(X_1)$ into connected open *noncritical regions*. Suppose that B_2 is constrained to move with its center C_2 remaining inside such a region R. Then the space $Q(X_1,X_2)$ of all free positions in $A_3'(X)$ of the center of B_3 decomposes into connected components K whose labelings $\lambda(K)$ (as defined in Definition 1.2) remain invariant throughout R. Furthermore, the component of $Q(X_1,X_2)$ having a given labeling (which, by Lemma 1.3, characterizes this component uniquely) varies continuously with $X_2\ \varepsilon\ R$.

(iv)   Next, we construct the connectivity graph $CG(X_1)$: Its nodes are of the form $[R,\lambda(K)]$, where R is a B_2-noncritical region and where $\lambda(K)$ is a labeling of some connected component of $Q(X_1,X_2)$ for any (hence every) $X_2 \in R$. An edge connects $[R,\Lambda]$ to $[R',\Lambda']$, if R and R' are adjacent regions having a portion β' of some B_2-critical curve as part of their common boundary and if X_2 can cross β' from R to R' in a way allowing a B_3-position X_3 to move continuously along with X_2 from a free position in some component K of $Q(X_1,X_2)$ for which $\lambda(K) = \Lambda$, $X_2 \in R$, to a free position in some component K' of $Q(X_1,X_2)$ with $X_2 \in R'$, $\lambda(K') = \Lambda'$. As shown earlier, with each critical curve segment β' of this sort there is associated a fixed crossing rule which is independent of the particular point on β' at which B_2 crosses from R to R'.

It follows from the preceding results that the number of connected components of the open manifold $P(X_1)$ is the same as the number of connected components of the graph $CG(X_1)$. Moreover, each component C of $CG(X_1)$ defines the following connected component $\mu(X_1,C)$ of $P(X_1)$:

$$\mu(X_1,C) = \{[X_2,X_3] :$$
$$[R,\Lambda] \in C,\ X_2 \in R,\ X_3 \in \psi_{X_1}(X_2,\Lambda)\ \}^- \cap P(X_1)$$

where $\psi_{X_1}(X_2,\Lambda)$ denotes the connected component of $Q(X_1,X_2)$ whose label is Λ. That is, the connected components of the finite graph $CG(X_1)$ can serve as discrete labels for the connected components of $P(X_1)$.

Of course, the noncritical regions R appearing in the preceding discussion depend on the position X_1 of the center of B_1, and hence they cannot be used directly to achieve a discrete labeling of the components of $P(X_1)$. However, since (by arguments analogous to Lemma 1.3) each such R is labeled uniquely by the circular sequence of displaced wall and critical curve sections constituting its boundary, and since (as will be shown below) only finitely many such sequences are possible, the regions R can themselves be given discrete labels. In what follows, we will label each B_2-noncritical region in this manner. Accordingly, given a position X_1 for the center of B_1, we can let $\tau(X_1,L)$ denote the B_2-noncritical region R labeled by L in the above sense.

The next step is to study the way in which $P(X_1)$ and $CG(X_1)$ depend on X_1. Adapting the strategy used in what has gone before, we proceed to define a collection of B_1-*critical curves* which collectively constitute the locus of all points X_1 such that if B_1 is placed with its center at X_1, then some discontinuity in the structure of $CG(X_1)$ can occur even if B_1 is moved only slightly. Such a discontinuity can only result if one of the following combinatorial events occurs.

(i)   The collection of labeled B_2-noncritical regions changes; that is, either one B_2-noncritical region splits into several subregions (or vice versa); or one such region shrinks to a point and then disappears, or the labeling of some non-critical region R changes. The latter situation can arise either when a boundary edge of R splits into subsegments, or when such an edge shrinks to a point and then disappears.

(ii)   The set of labels belonging to the collection of connected components of $Q(X_1,X_2)$ associated with each of the points X_2 belonging to some noncritical region R changes; that is, either one or more of the components of this set splits into subcomponents (or vice versa), or one component shrinks to a point and then disappears, or the labeling of one such component changes, again either because a boundary edge of $Q(X_1,X_2)$ splits into subedges, or because an edge of $Q(X_1,X_2)$ shrinks to a point and then disappears.

(iii)   The structure of the B_2-noncritical regions and of the B_3-connected components associated with them remains unchanged, but the graph $CG(X_1)$ changes due to the appearance or disappearance of one or more edges in it.

If none of the combinatorial changes (i), (ii), (iii) listed above occur, then each B_2-noncritical region having a given labeling varies continuously with X_1, and for X_2 moving continuously within one of those noncritical regions, the connected component of $Q(X_1,X_2)$ having a given labeling varies continuously with X_1 and X_2.

Indeed, suppose that the first assertion is false; then one could obtain a sequence X_{1n} $\rightarrow X_1$, where X_1 is a point at which none of the above combinatorial changes occurs, such that for some labeling L, the regions $\tau(X_{1n},L)$ converge to some set R^* which is different from $\tau(X_1,L)$. However, the boundary of $\tau(X_{1n},L)$ converges to a closed curve which is a concatenation of B_2-critical curves and bounding displaced walls and which encloses a region of B_2-noncritical points, whose label must be L. Thus both R^* and $\tau(X_1,L)$ are labeled L, which contradicts the fact that a given label attaches to just one noncritical region. The second assertion follows by similar arguments.

Note that a discontinuous event of one of these three types can only occur in consequence of the motion of some geometric element which appears in the analysis of the decomposition of $P(X_1)$ into its components and which moves with X_1. It is easy to enumerate all such elements, which are (a) the circle $\varrho_{12}(X_1)$ of radius r_1+r_2 and the circle $\varrho_{13}(X_1)$ of radius r_1+r_3 about X_1. These act as moving 'displaced walls,' which limit the motion of the centers of B_2 and B_3 respectively. (b) the circle $\varrho_{123}(X_1)$ of radius $r_1+r_2+2r_3$ about X_1. This appears as a moving type I critical curve (for the center of B_2) in the analysis of $P(X_1)$ into its components. It is the 'wall' B_1 displaced by the amount r_2+2r_3. (c) all loci of points p obtained by taking the intersection of the circle of radius r_1+r_3 about X_1 with a fixed wall or corner displaced by the amount r_3, and then by displacing such an intersection point by the amount r_2+r_3 in any direction. These are the moving type II critical curves (for the center of B_2) generated by the intersection of the displaced wall B_1 with a displaced fixed wall (cf. Figure 3.1 for a display of all these moving curves).

We call all of these moving curves *curves* (either boundaries or critical curves) *induced by B_1*.

Suppose that the center of B_1 moves slightly in a neighborhood N of some point X_1. Then the curves induced by B_1 and listed above move with it, and the points of their intersection with the other fixed displaced walls and critical curves change. As long as these changes are slight and quantitative, the combinatorial sturcture of the components of $P(X_1)$ will remain constant in N, and consequently X_1 will be a noncritical point. Thus, for X_1 to be critical the pattern of such intersections has to change qualitatively, and plainly this can happen only when either one of elements moving with X_1 becomes tangent to a displaced wall or critical curve, or when three such elements, at least one of which moves with X_1, meet at a point.

To assess the implications of this remark, we will begin by considering cases in which $\varrho_{13}(X_1)$ becomes tangent to some displaced wall or critical curve, or passes through the intersection of two other displaced walls or critical curves. Since $\varrho_{13}(X_1)$ does not count as a B_2-critical curve (see the definition of these curves for the two-circle problem, immediately preceding Lemma 1.5), this can never correspond to a discontinuity of the first type (i) listed above. Moreover, the only tangencies or triple intersections involving $\varrho_{13}(X_1)$ that can cause a discontinuity of type (ii) are those in which $\varrho_{13}(X_1)$ is tangent to a curve or passes through a corner which can form part of the boundary of a component of $Q(X_1,X_2)$, uniformly for all X_2 in some small open set, which is to say, tangent to an r_3-displaced wall or corner,

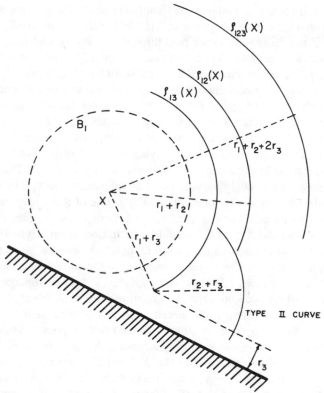

Figure 3.1 **Critical Curves and Displaced Walls Induced by B₁.**

or passes through the intersection of two r_3-displaced walls or corners. The locus of points X_1 at which this happens consists of the following curves:

(1)   walls and corners of V displaced by r_1+2r_3;
(2)   circles of radius r_1+r_3 about intersection points of r_3-displaced walls or corners.

These are our first two types of critical curves.

Next we consider extreme configurations of type (iii); but we will show that they do not exist. For this, suppose that X_1 is a point at which no critical configuration of type (i) or (ii) occurs. Then $\varrho_{13}(X_1)$ does not pass through the intersection E_0 of any two r_3-displaced walls or corners (if it did, some $Q(X_1',X_2)$ would change at $X_1' = X_1$, for every X_2 not within distance r_2+r_3 of E_0), and is not tangent to any r_3-displaced wall or corner (for similar reasons). Hence there exists a neighborhood U of X_1 such that for $X_1' \in U$, $\varrho_{13}(X_1')$ is not tangent to any r_3-displaced wall or corner and does not pass through the intersection of any two such walls. We can also take U small enough so that no type (i) or (ii) critical configuration occurs in U. Suppose that an edge connecting some cell $[L^a, T^a]$ to $[L^b, T^b]$ in $CG(X_1')$ exists for some $X_1' \in U$ but disappears as we pass through X_1. By the Corollary to Lemma 1.9, there exists a corner $E(X_1')$ at which two r_3-displaced walls (one of which may be the circle $\varrho_{13}(X_1')$ meet and bound an angle α such that for $X_2 \in L^a$(resp. $X_2 \in L^b$), the points of α lying near enough to its apex belong to a region whose exterior boundary has the label T^a (resp. T^b). Let X_1'' move continuously from X_1' to X_1 along a curve in U. Then the B_2-noncritical regions labeled L^a, L^b, and their bounding curves, will vary continuously, neither dividing into separate subparts, nor shrinking to points. Take points X_2^a, X_2^b lying in L^a, L^b respectively and varying continuously as those regions and their boundaries vary with X_1''. Then there is a uniquely defined continuously varying point of intersection $E(X_1'')$ of the two r_3-displaced walls which initially intersect at $E(X_1')$ (one of these may be $\varrho_{13}(X_1'')$). None of the continuously varying circles $\varrho_{23}(X_2^a)$ or $\varrho_{23}(X_2^b)$ pass through this point, since if either did X_2^a or X_2^b would by definition lie on a type II B_2-critical curve, rather than in the noncritical region L^a or L^b. Moreover, since no type (ii) critical configuration occurs in U, the boundaries of the components of $Q(X_1'', X_2^a)$ and $Q(X_1'', X_2^b)$ retain fixed labelings as X_1'', X_2^a, X_2^b vary continuously, and the boundary curves of these components vary continuously. It follows that the angle $\alpha = \alpha(X_1'')$ formed by the two r_3-displaced walls intersecting at the continuously varying point $E(X_1'')$ also varies continuously, and that all points in this angle and sufficiently near its apex remain in a component of $Q(X_1'', X_2^a)$ (resp. of $Q(X_1'', X_2^b)$) whose boundary labeling remains fixed as X_1'' and X_2^a (resp. X_2^b) vary continuously. This implies that for $X_1'' = X_1$ the points in the angle $\alpha(X_1)$ sufficiently near its apex belong to a component of $Q(X_1, X_2^a)$ (resp. $Q(X_1, X_2^b)$) with labeling T^a (resp. T^b). Used in its converse direction, the corollary to Lemma 1.9 now shows that an edge of $CG(X_1)$ does connect $[L^a, T^a]$ to $[L^b, T^b]$, contrary to assumption. This proves that configurations of type (iii) are impossible at points X_1 for which configurations of type (i) and (ii) do not occur.

Finally, we consider the more complex case of configurations of type (i). Here we need to consider all possible tangencies and triple intersections involving B_1-induced curves which can influence the structure or the labeling of B_2-noncritical regions. It is helpful to list all such interactions in a systematic table first, and then to give a more detailed description of the B_1-critical curves corresponding to each table entry. (Note, however, that some of the interactions appearing in the table shown can never actually arise because the geometric constraints that they impose are self-contradictory; these cases will be disposed of below.)

Table 3.1 is organized as follows: Each row has at most five entries: a serial number for convenient reference, two or three entries designating the nature of the curves involved in the critical configuration (two entries designate a tangency, whereas three entires designate a triple intersection), and a number referencing a paragraph in the detailed list of critical curves following this table, in which every type of curve is discussed. Each curve involved in a tangency or intersection is

Table 3.1. Critical Interactions of Curves.

Serial No.	1st Curve	2nd Curve	3rd Curve	Critical Curve No.
1	bd (B1)	bd		(3)
2	I (B1)	bd		(4)
3	II (B1)	bd		(20)
4	bd (B1)	I		(4)
5	I (B1)	I		(5)
6	II (B1)	I		(21)
7	bd (B1)	II		(6)
8	I (B1)	II		(7)
9	II (B1)	II		(22)
10	bd (B1)	II (B1)		—
11	I (B1)	II (B1)		—
12	II (B1)	II (B1)		—
13	bd (B1)	bd	bd	(8)
14	bd (B1)	bd	I	(10)
15	bd (B1)	bd	II	(12)
16	bd (B1)	I	I	(14)
17	bd (B1)	I	II	(16)
18	bd (B1)	II	II	(18)
19	I (B1)	bd	bd	(9)
20	I (B1)	bd	I	(11)
21	I (B1)	bd	II	(13)
22	I (B1)	I	I	(15)
23	I (B1)	I	II	(17)
24	I (B1)	II	II	(19)
25	II (B1)	bd	bd	(23)
26	II (B1)	bd	I	(24)
27	II (B1)	bd	II	(26)
28	II (B1)	I	I	(25)
29	II (B1)	I	II	(27)
30	II (B1)	II	II	(28)
31	bd (B1)	II (B1)	bd	(29)
32	bd (B1)	II (B1)	I	(30)
33	bd (B1)	II (B1)	II	(35)
34	I (B1)	II (B1)	bd	(31)
35	I (B1)	II (B1)	I	(32)
36	I (B1)	II (B1)	II	(36)
37	II (B1)	II (B1)	bd	(33)
38	II (B1)	II (B1)	I	(34)
39	II (B1)	II (B1)	II	(37)
40	bd (B1)	II (B1)	II (B1)	(38)
41	I (B1)	II (B1)	II (B1)	—
42	II (B1)	II (B1)	II (B1)	—

represented in the table by a mnemonic symbol which can be either 'bd', designating a boundary, i.e., a displaced wall limiting the motion of B_2, or 'I', designating a type I critical curve for B_2, or 'II', designating a type II critical curve for B_2. This mnemonic symbol always appears either by itself, designating a curve which does not dpeend on B_1, or is followed by '(B1)', designating a B_1-induced curve.

Concerning this table, note the following:

(i)   Since curves of type bd(B1) and type I(B1) always remain at the same distance from each other, no interactions between these curves are possible.

(ii)   Case (10) is impossible, since it would require a circle of radius r_2+r_3 about a point on a circle of radius r_1+r_3 about X_1 to be tangent to a circle of radius r_1+r_2 about X_1.

(iii) Case (11) describes an interaction which always takes place, namely that in which a circle of radius r_2+r_3 about a point on a circle of radius r_1+r_3 about X_1 is tangent to a circle of radius $r_1+r_2+2r_3$. Hence this condition does not generate any B_1-critical curve.

(iv) Case (12) is impossible because two B_1-induced type II B_2-critical curves cannot be tangent to each other at a free or semi-free position, because these curves are circular arcs of the same radius whose centers lie on the circle $\varrho_{13}(X_1)$, so that they can be tangent to each other only at a point interior to that circle, which is not a free position for either center of B_2 or B_3.

(v)   Case (41) is impossible since it would require some point to be at distance r_2+r_3 from two distinct points on the circle of radius r_1+r_3 about X_1, and also to be at distance $r_1+r_2+2r_3$ from X_1, which is plainly impossible.

(v)   Case (42) is impossible since it would require some point to be at the same distance r_2+r_3 from three distinct points on the circle of radius r_1+r_3 about X_1, contradicting the fact that two circles can intersect in at most two points.

Additional geometric details concerning the various possible types of critical interactions are given in Schwartz and Sharir (1983c). Here we will only note a few interesting cases. Note, for example, $\varrho_{12}(X_1)$ and $\varrho_{123}(X_1)$ are respectively the only displaced wall and the only type I critical curve generated by B_1 and affecting the motion of B_2. A first group of B_1-critical curves at which extreme configurations of type (i) arise is obtained by considering situations in which either $\varrho_{12}(X_1)$ or $\varrho_{123}(X_1)$ is tangent to another B_2-boundary or critical curve, or when $\varrho_{12}(X_1)$ or $\varrho_{123}(X_1)$ passes through an intersection of two B_2-boundary or critical curves. The resulting B_1-critical curves include the following:

(3)   (resp. (4)) $\varrho_{12}(X_1)$ (resp. $\varrho_{123}(X_1)$) is tangent to a B_2-boundary curve: The X_1-loci at which this happens are the walls and corners of V displaced by r_1+2r_2. (resp. $r_1+2r_2+2r_3$).

(12) (resp. (13)) $\varrho_{12}(X_1)$ (resp. $\varrho_{123}(X_1)$) passes thru an intersection point of a B_2-boundary curve with a type II B_2-critical curve: The X_1-loci at which this happens are the circles at radius r_1+r_2 (resp. $r_1+r_2+2r_3$) about the intersec-

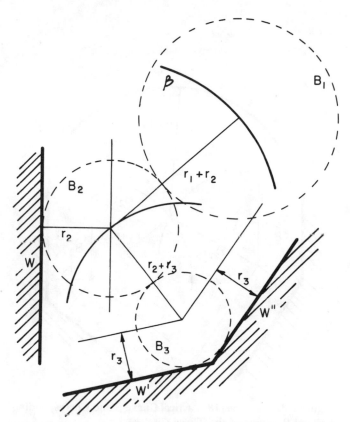

**Figure 3.2.   A Type (12) Critical Curve β, and Correspond-
ing Critical Positions of the Three Circles.**

tions of an r_2-displaced wall or corner W with the circle at radius r_2+r_3 about
an intersection of two r_3-displaced walls and corners W', W'' (cf. Figure 3.2).

(18)   (resp. (19)) $\varrho_{12}(X_1)$ (resp. $\varrho_{123}(X_1)$) passes thru an intersection point of two
type II B_2-critical curves: The X_1-loci at which this happens are the circles at
radius r_1+r_2 (resp. $r_1+r_2+2r_3$) about the intersection points of two circles at
radius r_2+r_3, each about an intersection of two r_3-displaced walls or corners
(cf. Figure 3.3).

Note that all these curves are either displaced walls, displaced corners, or circles
about one or another center.

A second group of B_1-critical curves arise from critical intersections of a B_2-type
II critical curve induced by B_1 with another B_2-boundary or critical curve.

These critical curves can all be seen to lie on circles of radius r_1+r_3 about one or
another center (see Schwartz & Sharir, 1983c).

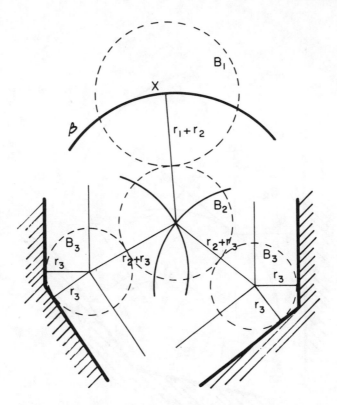

**Figure 3.3. A Type (18) Critical Curve β and Corresponding
Critical Positions of the Three Circles.**

A final group of B_2-critical curves is obtained by considering situations in which two or more B_2-critical curves, both induced by B_1, have a critical contact. As already noted in the remarks immediately following Table 3.1, these critical intersections can only arise from interaction of a B_1-induced type II critical curve with another B_1-induced boundary or critical curve and/or with stationary B_2-boundary or critical curves. Moreover, a B_1-induced type II B_2-critical curve can never be tangent to a B_1-induced B_2-boundary, and is always tangent to a type I B_2-critical curve (see Figure 3.4). Hence these potentially critical tangencies do not generate any B_1-critical point.

The remaining cases include the following B_1-critical curves:

(33) (resp. (34)) Two B_1-induced type II B_2-critical curves and a stationary B_2-boundary (resp. a type I B_2-critical curve) have a common intersection point: Consider a quadrangle $ABCD$, defined so that $|AB| = |AD| = r_1+r_3$, $|BC| = |CD| = r_2+r_3$, and assume it to be hinged at its vertices. Then the curves of type (33) (resp. (34)) are loci of points traversed by the vertex A as the

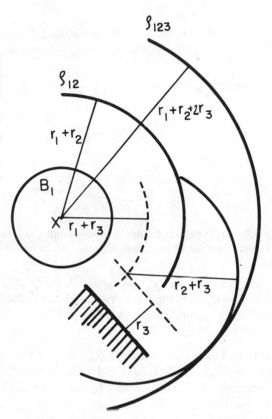

Figure 3.4.   Persistent and Impossible Touches Between B_1-induced B_2-critical Curves

quadrangle $ABCD$ moves in such a way that the vertices B and D glide along r_3-displaced walls or corners and the vertex C glides along an r_2-displaced (resp. (r_2+2r_3)-displaced) wall or corner (cf. Figure 3.5).

All critical curves of classes (29)–(38) are relatively simple algebraic curves. Curves of types (29), (30), (31), (32), (35), (36) and (38) are all 'glissetes' (cf. Lockwood & Prag, 1961) traversed by one vertex of a triangle or rigid quadrangle as its other vertices traverse a straight line or circle. Curves of types (33), (34) and (37) are produced by one vertex of a hinged quadrangle as its other three vertices slide along lines or circles.

We have now listed all the curves in the space of the variable X_1 along which there can occur configurations causing discontinuities in $CG(X_1)$. These curves partition the space A_1 of B_1-admissible positions into finitely many connected regions, which we will call B_1-*noncritical regions*. Our next task is to show that these regions possess properties analogous to those of noncritical regions of other

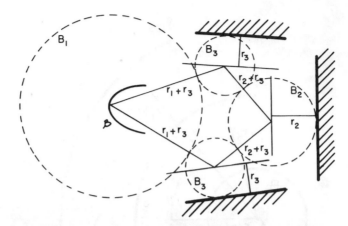

**Figure 3.5. A Type (33) Critical Curve β, and Correspond-
ing Critical Positions of the Three Circles.**

cases of the Movers' problem studied previously. A series of lemmas, given in
Schwartz and Sharir (1983c), accomplishes this, and leads to the following definitions and theorem, which generalize the principal definitions of Section 1, and
theorem 1.1.

DEFINITION 3.1:   Let the critical curve β be part of the boundary of a B_1-
noncritical region R, and let $C \in \sigma(R)$. Put $\zeta = [R,C]$. Then $\nu(X,\zeta)$ is the set of all
limit points of sequences $[Y_n,Z_n]$ such that $[Y_n,Z_n] \in \mu(X_n,C)$, taken over all
sequences $X_n \in R$ such that $X_n \to X$.

DEFINITION 3.2:   The connectivity graph CG_2 of an instance of the three-
circle movers' problem is an undirected graph whose nodes are all pairs of the form
$[R,C]$ where R is some connected B_1-noncritical region (bounded by B_1-critical
curves) and where $C \in s(R)$. The graph CG_2 contains an edge connecting $[R,C]$ and
$[R',C']$ if and only if the following conditions hold:
(1) R and R' are adjacent and meet along a B_1-critical curve β.
(2) For some one of the open connected portions $\beta' \phi'''' - \times o\phi \beta$ contained in the common
boundary of R and R' and not intersecting any other B_1-critical curve, and for some
(and hence every) point $X \in \beta'$ the sets $\text{int}(\nu(X,[R,C]))$ and $\text{int}(\nu(X,[R',C']))$ have a
non-null intersection.

THEOREM 3.1.   There exists a continuous motion c of B_1, B_2 and B_3 through
the space FP of free configurations from an initial configuration $[X,Y,Z]$ to a final
configuration $[X',Y',Z']$ if and only if the vertices $[R,C]$ and $[R',C']$ of the connectivity graph CG_2 introduced above can be connected by a path in CG_2, where R, R'
are the B_1-noncritical regions containing X, X' respectively, and where C (resp. C')
is the label of the connected component in $P(X)$ (resp. $P(X')$) containing $[Y,Z]$ (resp.
$[Y',Z']$).

To conclude this section, we will say a few words about the crossing rules associated with the various kinds of B_1-critical curves β listed above. As in the two-circle case, each of these rules describes a situation falling into one of the following three categories: either

(a) One or more components of $P(X_1)$ split, each into two components, as X_1 crosses β, which then pull away from each other as we move into the region lying on the other side of β. Conversely, various pairs of components may make contact and fuse together as β is crossed. A related possibility is that two otherwise disjoint portions ('lobes') of a single connected component should make contact as β is crossed, or conversely that a portion of one connected component should thin down to a point and then separate, but without connectivity being lost. Each of these latter cases represents a situation in which one component of $P(X_1)$ changes its label as β is crossed. In addition to these structural changes, some other components of $P(X_1)$ can change their labels at such a crossing, simply because some of the B_2-noncritical regions onto which these components project may change their labels, due to the appearance of a new boundary edge, or to the disappearance of an existing boundary edge (see below for details); or

(b) One or more components of $P(X_1)$ shrink, each to a point, and then disappear as X_1 crosses β (or vice versa); as before, some additional components of $P(X_1)$ may change their labels as β is crossed, for reasons similar to those mentioned above; or

(c) The number of connected components of $P(X_1)$ does not change at such a crossing, but some components change their labels as β is crossed.

Earlier, in preparing to describe the 38 kinds of B_1-critical curves, we noted that each critical position X_1 of the center of B_1 is associated either with a change in the collection of B_2-noncritical regions of $P(X_1)$, or (in the case of type (1) and type (2) B_1-critical curves β) with a change in the collection of B_3-components of $Q(X_1,X_2)$ which occurs uniformly for an entire region of positions X_2 of the circle B_2. Moreover, the combinatorial descriptors that have been associated with components of $P(X_1)$ are sets, namely sets of pairs $[L,T]$ comprising a single component of the connectivity graph $CG(X_1)$; here L describes some B_2-noncritical region of the set $A_2'(X_1)$ of all positions admissible for B_2 if the center of B_1 is placed at X_1, and T is an edge sequence which describes a connected component of $Q(X_1,X_2)$ for each X_2 in the region described by L.

Suppose first that β is not of type (1) or type (2), so that when X_1 crosses β the set of B_2-noncritical regions will change. This change occurs because when X_1 lies on β there occurs either a tangency between two B_2-boundary or critical curves, or a triple intersection of three such curves. Moreover, as we cross β at X_1 from one of the B_1-noncritical regions R adjacent to β to the region R' lying on the other side of β, one of the following phenomena will occur: Either

(a) Some B_2-noncritical region splits into two subregions which then pull away from each other (or conversely two such regions meet at a point and then fuse into one another); or

(b) Some B_2-noncritical region shrinks to a point and then either disappears or is replaced by another newly appearing B_2-noncritical region (or conversely some new B_2-noncritical region appears); or

(c) The label of some B_2-noncritical region changes.

It is important to realize that these changes may not always mean that the collection of connected components of $P(X_1)$ will change. Indeed, in order to determine the effect on the structure of $P(X_1)$ of such changes in the structure of B_2-noncritical regions, one first needs to analyze the manner in which the intersection of $P(X_1)$ with $U \times V$ changes, where U is a small neighborhood of the point Y at which the critical tangency or triple intersection of B_2-boundary or critical curves takes place. To see in more detail what this analysis will involve, suppose first that the crossing pattern at X_1 falls into category (a) above, i.e., that a B_2-noncrirical region L splits into two subregions L_1, L_2 locally at Y. This will cause each cell $[L,T]$ $\in CG(X')$, for X' lying on one side of β, to split into two subcells $[L_1,T]$ and $[L_2,T]$ as X' crosses β at X_1, and there will exist no edge linking these two cells directly, since the two noncritical regions L_1, L_2 will not be adjacent. However this does not necessarily imply that these cells have become disconnected from each other in $P(X_1)$, since it may still be the case that one can cross from positions in the subcell described by $[L_1,T]$ to positions in the subcell described by $[L_2,T]$ by passing through other cells, and in particular there may exist a strictly 'local' connection through a cell which projects onto the B_2-noncritical region that has just appeared between L_1 and L_2 and separated them. To find the cases in which this observation applies, one needs to analyze the geometric details of the neighborhood of the point at which L_1 and L_2 have pulled apart. (Note, however, that even if such a local analysis rules out relatively direct, local connections between cells $[L_1,T]$, $[L_2,T]$ of $CG(X')$, these cells may still be connected globally via some longer path in $CG(X')$. If this is the case, the structure of $P(X')$ for X' near X_1 will not change: only the way in which we label components by sets of pairs $[L,T]$ will change.)

Similarly, if the crossing at X_1 is of category (b), it may or may not allow one or more components of $P(X_1)$ to shrink and disappear. Some cases in which component disappearance is impossible will be revealed by local analysis of $CG(X')$, for X' near X_1, near the critical position Y of B_2 at which the disappearing B_2-noncritical region L vanishes. In particular, if such analysis shows that every pair $[L,T]$ is necessarily connected in $CG(X_1)$ to a pair $[L',T]$, where L' is a noncritical region adjacent to L which causes L to disappear by 'swallowing' it, then the set of components of $P(X_1)$ will not change even though L disappears; components will simply be renamed.

Finally, if the crossing at X_1 is of category (c), then the structure of the set of connected components of $P(X)$ will not change, though of course the sets labeling its components will generally change in this case also.

Note that the above considerations imply that when a B_1-critical curve β is crossed, several components of $P(X_1)$ (all of which contain cells $[L,T]$ which project onto the same B_2-noncritical region L) may simultaneously split, each into two subcomponents. Similarly, several components $[L,T]$ may shrink and disappear simultaneously if L shrinks and disappears. Additionally, the tangency or triple intersection of B_2-boundary or critical curves which occurs when X_1 comes to lie on a B_1-critical curve β will generally affect the labeling of all B_2-noncritical regions adjacent to the point Y at which a tangency or triple intersection occurs, usually by the appearance or disappearance of one of the boundary edges of these regions. This will cause changes in the levels of all connected components of $P(X_1)$ which contain cells $[L,T]$, for B_2-noncritical regions L adjacent to the point Y, above and beyond component relabelings which result from the splits or disappearances of some of these components.

These general principles underlie the information summarized in Table 3.2, which classifies the crossing rules that can be associated with each of the 38 different types of B_1-critical curves listed above. The crossing rules are labeled (a), (b) or (c), corresponding to the three possible changes in $P(X_1)$ listed above. As explained above, a renaming change (c) can always occur if a change of type (a) or (b) is possible.

To illustrate the statements concerning crossings represented in this table, we will consider a few representative types of B_1-critical curves, and analyze their associated crossing rules in more detail.

First consider crossing a B_1-critical curve of type (6). Recall that such a curve β consists of points X for which the circle γ_1 of radius r_1+r_2 about X becomes tangent to the circle γ_2 of radius r_2+r_3 about an intersection point D of two r_3-displaced walls or corners. Figure 3.6 shows the structure of B_2-noncritical regions in the neighborhood of the point Y of tangency, for two positions X of the center of B_1, one

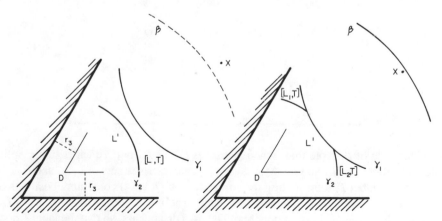

Figure 3.6.   Crossing a Type (6) B_1-critical Curve β; The Structure of B_2-noncritical Regions Near Critical Point of Tangency.

Table 3.2.   Critical Curves and their Crossing Rules.

Type of Critical Curve	Associated Crossing Rules
1	(a) or (c)
2	(b) or (c)
3	(a) or (c)
4	(c)
5	(c)
6	(a) or (c)
7	(c)
8	(b) or (c)
9	(a) or (c)
10	(a) or (c)
11	(c)
12	(b) or (c)
13	(c)
14	(c)
15	(c)
16	(c)
17	(c)
18	(c)
19	(c)
20	(a) or (c)
21	(c)
22	(c)
23	(b) or (c)
24	(c)
25	(c)
26	(c)
27	(c)
28	(c)
29	(b) or (c)
30	(c)
31	(c)
32	(c)
33	(c)
34	(c)
35	(c)
36	(c)
37	(c)
38	(c)

on either side of β. Note that, when X crosses β, the B_2-noncritical region L splits into two subregions L_1 and L_2. Moreover, the characteristic sets of L, L_1 and L_2 will all contain a label T, representing the component of $Q(X,X_2)$ containing positions of the center of B_3 which lie in an angular neighborhood near D, for $X_2 \in L$, L_1 or L_2; however, for $X_2 \in L'$ this component of $Q(X,X_2)$ disappears, so that the label T does not appear in the characteristic of L'. Since the center of B_2 cannot cross γ_1, it

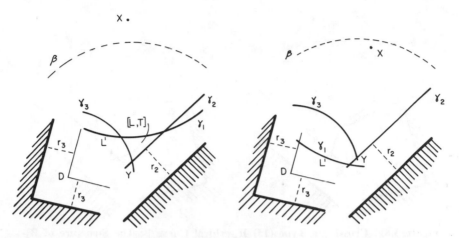

Figure 3.7. Crossing a Type (12) B_1-critical Curve β; The Structure of B_2-non-critical Regions Near Critical Point of Intersection.

follows that, if X lies on the inner side of β, there is no way for B_2 and B_3 to cross from the cell labeled $[L_1,T]$ to the cell labeled $[L_2,T]$, with the center of B_2 remaining near Y. On the other hand, if X lies on the outer side of β, the two cells $[L_1,T]$, $[L_2,T]$ merge into a cell labeled $[L,T]$. It follows that, unless B_2 and B_3 can cross from $[L_1,T]$ to $[L_2,T]$ along a path involving more global motions, e.g., by moving B_2 around B_1, the crossing rule associated with β is of type (a). Of course, as noted above, it is possible that this locally type (a) crossing really is of type (c) because of global connections ignored by the purely local analysis just given.

Next consider the situation which occurs when we cross a B_1-critical curve β of type (12). Recall that such a curve is the locus of points X for which the circle γ_1 of radius r_1+r_2 about X passes through the intersection point Y of an r_2-displaced wall or corner γ_2 and a circle γ_3 of radius r_2+r_3 about an intersection point D of two r_3-displaced walls or corners. Figure 3.7 shows the B_2-noncritical regions in the vicinity of Y, for X lying on the inner side of β, and also for X lying on the outer side of β. Note that, as X crosses β from its outer side to its inner side, a small B_2-noncritical region L shrinks to the point Y and then disappears. Moreover, the characteristic set of L contains a label T corresponding to the connected component of $Q(X,X_2)$, for $X_2 \in L$, which includes all admissible positions for the center of B_3 which lie in a neighborhood of D. Finally, T does not appear in the (sole) B_2-noncritical region L' adjacent to L, since, when B_2 moves into L', the component of $Q(X,X_2)$ labeled T will itself shrink to a point and disappear. It follows that, in this case, the crossing rule at β is of type (b), since the cell of $P(X)$ labeled $[L,T]$ shrinks to a point and disappears as β is crossed, and since in the example $[L,T]$ has no connections to other cells of $P(X)$. However, if the geometry of the walls generating γ_2 and γ_3 were different, so that when B_1 is placed at X and B_2 is placed at Y, B_3 is

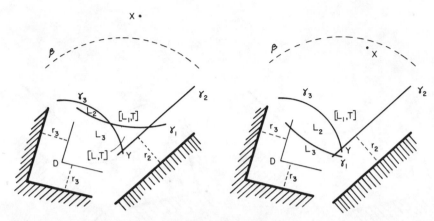

Figure 3.8.   Crossing a Type (13) B_1-critical Curve β; The Structure of B_2-noncritical Regions Near a Critical Point of Intersection.

not 'stuck' at D, but can still move freely near D, then the crossing rule would be of type (c).

Finally, consider the situation which occurs when we cross a B_1-critical curve β of type (13). Such a curve is defined in much the same way as type (12) critical curves, except that the circle γ_1 about $X \in \beta$ now has radius $r_1+r_2+2r_3$. Figure 3.8 shows the B_2-noncritical regions for two positions of X, one on either side of β, in the vicinity of the point Y in which the three curves $\gamma_j, j=1, \ldots, 3$, intersect when $X \in \beta$. Note that, as X crosses β, the small B_2-noncritical region L shrinks to a point and then disappears as in the preceding case, but this time it has a neighboring region L_1 whose characteristic set also contains the label T, and in the graph $CG(X)$ an edge connects $[L,T]$ to $[L_1,T]$ for X lying on the outer side of β. Thus the disappearance of L does not affect the overall structure of $P(X)$, but only causes a change in the sets which label certain of the components of $P(X)$. More specifically, the component of $CG(X)$ which contains the pairs $[L,T]$ and $[L_1,T]$ will no longer contain $[L,T]$ once β has been crossed from outside to inside, and furthermore, the labeling of the B_2-noncritical regions L_1,L_2 and L_3 will change due to the disappearance or addition of a boundary curve in each of them. Therefore the crossing rule applicable to curves of this kind is necessarily of type (c).

The crossing rules which apply to the other 35 types of B_1-critical curves can be derived in much the same way; a summary has been given in Table 3.2 above. We leave it to the reader to work out these rules along the lines of the few examples we have developed in fuller detail.

The complexity of the algorithm for the 3-circle movers' problem that we have described is bounded below by the size of the connectivity graph which it has to search. We can give the following very crude estimate of this size. First note that the total number of B_1-boundary and critical curves is $O(n^5)$ (there can exist $O(n^5)$ critical curves of type (28) in the worst case). Since these are all algebraic curves of some fixed low degree, they will intersect in at most $O(n^{10})$ points, so that the number of B_1-noncritical regions is also $O(n^{10})$. For each such noncritical region R, the number of connected components of $P(X)$ for any $X \in R$ is bounded by the size of the (two-circle) connectivity graph $CG(X)$ which, as shown in Section 2, can have at most $O(n^3)$ nodes. Thus the size of the three-circle connectivity graph is at most $O(n^{13})$, and, using techniques similar to those outlined in Section 2 for the 2-circle problem, one can construct this graph, and then search it in time no more than $O(n^{13})$.

4. THE CASE OF ARBITRARILY MANY CIRCULAR BODIES

The treatment of three moving circles in the preceding section used the solution of the two-circle problem repeatedly, both in order to obtain labels for the connected components of $P(X)$, and to construct a path between two specified configurations when such a path exists. This suggests a recursive approach to the motion-planning problem for an arbitrary number of circular bodies moving amidst polygonal barriers. In this section we will give a brief account of the general recursive approach that we propose, omitting most detail, and using informal arguments mainly. Note, however, that, as can be seen comparing the analysis required in the three-circle case to that which suffices in the two-circle case, the complexity of detail that can be expected to appear in a full treatment by the method to be sketched will increase rapidly with the number of circles.

Let $B_1, \ldots, B_k$ be k circular bodies with centers $C_1, \ldots, C_k$ and radii $r_1 \geq \ldots \geq r_k$ respectively. We assume that these circles are free to move in a polygonal region V, but that none of these circles may touch or penetrate any wall or other circle. Then the space FP of *free configurations* $[X_1, \ldots, X_k]$ of the centers of these circles forms a $2k$-dimensional open manifold, and our problem is to decompose this space into its connected components.

To achieve such a decomposition we can project FP into the 2-dimensional space A_1 of the positions available for the center C_1 of B_1. For each such fixed position X_1, consider the 'fiber' space $P(X_1)$ of all configurations $[X_2, \ldots, X_k]$ of the centers of the remaining circles such that $[X_1, \ldots, X_k] \in FP$. We can decompose $P(X_1)$ into its components by noting that it represents the space of free configurations of the remaining $k-1$ circles confined to move in the space $V(X_1)$ obtained by adding B_1 as an additional barrier in V. Although B_1 is not polygonal, it can be regarded as a displaced point, and the methods for handling $k-1$ circles can be adapted to handle displaced walls of this form. Thus, using the algorithm for $k-1$ circles, we can compute the corresponding connectivity graph $CG(X_1)$, and use each of its connected components to label a corresponding connected component of $P(X_1)$ in a $1-1$ manner.

We then divide the points X_1 into critical and noncritical points, where X_1 is critical if the connected components of $P(X_1)$ change discontinuously as X_1 is moved slightly; otherwise X_1 is noncritical. One can show that the critical points lie on finitely many critical curves (although their number increases exponentially with k), which partition A_1 into finitely many noncritical regions. The next step is to generalize Lemmas 3.1–3.4 to this case. That is, one must first show that there exists a continuous motion between two configurations $[X_1, \ldots, X_k]$ and $[X_1', \ldots, X_k']$, such that X_1, X_1' both belong to some noncritical subregion R, and such that during that motion the first circle moves with its center remaining in R, if and only if the label of the connected component of $P(X_1)$ containing $[X_2, \ldots, X_k]$ and of the component of $P(X_1')$ containing $[X_2', \ldots, X_k']$ are identical. Then one wants to show that for each label C of a connected component $\mu(X,C)$ of $P(X)$ for X in some noncritical region R, the set $\mu(X,C)$ and its interior vary continuously (in the Hausdorff topology of sets) with $X \in R$, and admits a continuous extension to the closure of R. Continuing in analogy with the treatment of the two- and three-circle cases, the next aim is to show that a continuous motion in which the center of B_1 crosses a critical curve β separating between two noncritical regions R, R' can take place if and only if the initial (resp. final) configuration $[X_1, \ldots, X_k]$ (resp. $[X_1', \ldots, X_k']$) are such that the labels C (resp. C') of the connected component of $P(X_1)$ (resp. $P(X_1')$) containing $[X_2, \ldots, X_k]$ (resp. $[X_2', \ldots, X_k']$) have the property that the limits of $\mu(X,C)$ (resp. $\mu(X,C')$) as X approaches the common boundary β from the R-side (resp. the R'-side) have overlapping interiors. Moreover, one wants to show that this condition does not depend on the particular point on β at which B_1 crosses from R to R'.

All these results would enable us to define a finite connectivity graph, in much the same way as was done in the three-circle case, and to reduce the problem to a combinatorial path-searching through that graph; this would yield a recursive solution to the k-circle problem.

CHAPTER 4

On the Piano Movers' Problem: IV. Various Decomposable Two-Dimensional Motion Planning Problems

MICHA SHARIR
ELKA ARIEL-SHEFFI

Department of Mathematical Sciences
University of Tel Aviv

Various special motion planning problems involving arbitrarily many degrees of free-dom are shown to admit relatively simple solutions by techniques based on the connec-tivity graph approach described by Schwartz and Sharir. The solutions exploit the particularly simple configuration space structure of the robot systems considered. A typical problem is that of planning motions for a 2-D robot system consisting of several arms all jointed at one common endpoint and free to rotate past each other. The algorithm given for solving this problem runs in time $O(n^{k+4})$, where k is the number of arms.

0. INTRODUCTION

The piano movers' problem (see Reif, 1979; Lozano-Perez & Wesley, 1979) is that of finding a continuous motion that will take a given body (which may consist of several rigid subparts jointed together) from a given initial position to a desired final position, but which is subject to certain geometric constraints during the motion. These constraints forbid the body to come in contact with certain obstacles or 'walls.' Earlier papers (Schwartz & Sharir, 1983a,c) have described a general technique for solving such problems, which is based on decomposition of the space *FP* of free configurations of the robot system into connected subcells, and an analysis of adjacency relations between these cells. This approach results in the construction of a discrete *connectivity graph* whose nodes represent the cells in the decomposition of *FP,* and whose edges connect nodes corresponding to adjacent cells. Search for a continuous motion between two specified system configurations then reduces to searching for a path in the connectivity graph connecting the two nodes corresponding to the cells containing the initial and final configurations.

FP is a k-dimensional manifold, where k is the number of degrees of freedom of the system. To decompose *FP,* one projects it onto a subspace *A* of lower dimen-

---

This work has been supported in part by ONR Grant N00014-82-K-0381, by a grant from the U.S.-Israeli Binational Science Foundation, and by a grant from the Bat-Sheva Fund.

sion, and then partitions A into connected regions, each such region R having the property that connected components of the 'fiber space' $P(X)$, of points in FP projecting into X, vary continuously as X varies in R, and remain qualitatively constant. It follows that each such region R gives rise to a fixed number of nonadjacent cells of FP, all projecting onto R. To construct the connectivity graph, one must calculate adjacency relationships between cells lying above R and cells lying above R' for every pair of regions, R,R' which are adjacent in A.

The advantage of this technique is that it reduces a k-dimensional problem into decomposition problems involving manifolds of lower dimensions. For systems with $k=3,4$ the technique has been used to obtain motion planning algorithms for a line segment moving in 2-D space amidst polygonal barriers (here $k=3$), and for two independent circular bodies moving in the same environment ($k=4$ for this problem). The main problem one faces in applying this technique to systems with a larger number of degrees of freedom is that normally application of this projection method reduces the problem to similar problem for a smaller, but still large k. In such cases, it is often possible to apply the technique recursively to the reduced problems, repeating this process until k is reduced to 1 or 2; for the low-dimensional problems which remain, decomposition can then be performed in a straightforward manner. However, each step of projection complicates significantly the geometric structure of the manifold to be analyzed, and so recursive application of the technique grows difficult after two or three levels of recursion. The motion-planning algorithm for three independent circular bodies presented in [Schwartz & Sharir, 1983c] which involves two levels of recursive decomposition, illustrates this point.

The point of this paper is to observe that, for certain special robot systems, the decomposition of the configuration space FP can be accomplished using just one projection, even though the number of degrees of freedom involved may be arbitrarily large. This is because the resulting fiber spaces $P(X)$ are decomposable as Cartesian products of many simple subspaces of low dimension. This observation allows us to derive relatively simple motion planning algorithms for these systems (although these algorithms have complexity which grows exponentially with k). In Section 1, we present such an algorithm for the case of a system of k straight segments ('arms'), jointed at one common endpoint, but able to rotate freely about that point, which unimpeded motion of one arm past the other. Motion planning for such a system can be accomplished using a simple generalization of the motion-planning techniques of Schwartz and Sharir (1983a) for a single straight segment (a 'ladder'). In section 2, we comment on various other systems involving arbitrarily many degrees of freedom which are amenable to similarly simple solution techniques.

1.   MOTION PLANNING ALGORITHM FOR A 'SPIDER'

In this section we generalize the 'ladder' case of the Piano-movers' problem discussed in Schwartz and Sharir (1983a) to a *spider B with k arms*. This system consists of k straight segments $PQ_1, PQ_2, \ldots, PQ_k$, all having a common end-

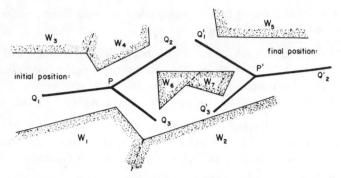

Figure 1.1. An instance of the 'spider' motion planning problem

point P about which each segment can rotate freely independently of the positions of the remaining segments. (In particular, two segments can rotate past one another.) We will refer to the segment PQ_i as the i-th arm of the spider, and denote it simply by its index i. The lengths $d_i = |PQ_i|$, $i=1, \ldots, k$ need not be equal, and in fact are assumed to be distinct from each other.

The region V in which B is free to move is assumed to be bounded by polygonal walls, having the same properties as in Schwartz and Sharir (1983a). Each configuration of B will be denoted as $[X,\theta_1, \ldots ,\theta_k]$, where X is the position of P and where θ_i is the orientation of the i-th arm of B. Figure 1.1 shows an instance of the 'spider' motion-planning problem.

Assume that B is to be placed in V with P at some point X. X is called an *admissible point* if there exists at least one orientation tuple $\theta=[\theta_1, \ldots ,\theta_k]$ such that, at position $[X,\theta]$, B does not touch any wall. We will call a position of B at which it does not touch any wall a *free position*, and a position of B at which it does not cross thru any wall a *semi-free position*. (Thus, a semi-free position is either a free position or a position at which B touches a point of some wall, but does not enter the interior of the wall region.) The set FP of all free positions of B is plainly an open $k+2$-dimensional manifold, and the set SFP of all semi-free positions is closed.

Similarly, for each $i=1, \ldots , k$, we call a position $[X,\theta_i]$ of the i-th arm of B a *free position of i*, if at this position i does not touch any wall. FP_i denotes the 3-D set of all free positions of i. Likewise, $[X,\theta_i]$ is a *semi-free position for i* if at this position i does not cross through any wall, and SFP_i denotes the 3-D set of all semi-free positions for i. Note that FP_i (resp. SFP_i) is the set of free (resp. semi-free) positions of a single straight segment of length d_i, as defined in Schwartz and Sharir (1983a).

To analyze the irregularly shaped $k+2$-dimensional manifold SFP, we project it into the 2-dimensional region in which the point P is free to move, and then consider the set of orientation tuples available to B for each fixed position of P. As in Schwartz and Sharir (1983a), we have

LEMMA 1.1: For each fixed admissible position X of P, and for each $i=1, \ldots ,k$, the set $O_i(X)$ of all orientations representing free positions of i is a finite collection of open angular sectors, whose endpoints represent semi-free orientations of i. (Note that orientations are represented by angles θ_i on the unit circle.)

Lemma 1.1 implies that the set $O(X)$ of all orientation tuples $\theta = [\theta_1, \ldots ,\theta_k]$ for which $[X,\theta] \in FP$, is given by $O(X) = \Pi\, O_i(X)$, and is an open set consisting of a finite collection of connected cells, each of which is a Cartesian product of k angular sectors.

We will use also the following notations. For each $i=1, \ldots ,k$ we denote by $O_i^*(X)$ the set of all semi-free orientations of the i-th arm of B (when P is fixed at X). The Cartesian product $O^*(X) = \Pi\, O_i^*(X)$ is the set of all orientation tuples θ for which $[X,\theta]\ \varepsilon\ SFP$. For each $i=1, \ldots ,k$ we denote by $T_i(X)$ the set $O_i^*(X) - O_i(X)$ of *touches* of the i-th arm of B. Then, plainly, for each orientation tuple $\theta \in O^*(X) - O(X)$ there exists at least one i such that $\theta_i \in T_i(X)$.

The following lemma and corollary follow by easy adaptation of the arguments used to prove similar statements in Schwartz and Sharir (1983a).

LEMMA 1.2: Let X be an admissible point. Then for each $\varepsilon > O$ and each $i=1, \ldots ,k$ there exists a neighborhood N of X such that for all $Y \in N$, each point of $O_i^*(Y)$ lies within a distance ε of some point of $O_i^*(X)$, and conversely each nonisolated point of $O_i^*(X)$ lies within a distance ε of some point of $O_i^*(Y)$.

COROLLARY: If X is an admissible point, then for each compact subinterval K of $O_i(X)$ there exists a neighborhood N of X such that for each $Y \in N$, $O_i(Y)$ contains K.

Each point in the set $T_i(X) = O_i^*(X) - O_i(X)$ introduced above represents an orientation θ_i for which PQ_i touches a wall section W, either at its extremity Q_i or at a point intermediate between P and Q_i. In what follows we shall label each such point θ_i (in $T_i(X)$) with the set $s_i(\theta_i,X)$ of wall sections W touched by the i-th arm of B when this arm is given the orientation θ_i with P at X.

DEFINITION 1.1:   (a) Let $1 \leq i \leq k$, and let $[X,\theta_i] \in SFP_i$. Give PQ_i the position/orientation $[X,\theta_i]$, and extend the segment PQ_i to a line L. Let S be the side of L that PQ_i will enter if turned slightly in the clockwise (resp. counterclockwise) direction. Let W be a wall section, and suppose that when PQ_i is given the position/orientation $[X,\theta_i]$ it touches W. If the intersection of W with the circle of center X and radius equal to the length of PQ_i lies entirely within S, then $[X,\theta_i]$ is called a *clockwise* (resp. *counterclockwise*) stop of i against W.
(b) For each $i=1, \ldots ,k$, the set of pairs

$$\{[s_i(\theta_i,X), s_i(\theta_i',X)] : \theta_i,\ \theta_i'\ \varepsilon\ T_i(X) \mid$$

θ_i is a clockwise stop, θ_i' is a counterclockwise stop, and

the whole interval between θ_i and θ_i' belongs to $O_i(X)$ }

is called the *i-th characteristic* of X and will be written as $\sigma_i(X)$.
(c) The *characteristic* $\sigma(X)$ of X is defined to be the Cartesian product $\Pi \, \sigma_i(X)$.

It is easy to prove the following two lemmas and corollary by adapting the arguments used in Schwartz and Sharir (1983a) to prove corresponding statements for the case of a single ladder.

LEMMA 1.3: Let $1 \leqslant i \leqslant k$. Let $X_n \rightarrow X$ and suppose that a wall section W belongs to all the sets $s_i(\theta_{in}, X)$ for all n and for some $\theta_{in} \in T_i(X_n)$. Then the limit orientation θ_i of any convergent subsequence of the θ_{in} belongs to $T_i(X)$, and $W \in s_i(\theta_i, X)$. Furthermore, if for all n, $[X_n, \theta_{in}]$ is a clockwise (resp. a counterclockwise) stop of i against W, then $[X, \theta_i]$ is also a clockwise (resp. counterclockwise) stop of i against W.

LEMMA 1.4: Let $1 \leqslant i \leqslant k$. (a) If $[X, \theta_i]$ is the clockwise (resp. counterclockwise) endpoint of an interval of $O_i^*(X)$, it is a clockwise (resp. counterclockwise) stop of i against W.
(b) If $[X, \theta_i]$ and $[X, \theta_i']$ are both clockwise (resp. counterclockwise) stops of i against W, then $\theta_i = \theta_i'$.
(c) If $[X, \theta_i]$ is a position/orientation of PQ_i such that $\theta_i \in T_i(X)$ and such that PQ_i touches a convex wall section W, then it is either a clockwise stop or counterclockwise stop of i against W, or both.

COROLLARY: For each $i = 1, \ldots, k$, each wall section W and each admissible point X, there are at most two orientations $\theta_i, \theta_i' \in T_i(X)$ at which PQ_i touches W.

DEFINITION 1.2: Let $1 \leqslant i \leqslant k$. and let X be an admissible point. For each wall section W which labels a clockwise stop (resp. a counterclockwise stop) in $T_i(X)$, we let $\psi_i(X, W)$ (resp. $\psi_i'(X, W)$) denote the unique clockwise (resp. counterclockwise) stop θ_i such that $W \in s_i(\theta_i, X)$.

If $\sigma_i(X)$ is null we will find it technically convenient in what follows to introduce a nominal element Ω of $\sigma_i(X)$ and to define $\psi_i(X, \Omega)$ to be the angle 0, and $\psi_i'(X, \Omega)$ to be the angle 360.

We continue to argue in a manner paralleling Schwartz and Sharir (1983a): The next three lemmas follow by straightforward adaptation of the arguments given in that paper.

LEMMA 1.5: Suppose that P lies at an admissible point X of V which has the following properties:
(*) For each $i = 1, \ldots, k$ there does not exist a semi-free orientation of i (with P at X) such that PQ_i touches the boundary of V' in more than one point.

(**) For each $i=1, \ldots ,k$ there does not exist a semi-free orientation of i (with P at X) in which PQ_i touches a wall perpendicularly at Q, or in which the extremity Q_i of PQ_i touches a corner or end of a wall section.

Let $\theta \in T(X)$. Then for each $i=1, \ldots ,k$ for which the orientation θ_i is a touch, θ_i is not an isolated point of $O_i^*(X)$, and is not the end-point of more than one arc of $O_i^*(X)$. Moreover, $s_i(\theta_i,X)$ contains just one wall section W.

LEMMA 1.6: The set of points satisfying the hypotheses (*) and (**) of Lemma 1.5 is open.

LEMMA 1.7: Let X satisfy the hypotheses of Lemma 1.5. Then there exists an open neighborhood N of X such that for $Y \in N$ we have $\sigma(Y) = \sigma(X)$, and for each $i=1, \ldots ,k$, $O_i^*(Y)$ consists of exactly as many arcs as $O_i^*(X)$. Moreover, if $[W,W'] \in \sigma_i(X) = \sigma_i(Y)$, then $\psi_i(Y,W)$ (resp. $\psi_i'(Y,W')$) depends continuously on Y for $Y \in N$.

Let V be the locus of all points violating one of the conditions (*), (**) of Lemma 1.5. We will see below that V is the union of a finite collection of curves, which we call the *critical curves* of our case of the mover's problem. Removal of these critical curves divides the two dimensional space of all admissible points into a finite collection of connected open regions R, which we call the *noncritical regions* of our problem. The following corollary is an immediate consequence of what has already been shown:

COROLLARY: The set $\sigma(X)$ is constant on each connected subregion R of the set A of admissible positions for which R contains no critical curve.

DEFINITION 1.3: For each such R we put $\sigma(R) = \sigma(X)$, where X is a point chosen arbitrarily from R.

The critical curves of our problem fall into the four following categories.

Type I: For each convex wall section W and for each $i=1, \ldots ,k$, the locus of all points at distance d_i from W;

Type II: For each common endpoint Z of a pair of neighboring wall sections and for each $i=1, \ldots ,k$, the circle of radius d_i about Z;

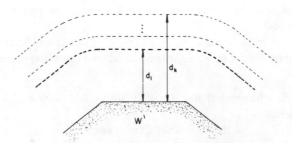

Figure 1.2.   Type I critical curves

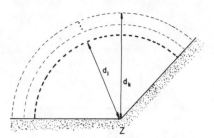

Figure 1.3.   Type II critical curves

Type III: For each wall section W_1 and each corner C of a different wall section W_2, the set of all points traced by P as the i-th arm PQ_i of B moves touching W_1 and C. (See Figs. 1.4(a), 1.4(b))

Type IV: For each wall section W and each $i=1, \ldots ,k$, the set of all points traced by P as PQ_i slides along an edge of W. (See Figure 1.4(c).)

These four types of critical curves coincide with the critical curves arising in the case of a single segment, as described and analyzed in detail in Schwartz and Sharir (1983a). We will refer to a critical curve consisting of positions at which the i-th arm can have a critical wall contact, a *critical curve for i*. Note that there are $O(kn^2)$ critical curves for a k-armed spider.

LEMMA 1.8:   Let R be a connected open noncritical subregion of the set A of all accessible points. Then one can move continuously through FP from a given free position and orientation tuple $[X,\theta]$ to another such $[X',\theta']$ where both $X, X' \in R$, via a motion during which P remains in R, it and only if for each $i=1, \ldots ,k$ the clockwise endpoint $\gamma_i(\theta,X)$ of the arc of $O_i^*(X)$ to which θ_i belongs has a 'marking' $s_i(\gamma_i(\theta_i,X),X)$ equal to the 'marking' $s_i(\gamma_i(\theta',X'),X')$ of the endpoint $\gamma_i(\theta_i',X')$.

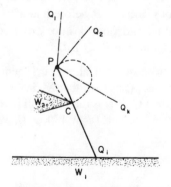

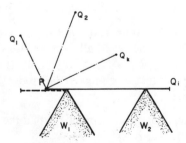

Figure 1.4(a). PQ_i in contact with two walls. Q_i lies along a wall.

Figure 1.4(b). PQ_i in contact with two walls. Q_i does not lie along a wall.

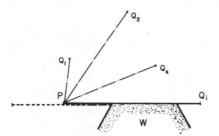

Figure 1.4(c).   PQ$_i$ in contact with two points of the same wall.

Proof: It follows readily from the proof of Lemma 1.7 that, for each $i=1, \ldots ,k$, the endpoint $\gamma_i(\theta_i, Y)$ changes continuously as $[Y,\theta]$ moves continuously through FP with Y remaining in R, and that the marking $s_i(\gamma_i(\theta_i, Y), Y)$ cannot change during such a motion. This proves the 'only if' part of the present lemma.

For the converse, take a curve $c(t)$, $0 \leq t \leq 1$, that connects X to X' in R. Fix $1 \leq i \leq k$, and let W_i be the unique element of $s_i(\gamma_i(\theta_i, X), X) = s_i(\gamma_i(\theta_i', X'), X')$. Let $\gamma_i^{cc} = \gamma_i'(\theta_i, X)$ denote the counterclockwise end of the interval of $O_i{}^*(X)$ containing θ_i, and let W_i' be the unique element of $s_i(\gamma_i^{cc}, X)$. Note that, by Lemma 1.7, W_i' is also the unique element in $s_i(\gamma_i'(\theta_i', X'), X')$. Then, plainly, θ_i and θ_i' belong to the open angular intervals $(\psi_i(X, W_i), \psi_i'(X, W_i'))$ and $(\psi_i(X', W_i), \psi_i'(X', W_i'))$ respectively. By Lemma 1.7, the mapping

$$t \to [c(t), (\psi_1(c(t), W_1) + \psi_1'(c(t), W_1'))/2, \ldots ,$$

$$(\psi_k(c(t), W_k) + \psi_k'(c(t), W_k'))/2 \,]$$

defines a continuous free motion of B through R. Since for each $i=1, \ldots ,k$ we have $\theta_i \in (\psi_i(X, W_i), \psi_i'(X, W_i'))$ and $\theta_i' \in (\psi_i(X', W_i), \psi_i'(X', W_i'))$, it is clear that, by performing approriate rotations of the various arms of B before and after this motion, we can pass between the specified initial and final positions of B via a continuous motion of B through FP with P remaining in R. Q.E.D.

DEFINITION 1.4:   Let R be a connected open noncritical region. Then
(a) $C(R)$ is the set of all free positions/orientations $[X,\theta]$ such that $X \in R$.
(b) If $\xi = [[W_1, W_1'], \ldots ,[W_k, W_k']] \in \sigma(R)$, then $C(\xi, R)$ is the set of all $[X,\theta] \in C(R)$ such that θ belongs to the product of the open intervals $(\psi_i(X, W_i), \psi_i'(X, W_i'))$, $i=1, \ldots ,k$.

It is obvious from Lemma 1.8 that the connected components of $C(R)$ are the sets $C(\xi, R)$, $\xi \in \sigma(R)$.

Next we consider what happens when B crosses between regions R_1, R_2 separated by a critical curve. The following simple lemma generalizes a similar lemma given in Schwartz and Sharir (1983a), and rules out extreme cases that would otherwise be troublesome.

LEMMA 1.9:   Let $p(t)=[x(t),\theta(t)]$ be a continuous curve in the open $(k+2)$-dimensional manifold FP of free positions of B. Suppose that the end-points $[X,\theta]$, $[X',\theta']$ of p are specified. Let $\{X_1 \ldots X_n\}$ be any finite collection of points in the 2-dimensional space V not containing either X or X'. Then by moving p slightly we can assume that, during the motion described by p, P never passes through any of the points $X_1 \ldots X_n$.

Proof: See the proof of Lemma 1.9 of Schwartz and Sharir (1983a).

Remark: As in Schwartz and Sharir (1983a), a similar argument, based on Sard's lemma, shows that, by modifying any given free motion very slightly, we can always ensure that the curve $x(t)$ traced out by P during the motion p has a non-vanishing tangent everywhere along its length, and that, given any finite set β_1, $\ldots ,\beta_n$ of smooth curves in two-dimensional space, we can assume that the tangent to $x(t)$ lies transversal to β_j at any point in which $x(t)$ intersects β_j (see Schwartz, 1969). Moreover, we can assume that $\theta(t)$ is constant and $x(t)$ is linear in t for all points along p lying in a sufficiently small neighborhood of each such intersection.

It follows from Schwartz and Sharir (1983a) that the critical curves of the case of the mover's problem considered here are always smooth, and that two critical curves can have only finitely many intersections. Thus, to characterise the connected components of the $(k+2)$ dimensional manifold FP, it is sufficient to analyse what happens as $P = x(t)$ crosses between regions R_1,R_2 along a line L transversal to a critical curve β separating these two regions, such that L does not pass through any point common to two critical curves. Moreover, we can suppose that each θ_i maintains a constant orientation in the neighborhood of each such crossing.

LEMMA 1.10:   Suppose that (a portion of) the critical curve β forms part of the boundary of a noncritical region R, and that $\xi = [[W_1,W_1'], \ldots ,[W_k,W_k']] \in \sigma(R)$. Let $X \in \beta$, and let $Y_n \in R$, and $Y_n \to X$. Then for each $i=1, \ldots ,k$ the sequence $\psi_i(Y_n,W_i)$ (resp. $\psi_i'(Y_n,W_i')$) converges to $\psi_i(X,W_i)$ (resp. $\psi_i'(X,W_i')$). Moreover, the function $\psi_i(X,W_i)$ (resp. $\psi_i'(X,W_i')$) depends continuously on X for $X \in \beta$. Finally, the entire (counterclockwise) open angular sector $(\psi_i(X,W_i), \psi_i'(X,W_i'))$ belongs to $O_i^*(X)$.

Proof: Similar to the proof of Lemma 1.10 of Schwartz and Sharir (1983).

LEMMA 1.11:   Suppose that (a portion of) a smooth critical curve β separates two connected noncritical regions R_1,R_2 and that $R_1 \cup R_2 \cup \beta$ is open. Let $\xi_1 = [[W_{11},W_{11}'], \ldots , [W_{1k},W_{1k}']] \in \sigma(R_1)$ (resp. $\xi_2 = [[W_{21},W_{21}'], \ldots , [W_{2k},W_{2k}']] \in \sigma(R_2)$), and let $C_1 = C(\xi_1,R_1)$, and $C_2 = C(\xi_2,R_2)$. Then the following two sets of conditions are equivalent:

Condition A: There exists a point $X \in \beta$ such that for each $i=1, \ldots ,k$ the open angular intervals $(\psi_i(X,W_{1i}),\psi_i'(X,W_{1i}'))$ and $(\psi_i(X,W_{2i}), \psi_i'(X,W_{2i}'))$ are subsets of $O_i^*(X)$ and have a non-null intersection.

Condition B: There exists a smooth path $c(t) = [x(t),\theta(t)] \in FP$ which has the three following properties:

(i) $c(0) \in C_1$, $c(1) \in C_2$;

(ii) $x(t) \in R_1 \cup R_2 \cup \beta$ for all $0 \leqslant t \leqslant 1$;

(iii) $x(t)$ crosses β just once, transversally, when $t=t_0$, $0<t_0<1$, and $\theta(t)$ is constant for t in the vicinity of t_0.

Note that condition B amounts to saying that C_1 and C_2 lie in the same arcwise connected (and hence connected) component of FP (see also the remarks made in Definition 1.4).

   Proof: The proof of Lemma 1.11 of Schwartz and Sharir (1983a) applies here as well.

   Next we show that, if β is an (open) critical curve section not intersected by any other critical curve, and if condition A of Lemma 1.11 holds for one point X along β, then it holds for all X along β. This makes it easy to calculate the relationships of connectivity in which we are interested by applying Lemma 1.11 to an arbitrarily selected point of β.

   LEMMA 1.12:   Let the smooth critical curve β separate the two noncritical regions R_1, R_2. Let β' be a connected open segment of β not intersecting any other critical curve, and suppose that $\beta' \cup R_1 \cup R_2$ is open. Let $\xi_1 = [[W_{11}, W_{11}'], \ldots, [W_{1k}, W_{1k}']] \in \sigma(R_1)$ (resp. $\xi_2 = [[W_{21}, W_{21}'], \ldots, [W_{2k}, W_{2k}']] \in \sigma(R_2)$). Then the set of $X \in \beta'$ for which for each $i=1, \ldots, k$ the open clockwise angular sectors $(\psi_i(X, W_{1i}), \psi_i'(X, W_{1i}))$ and $(\psi_i(X, W_{2i}), \psi_i'(X, W_{2i}'))$ are subsets of $O_i^*(X)$ and overlap is either all of β' or is empty.

   Proof: The proof of Lemma 1.12 of Schwartz and Sharir (1983a) can be applied here.

   Remark: If one assumes that no two critical curves coincide with one another (which assumption is discussed below), then the statements of the three preceding lemmas can be simplified. Indeed, suppose that the curve section β appearing in these lemmas is a critical curve for the i-th arm. Then as β is crossed the characteristics $\sigma_j(X)$ remain unchanged, and each connected sector of $O_i^*(X)$ varies continuously, for all $j \neq i$. Discontinuous changes can occur only for the configuration space features connected with the i-th arm. This remark makes the connection between the three preceding lemmas and their counterparts in the case of a single ladder even more obvious.

   As in Schwartz and Sharir (1983a), the chain of lemmas described so far enables us to reduce the case of the movers' problem considered here to a finite combinatorial search:

   DEFINITION 1.5:   The *connectivity graph CG* of an instance of our case of the movers' problem is an undirected graph whose nodes are all the pairs $[R, \xi]$, where R is a connected open noncritical subregion of V and where $\xi \in \sigma(R)$. An edge connects two nodes $[R_1, \xi_1]$ and $[R_2, \xi_2]$ in CG if and only if the following conditions hold:

(1) R_1 and R_2 are adjacent and meet along a critical curve β.

(2) There exists a (maximal) open portion of β' of β contained in the common

boundary of R_1 and R_2 and not intersecting any other critical curve, such that for some (hence every) point X on β' the open angular sectors $(\psi_i(X,W_{1i}),\psi_i'(X,W_{1i}'))$ and $(\psi_i(X,W_{2i}),\psi_i'(X,W_{2i}'))$ overlap for each $i=1, \ldots ,k$, where $\xi_1 = [[W_{11},W_{11}'], \ldots , [W_{1k},W_{1k}']]$ and where $\xi_2 = [[W_{21},W_{21}'], \ldots , [W_{2k},W_{2k}']]$.

We are now in position to state our main theorem:

THEOREM 1.1:   There exists a motion c of B through the space FP of free positions from an initial $[X,\theta]$ to a final $[X',\theta']$ if and only if the vertices $[R,S]$ and $[R',S']$ of the connectivity graph CG introduced above can be connected by a path in CG, where R (resp. R') is the noncritical region containing X (resp. X'), and where

$$S = [\ [s_1(\gamma_1(\theta_1,X),X) \ , \ s_1(\gamma_1'(\theta_1,X),X)], \ldots ,$$

$$[s_k(\gamma_k(\theta_k,X),X) \ , \ s_k(\gamma_k'(\theta_k,X),X)] \ ],$$

and

$$S' = [\ [s_1(\gamma_1(\theta_1',X'),X') \ , \ s_1(\gamma_1'(\theta_1',X'),X')], \ldots ,$$

$$[s_k(\gamma_k(\theta_k',X'),X'), \ s_k(\gamma_k'(\theta_k',X'),X')] \ ],$$

Remark: For this theorem to apply, X and X' should not lie on a critical curve. Thus, if X lies on a critical curve, we first move B a little so as to change X to a point inside some noncritical region, and then apply the theorem as stated above.

Proof: Same as the proof of Theorem 1.1 of Schwartz and Sharir (1983a).

If we assume that no two critical curves overlap, it follows that, if β is a critical curve generated by the i-th arm of B, then, as we cross β from a region R_1 to a region R_2, precisely one interval (or a pair of related intervals) of $O_i(X)$ or its 'marking' changes discontinuously as X crosses β, in one of the three following ways:

(i)   One interval of $O_i(X)$ may shrink to a point and then disappear (or vice-versa one new interval may appear).

(ii)   As we leave R_1 a new stop of i may appear within some interval I of $O_i(X)$, dividing I into two parts which then pull apart as we move into R_2.

(iii) As we cross β the 'marking' S' of one endpoint of some interval may change to S''.

Furthermore, for each $j \neq i$, the structure and labeling of the components of $O_j(X)$ remain unchanged as β is crossed. It follows that the structure of $O(X)$ itself changes in a way very simply related to the manner in which $O_i(X)$ changes.

Suppose for specificity that $\sigma(R_1)$ always contains as many points as $\sigma(R_2)$. Then, in case (i), $\sigma(R_1)$ contains $\sigma(R_2)$, and $\sigma(R_1) - \sigma(R_2)$ contains several elements, all having the same i-th component τ. In this case, we connect each node $[R_2,S]$, $S \in \sigma(R_2)$ such that $S_i \neq \tau$, of the connectivity graph CG to $[R_1,S]$, but leave each node $[R_1,S]$ for which $S_i = \tau$, unconnected to any node $[R_2,S]$.

In case (ii), all the elements of $\sigma(R_2) - \sigma(R_1)$ have the same i-th component $\tau =$

$[W_1,W_2]$, and all the elements of $\sigma(R_1) - \sigma(R_2)$ have as their i-th component one of the pairs $\tau_1 = [W_1,W]$, $\tau_2 = [W,W_2]$. Here we connect each node $[R_1,S]$ to $[R_2,S]$ for those S satisfying $S_i \neq \tau,\tau_1,\tau_2$, but connect each node $[R_2,S]$ to both $[R_1,S']$ and $[R_1,S'']$, where $S_i=\tau$, $S_i'=\tau_1$, $S_i''=\tau_2$, and where $S_j=S_j'=S_j''$ for each $j \neq i$.

In case (iii), all elements in $\sigma(R_1) - \sigma(R_2)$ have the same i-th component τ, and all elements of $\sigma(R_2) - \sigma(R_1)$ have the same i-th component τ'. In this case, we connect $[R_1,S]$ to $[R_2,S]$ for each S satisfying $S_i \neq \tau,\tau'$, but connect $[R_1,S]$ to $[R_2,S']$, for each S,S' such that $S_i=\tau$, $S_i'=\tau'$, and $S_j=S_j'$ for each $j \neq i$.

If two or more critical curves coincide, we can separate between them by an infinitesimal displacement of the walls or arm-lengths defining them. This creates an infinitesimally narrow strip between the displaced curves. Then, by regarding this strip as a noncritical region, we can obtain the crossing rules of the originally coinciding curves as combinations of the crossing rules applying to each of the individual curves. This technique, described in detail in Schwartz and Sharir (1983a), can be easily adapted to the present situation.

Sketch and Analysis of the Motion-Planning Algorithm for a Spider
The preceding analysis can be easily developed into a motion-planning algorithm for a spider. The algorithm is very similar to the algorithm developed for the case of a ladder in Schwartz and Sharir (1983a). Indeed, partitioning of the space V into noncritical regions can proceed exactly as in the case of a ladder, with the sole difference that the case of a spider involves more critical curves (although these have the same types as in Schwartz and Sharir (1983a). Note that the set of critical curves for the present 'spider' problem is simply the union of the sets of critical curves for the motion of each of the spider's arms, and that the crossing rule applicable to each of these critical curves relates in the simple combinatorial manner described above to the crossing rules governing the single-arm critical curves.)

After using the critical curves to partition the space V, we go on to construct the connectivity graph. For each noncritical region R and for each $i=1, \ldots ,k$, we can compute $\sigma_i(R)$ by picking an arbitrary point $X \in R$ and by computing $\sigma_i(X)$, applying the technique of Schwartz and Sharir (1983a) to the i-th arm of B. The characteristic $\sigma(R)$ is obtained as the Cartesian product of all the characteristics $\sigma_i(X)$. We can then build the connectivity graph CG, using the crossing rules stated above in a straightforward manner, and finally search through this graph to find a path connecting the two nodes corresponding to the initial and final given configurations $[X,\theta_1, \ldots ,\theta_k]$ and $[X',\theta_1', \ldots ,\theta_k']$ of the spider. (To find these nodes, we first find the connected noncritical regions R, R' containing X, X' respectively, and then find, for each i, the labels of the endpoints of the angular sectors containing θ_i, θ_i', respectively. These labels and regions define the required nodes of CG.)

To analyze the complexity of the algorithm, we have only to adapt Lemma 3.1 of Schwartz and Sharir (1983a), obtaining

LEMMA 1.13: The total number of critical curve sections and noncritical regions is $O(k^2 n^4)$. Also, for each admissible point X, the total number of components of $O(X)$ (that is, the size of $\sigma(X)$) is $O(n^k)$.

Proof: The first statement follows as in Lemma 3.1 of Schwartz and Sharir (1983a) because there are $O(kn^2)$ critical curves. The second statement follows because $O(X)$ is a Cartesian product of k subspaces, each of which can have at most $O(n)$ components, again by Lemma 3.1 of Schwartz and Sharir (1983a). Q.E.D.

From this last lemma, the total number of nodes and vertices in CG is seen to be $O(k^2n^{4+k})$, and it is also easy to construct CG within this time bound, using the technique sketched above. Thus we have

PROPOSITION 1.1:   The running time of our algorithm is $O(k^2n^{k+4})$, where k is the number of the spider's arms, and where n is the number of wall edges.

2. POSSIBLE EXTENSIONS

The technique described in the preceding section can be applied to other robot systems having similar decomposability properties. As a first example, consider the system B consisting of a rigid 2-D cart to which there are affixed k straight arms at various points, each free to rotate about the point of contact with the cart, unimpeded by the positions of the other arms. Suppose first that the motion of the cart itself is restricted, e.g., that it can only translate without rotations, and that the arms attached to it move in a space confined by polygonal barriers as in the preceding section. Then we can plan the motion of this system using a technique quite similar to that described above. More specifically, we project the space FP of free positions of B onto the one-dimensional space of the positions of the cart itself. For each such position X, the space $O(X)$ of the tuples of orientations of the arms of B is decomposable into the Cartesian product of the spaces $O_i(X)$ of free orientations of each of the arms, and the technique of the preceding section can then be used to define critical positions of the cart, partition the 2-D space of positions of the cart into connected noncritical regions, and build a connectivity graph in much the same way as before.

If the motion of the cart is unrestricted, then the above technique can still be applied, but now we must project FP onto the 3-D space of free positions of the cart. This leads to a more complicated problem, since we will have to partition this 3-D space into noncritical regions. Although more difficult, this task can still be attacked by methods similar to those developed in Schwartz and Sharir (1983c).

All the preceding generalizations, including the algorithm developed in Section 1, are based on the availability of an algorithm for planning the motion of a ladder. As similar algorithms are developed for more complex systems, it will generally be possible to use these algorithms as a basis for developing motion-planning algorithms for systems involving special decomposable combinations of arbitrarily many such "basic" systems.

CHAPTER 5

On the Piano Movers' Problem: V. The Case of a Rod Moving in Three-dimensional Space Amidst Polyhedral Obstacles

JACOB T. SCHWARTZ

Courant Institute of Mathematical Sciences
New York University

MICHA SHARIR

Department of Mathematical Sciences
Tel Aviv University

This paper, a fifth in a series, solves some additional 3-D special cases of the piano movers' problem, which arises in robotics. The main problem solved in this paper is that of planning the motion of a rod moving amidst polyhedral obstacles. We present polynomial-time motion-planning algorithms for this case, using the connectivity-graph technique described in the preceding papers. We also study certain more general polyhedral problems, which arise in the motion planning problem considered here but have application to other similar problems. Application of these technique to the problem of planning the motion of a general polyhedral body moving in 3-space amidst polyhedral obstacles is also described.

1. INTRODUCTION

The strategy used in the analysis of the piano movers' problem begun in Schwartz and Sharir (1983a) and continued in Schwartz and Sharir (1983c) can be regarded as an optimization of the general but catastrophically inefficient algorithm described in Schwartz and Sharir (1983b); specifically we optimize by treating 'easy' dimensions in a special, direct manner. This strategy can be described in the following general terms. We are given an n-dimensional algebraic manifold FP, representing the free positions of a body (or group of related bodies) B constrained to move freely in a two- or three-dimensional space bounded by certain obstacles. Our task is to find the connected components of FP, or, what is much the same thing, to determine for two given positions p_1, p_2 of B whether or not they lie in the same (arcwise) connected component of FP. As in Schwartz and Sharir (1983a), we can proceed by fixing the values of k of the n parameters specifying a position in FP. Let X denote the value of the parameters thereby fixed, and consider the corresponding 'fibers'

This research has been supported by ONR Grants N00014-75-C-0571 and N00014-82-K-0381, and by a grant from the U.S.-Israeli Binational Science Foundation.

$P(X)$ consisting of all allowed values of the remaining $n-k$ parameters (call them Y). In the special case analysed in Schwartz and Sharir (1983a), namely that of a rigid polygonal body moving in 2-space amidst polygonal barriers, it is seen that unless X lies in a certain 1-dimensional 'critical' curve of the 2-dimensional space over which it varies, each of the connected components of $P(X)$ can be given a discrete combinatorial characterization which remains invariant under a continuous change of X, provided that X avoids the aforementioned critical submanifolds. However, as X crosses a critical submanifold β in 2-space, certain typical changes, generally of one of the following three kinds, will occur:

(i)   A component of $P(X)$ can disappear (or appear);
(ii)   Two components of $P(X)$ can join together into one component, or conversely one component may split into two disconnected subparts.
(iii) The combinatorial description (or 'marking') of a component of $P(X)$ may change, even though the topological structure of $P(X)$ does not change, and even though the connected components of $P(X)$ move continuously as β is crossed.

In the elementary special case studied in Schwartz and Sharir (1983a), the critical submanifolds β are easily characterized in terms of the geometric parameters of the problem (i.e., the shape of the obstacles and of the body). Each critical curve is shown to be of a lower dimension than that of the 2-dimensional space in which X varies, and "crossing rules" are established for each critical submanifold β. These rules describe how $P(X)$ changes as X crosses β from one side to the other. The crossing rules are seen to remain invariant over any connected subset of β not containing any singular point. Once this is established, we find the connected components of the set of all noncritical values of X; since k is less than n, this is a significantly easier problem than the one with which we began. The number of these components is easily seen to be finite, and they can be identified by the collection of 'critical surface patches' forming their boundary. These observations reduce our original problem to a purely combinatorial one. More specifically, we can construct a 'connectivity graph' CG whose nodes are pairs of the form $[R,S]$, where R is a connected component of the noncritical subpart of the region in which X varies, and where S is the combinatorial "marking" of some connected component of $P(X)$ for any (hence every) $X \in R$. An edge will connect $[R_1,S_1]$ to $[R_2,S_2]$ in CG if R_1,R_2 are two noncritical regions having a common boundary β, and if the $P(X)$ component marked S_2 on the R_1 side of β makes contact with the component marked S_1 on the R_2 side of β. Finally, let $[X_1,Y_1],[X_2,Y_2]$ be two given configurations of the body (or bodies). Then we can map $[X_1,Y_1]$ and $[X_2,Y_2]$ to nodes $[R_1,S_1]$ and $[R_2,S_2] \in CG$, where R_j is the noncritical component of k-space containing X_j, and where S_j is the marking of the component $P(X_j)$ containing Y_j, for $j=1,2$. Our original problem is thereby reduced to a simple combinatorial search in CG to check whether $[R_1,S_1]$ and $[R_2,S_2]$ belong to the same component of the finite graph CG.

We also note that there is no need to identify the noncritical regions R explicitly.

This fact appears clearly in the algorithm presented in Schwartz and Sharir (1983a). Instead of explicitly identifying these regions, we can proceed in the following simpler manner: First identify all 'critical patches' formed by dividing the critical submanifolds into pieces along lines formed by the intersection of the critical submanifolds. With each such patch β associate two 'sides' $r(\beta)$ and $l(\beta)$, and determine the crossing applicable to β. Then, given a side R_1 of some critical surface patch and another side R_2 of another patch, determine whether these two sides are connected to each other, that is, whether a point X_1 on the R_1 side of β_1 (sufficiently near β_1) can be connected to a point X_2 on the R_2 side of β_2 without going through any critical submanifold. Using any geometric procedure able to make this decision, we can build a modified connectivity graph replacing critical regions by sides of critical surface patches, and connecting $[R_1, S_1]$ to $[R_2, S_2]$ if either $S_1 = S_2$, and R_1 and R_2 can be connected to each other, without crossing any critical submanifold, and also if R_1 and R_2 are the two opposite sides of the same critical surface patch and the marking S_1 would be changed to S_2 when we apply the crossing rules applicable for the patch β.

It is also of significance that our approach is capable of being used recursively. Our initial problem can be posed as follows: Given the Euclidean n-space E^n, and a collection of critical manifolds of dimension $n-1$ (these manifolds constitute the boundary of the manifold FP of free positions), find an effective procedure that will determine for any two given points Z_1, Z_2 whether they belong to the same connected component of $E^n - V$. Our strategy breaks this problem into two subproblems of size k and $n-k$ by projecting onto E^k. First we have to identify connected components of $P(X)$ for each $X \in E^k$ and mark them in some discrete manner. Then we need to identify connected components of $E^k - V_k$, where V_k is the collection of critical submanifolds in the X-space. If the dimension k is too large, we can repeat this procedure recursively several times, until k reduces to a manageable size (usually 1 or 2). However, each level of application of the strategy outlined complicates the geometry with which we have to deal, so that in practical terms it is not always possible to apply the proposed approach efficiently. In Schwartz and Sharir (1983c) the general recursive principle which we have just sketched is illustrated in the solution of the motion-planning problem for three circles moving in 2-space. Note also that, as shown in Schwartz and Sharir (1983b), the motion-planning problem can be solved in general using a systematic projection/decomposition technique due to Collins; however, this procedure is very inefficient.

In this paper we continue our study of the mover's problem, by attacking a special case of it having a larger number of degrees of freedom than that considered in Schwartz and Sharir (1983a). Specifically, we consider the case of a rigid rod B moving in 3-dimensional space V bounded by polyhedral walls. This motion-planning problem involves 5 degrees of freedom, namely 3 translational and 2 rotational parameters. Although problems of 6 degrees of freedom have already been attacked in Schwartz and Sharir (1983c), the geometry involved in the case of a moving rod is a bit more complicated. We will show that this problem can be handled by the 'projection' approach outlined above. For this, we begin by projecting FP into the

2-space A of all possible orientations of B (we will identify A with the unit 3-sphere). Thus, for each orientation θ of B, $P(\theta)$ consists of all translations X of B for which B remains free of intersection with the walls when given position/orientation $[X, \theta]$. Section 2 shows how to describe and label the components of $P(\theta)$ by studying the simpler case of a single point moving amidst polyhedral walls in 3-space. This simplified problem involves only 3 degrees of freedom, but its solution uses methods which generalize to similar problems in an arbitrary number of degrees of freedom, which involve only linear constraints.

After these 'purely polyhedral' questions have been treated, we return in Section 3 to our original analysis of a rigid rod. There we describe the geometry of the noncritical regions in the θ-space, develop crossing rules for the critical curves which separate these regions, and finally reduce the problem to a purely combinatorial affair involving an appropriate 'connectivity graph'. Since θ-space is only 2-dimensional, the geometric details of its decomposition into noncritical regions are similar to those pertaining to the decomposition described in Schwartz and Sharir (1983a).

2. PROJECTING ONTO A PURELY POLYHEDRAL SUBPROBLEM

Our problem is to plan a continuous motion of a rigid rod B free to move (i.e., translate and rotate) in 3-dimensional space bounded by a finite collection of polyhedra. Let P be a designated endpoint of B, and let Q be the other endpoint. In what follows it will be convenient to specify a general position of B by a quintuple $[x, y, z, \phi, \psi]$, where $X = [x, y, z]$ denotes the Cartesian position of P, and where the spherical coordinates $\theta = [\phi, \psi]$ represent the orientation of B.

We first wish to eliminate the three translational parameters from the problem, and so project FP onto the 2-dimensional θ-space. To do this, let θ be a fixed orientation of B, and let B_0 denote the segment occupied by the rod B when it is placed at orientation θ with P at the origin. If we constrain B to move without changing this orientation, its motions are all purely translational, and so have just 3 degrees of freedom. It is easy to see that the set $P(\theta)$ of all points X to which P can move while θ is held fixed is the set of all X such that $(B_0 + X) \cup V^c = \varnothing$, where V^c is the wall region, i.e., the complement of the free space V. That is $P(\theta) = (V^c - B_0)^c$. (Here $+$ and $-$ denote pointwise vector addition and subtraction respectively.) Assume that V^c can be cut into finitely many convex polyhedra W_1, $\ldots$, W_n which intersect only at faces common to two such parts. Then we can write

$$P(\theta) = (\cup_i (W_i - B_0))^c$$

Let us agree to write $ext(S)$ for the collection of extreme points of the set S, and $conv(T)$ for the convex hull of the set of points T. Put $EB_0 = ext(B_0) = \{O, L\theta\}$ (where L is the length of B), $EW_i = ext(W_i)$. Then it is easy to see that

$$W_i - B_0 = conv(EW_i - EB_0) = conv(EW_i \cup (EW_i - L\theta))$$

and

$$P(\theta) = (\cup_{i} conv(EW_i - EB_0))^c$$

The problem of characterising $P(\theta)$ and its connected components is therefore a special case of the following more general problem.

Problem: Given convex polyhedra $K_1, K_2, \ldots, K_m$, all of which are n-dimensional and contained in another n-dimensional convex polyhedron K_0, find all the connected components of $L = K_0 \cap (\cup_i K_i)^c$ and label them unambiguously.

In the following paragraphs we will describe a recursive procedure to solve this problem for arbitrary dimensions n.

To solve the polyhedral problem just stated we can proceed in the following recursive manner. We are given m convex n-dimensional polyhedra $K_1, K_2, \ldots, K_m$ (some of which may overlap) in Euclidean n-space, such that all are contained in the interior of another convex polyhedron K_0. Our aim is to compute, and give discrete labelings to, the connected components of the complement L of $\cup K_i$ in K_0. We will refer to an m-dimensional face of a polyhedron K as an $m-face$ of K; thus faces of maximal dimension are $(n-1)$-faces, and for each $j < n-1$ every j-face is a subface of a $(j+1)$-face. An m-face will also be called a face of *codimension $n-m$*. By the *interior* of a k-face F we mean its interior in the k-plane which contains F.

The procedure that we are about to present assumes that the polyhedra with which we deal are in 'general position'. To define this property rigorously, we will assume that each of the polyhedra K_i (and K_0) are defined as an intersection of a given collection of half-spaces. We then say that our polyhedra are in general position if the following condition holds:

(*) Any collection of $n+1$ distinct bounding half-spaces of the given polyhedra have an empty intersection.

Later in subsequent development of the more general analysis sketched in the introduction, we will say that θ is a *critical orientation* if the polyhedra $K_i = K_i(\theta)$ appearing for this value of θ do not satisfy (*). In some cases this violation of (*) may occur because some of the bounding half-spaces happen to coincide with each other, or because such a half-space is about to become redundant, i.e., touches the polyhedron that it bounds at a lower-dimensional face, and so forth. Note therefore that if the polyhedra are in general position then any collection of $n+1$ distinct faces of the given polyhedra have a null intersection, but that this latter condition does not imply condition (*) if degenerate coincidences or redundancies of the bounding half-spaces do occur.

We prepare for what follows by proving a number of easy auxiliary lemmas.

LEMMA 1: Suppose that a collection K_i of convex polyhedra is in general position in the sense defined above, and that $F_1, \ldots, F_m$ are a set of faces of these polyhedra, of respective codimensions $c_1, \ldots, c_m$. Then if the interiors of these faces intersect, we have $c_1 + \cdots + c_m \leq n$, and the dimension of the intersection is precisely $n - c_1 - \cdots - c_m$.

Proof: Suppose that the K_i's are in general position, and let $F_1, \ldots, F_m$ have codimensions $c_1, \ldots, c_m$. Each face F_j is the intersection of the c_j $(n-1)$-faces of the polyhedron K_i on which it lies; thus if $c_1 + \cdots + c_m > n$ we must have $F_1 \cap \cdots \cap F_m = \varnothing$. Hence we need only consider the case $d \equiv n - c_1 - \cdots - c_m \geq 0$. Here, let G_j be the interior of F_j for $j=1, \ldots, m$ and suppose that these interiors intersect. Since the interiors of any two distinct faces lying on the same polyhedron are disjoint from one another, it follows that no two faces F_j lie on the same polyhedron. The intersection I of the G_j's contains a relatively open set in the intersection of all the $c_1 + \cdots + c_m$ hyperplanes (of codimension 1) containing an $(n-1)$-face which contains one of the G_j. Hence the dimension of the intersection of the G_j is at least d.

Next suppose that the closed convex set I has dimension d' greater than d, and let H be the hyperplane of smallest dimension containing I. Since the boundary of I is plainly contained in the intersection of H with the union of the boundaries of the G_j, and since H, being of dimension d' cannot be separated by a union of sets of dimension less than $d'-1$, one of those G_j, say for definiteness G_1, has an (open) face G'_1 which intersects I in a set of dimension at least $d' - 1 \geq d$. Thus if we replace G_1 by G'_1, we raise c_1 by 1 and lower the dimension of the intersection by at most 1. This step can be repeated till $c_1 + \cdots + c_m = n+1$, at which point a non-null intersection (at least zero-dimensional) must remain. But this is plainly impossible in view of the remarks made in the preceding paragraph. This proves our lemma. Q.E.D.

LEMMA 2: Let a set of convex polyhedra K_i be in general position, and let F be an $(n-1)$-face of K_1. Then the set of all vertices (i.e., extreme points) of $K_1 \cap K_2$ is the set of all intersections of m-faces of K_1 with $(n-m)$-faces of K_2, $1 \leq m \leq n-1$; and similarly, the set of all vertices of $F \cap K_2$ is the set of all intersections of m-faces of F with $(n-m)$-faces of K_2, $1 \leq m \leq n-2$.

Proof: Since by Lemma 1 no $(m-1)$-face of K_1 intersects any $(n-m)$-face of K_2, and no $(n-m-1)$-face of K_2 intersects an m-face of K_1, it follows that the unique point p of intersection of an m-face F_1 of K_1 and an $(n-m)$-face F_2 of K_2 must be interior to both the intersecting faces. Thus linear coordinates can be established near p in which K_1 (resp. F_1) appears locally as the set $x_1 \geq 0, \ldots, x_{n-m} \geq 0$ *(resp. $x_1 = 0, \ldots, x_{n-m} = 0$)*, and K_2 (resp. F_2) appears locally as $x_{n-m+1} \geq 0, \ldots, x_n \geq 0$ (resp. $x_{n-m+1} = 0, \ldots, x_n = 0$). In these coordinates p appears as $(0, \ldots, 0)$, and $K_1 \cap K_2$ appears locally as $x_i \geq 0$, $i=1, \ldots, n$. This makes it plain that p is an extreme point, i.e., a vertex, of $K_1 \cap K_2$.

Conversely, take any $p \in K_1 \cap K_2$, and suppose that it lies in faces F_1, F_2 of K_1, K_2 having (largest possible) codimensions c_1, c_2 respectively. As noted in the proof of the preceding lemma, p must belong to the intersection of the interiors of F_1 and F_2. Since the K_i's are in general position, we must have $c_1 + c_2 \leq n$, and the intersection of these interiors is of dimension $n - c_1 - c_2$. If this dimension is not zero, then p lies in the interior of this intersection, and consequently cannot be an extreme point of $K_1 \cap K_2$.

This proves the first assertion of lemma 2; The proof of the second is similar and is left to the reader. Q.E.D.

Now we can describe a procedure for computing and labeling the connected components of the set L of the problem mentioned at the end of the preceding section. The input to this procedure is a collection of polyhedra $K_0, K_1, \ldots, K_m$. It will be convenient to assume that each polyhedron K_i is represented by a (finite) collection of half spaces whose intersection is equal to K_i. The output of the procedure is the set of all connected components of L. Each such component is represented as the set of its $(n-1)$-faces; each such face is in turn represented as the set of all its $(n-2)$-faces, and so on, till eventually we descend to 0-faces, i.e., points, each represented by its coordinates.

Our procedure uses recursion and proceeds downward through successive dimensions n. If $n = 1$ the task is trivial, since K_0 is an interval, and $K_1, \ldots, K_m$ are subintervals of K_0. In this case, each connected component of L is also an open subinterval of K_0, and is represented by its endpoints.

Next suppose that $n > 1$. Fix $i \geq 0$, and let F be an $(n-1)$-dimensional face of K_i. Compute $F_j = K_j \cap F$ for each $j \geq 1$, $j \neq i$. F_j is represented by the set U of all the half spaces defining K_j and by all half spaces defining K_i except for the half space H whose boundary plane M contains F; H is replaced in U by M. Condition (*) ensures that each F_j is either empty or is an $(n-1)$-dimensional convex polyhedral subset of F. The following lemma shows that condition (*) is hereditary, in the sense that it also holds for these new sets.

LEMMA 3: Let K_i, F and F_j be as above, and let $G_1, \ldots, G_n$ be n distinct $(n-2)$-dimensional hyperplanes contained in the plane of the face F and drawn through the various $(n-2)$-dimensional faces of the convex subsets F_j of F. Then the intersection of the G_i's is empty.

Proof: Suppose the contrary, i.e., let x be a point in the intersection of G_1, $\ldots, G_n$. Each G_i is the intersection of the hyperplane M containing F with a hyperplane M_i which defines a face J_i of one of the polyhedra K_j, such that $x \in J_i$. By (*) it also follows that all the M_i's are distinct and that none of them coincides with M. But this contradicts (*), because the $n+1$ distinct planes $M, M_1, \ldots, M_n$ intersect at X. This establishes our lemma. Q.E.D.

LEMMA 4: Let G be a connected component of the set

$$F - \bigcup_{j \geq 1, j \neq i} K_j.$$

where $i \geq 0$ and F is an $(n-1)$-dimensional face of K_i. Let p be an interior point of G. Then there exists a neighborhood U of P whose intersection with the side of G outside K_i either lies outside K_0 or is contained in L.

Proof: If $i=0$, the claim is obvious. Otherwise, p does not belong to any K_j, $j \geq 1$, $j \neq i$, and it follows that there exists a neighborhood U of t which is disjoint from each of these K_j's. If near p the exterior side E of K_i is contained in K_0, then it is clear that $U \cap E$ has the property asserted. Q.E.D.

We now proceed by considering all faces F of all the polyhedra $K_0, \ldots, K_m$ in turn and decomposing $F - \bigcup_{j \neq l} K_j$ into its connected components. This gives a set

H of connected $(n-1)$-dimensional face components, each of which will have been labeled in the manner explained above.

Next we define a relation Ξ_0 on H as follows: Two face components $G, G' \in H$ are related by Ξ_0 if they have a common $(n-2)$-dimensional subface. (The representation of connected face components that our algorithm provides makes it easy to check this condition). Finally, we take the reflexive and transitive closure of Ξ_0 to obtain an equivalence relation Ξ_1 on H. The following lemma shows that the equivalence classes of this relation approximate a representation of the connected components of L that we seek:

LEMMA 5: Let K_j, F_j, L and H be as above, and suppose as before that the K_j are in general position. Then all the faces belonging to the same connected component of the boundary of a connected component of L are equivalent under Ξ_1; conversely, each equivalence class of the equivalence relation Ξ_1 consists of all faces belonging to some connected component of the boundary of some connected component of L.

Proof: Let S be a connected component of L, and let T be its closure. Let E be the intersection of T with some face F of the polyhedra K_i, and let G be a connected component of E. G is plainly a connected subset of

$$F - \bigcup_{j \geq 1, j \neq i} K_j,$$

and is in fact a maximal connected such subset; hence $G \in H$. Conversely, by Lemma 4 each $G \in H$ is contained in the closure of some connected component S of L, and by maximality of G it is easy to see that G coincides with a connected component of the intersection of T with the face containing G. Hence the set H is the set of all faces of connected components of L.

Next let G, G' be two connected face components such that $[G, G'] \in \Xi_0$; let I be their common $(n-2)$-dimensional subface. Take an internal point p of I, and a neighborhood of p not intersecting any other $(n-1)$-faces of L. We can plainly introduce coordinates near p which make G, G' appear locally as two coordinate half-planes, and L appear locally as the 'wedge' of Euclidean space which these half-planes bound. Note that since all half-spaces bounding the polyhedra K_i are linearly independent (as follows from Lemma 1) no other face passes through p, or else p would be a boundary point rather than an interior point of I. This makes it clear that G and G' are faces of the same connected component of L. Hence, for any pair $[G,G'] \in \Xi_1$, G and G' are faces belonging to the same connected component of the boundary of a connected component of L. Conversely, suppose that G and G' $\in H$ are faces belonging to the same connected component of the closure S of a connected component of L. Then, since the boundary of L is a piecewise flat manifold of dimension $n-1$, one can find a sequence $G=G_1,G_2, \ldots ,G_j=G'$ of faces of S such that for each $j<l$ G_j and G_{j+1} meet at some $(n-2)$-dimensional subface of S. Then by definition $[G_j,G_{j+1}] \in \Xi_0$, so that $[G,G'] \in \Xi_1$ by transitivity. Q.E.D.

To complete the procedure for finding and labeling connected components of L, it remains to determine which connected boundary components bound the same component of L. For this, the following simple technique, similar to that described in Ocken, Schwartz, and Sharir (1983), is available. For each boundary component Γ choose any point $X_\Gamma \in \Gamma$. If Γ is an interior boundary component of some connected component C of L, then any straight ray emerging from X_Γ upward in some standard vertical direction, must eventually intersect the exterior boundary of C, and before the first such intersection it can intersect only interior boundary components of C. We therefore draw such a ray from X_Γ, find all its points of intersection with boundary components, and order these points along the ray. These points divide the ray into intervals which alternately lie either in L or in its complement. We then equivalence every pair of boundary components which intersect our ray in two endpoints of a ray interval all of whose interior points belong to the open set L. We repeat this procedure for all components Γ. It then follows that the extended equivalence relation Ξ_2 of Ξ_1 thereby obtained has the property that all the faces of each connected component of L are equivalent under Ξ_2, and, conversely, each equivalence class of Ξ_2 consists of all faces of some connected component of L. This completes the description of our recursive procedure.

Next we must discuss situations for which condition (*) is violated. Each X at which (*) is violated will lie in the intersection of $n+1$ faces of our set of polyhedra. This can be expressed as an intersection of finitely many half-spaces with $n+1$ planes of dimension $n-1$. To test for the existence of such a point X we can in principle simply intersect all possible combinations of $n+1$ face planes, and then check whether their intersection has a non-null intersection with all the additional half spaces defining faces in their respective planes. Note that the first part of this test can be expressed algebraically by saying that the coefficients (a,b) of the $n+1$ planes (which we write as $ax + b = 0$) are linearly dependent. When we do have $n+1$ intersecting faces, then by grouping these faces according to the respective polyhedra that contain them, we can re-express the fact that the $n+1$ planes considered have a non-null intersection, as follows: There exist t polyhedra K'_1,K'_2, $\ldots ,K'_t$ among the given polyhedra, and a sequence of positive integers $s_1, \ldots ,s_t$ whose sum is $n+1$, and for each $i \leq t$ there exists an $(n-s_i)$-face of K'_i, such that all these faces have a nonempty intersection.

The sequence of codimensions $s_1, \ldots ,s_t$ appearing in the preceding sentences defines a useful characteristic of the critical situations that we will need to analyze in what follows. Let us illustrate this classification for dimensions $n = 2$ and 3. First suppose $n = 2$. Since there are only two partitions of 3, only the following two critical two-dimensional configurations are possible:

(i)   A corner of some polygon K touches an edge of another polygon K' (this corresponds to the partition $3 = 2+1$).

(ii)   Three edges, each of a different polygon, intersect at a point (this corresponds to the partition $3 = 1+1+1$).

When $n = 3$, there are four possible partitions of 4. We can list them, each with the corresponding intersection conditions, as follows.

(i)   Two edges, each of a different polyhedron, meet at a point (corresponding to $4 = 2+2$).

(ii)   A corner of one polyhedron touches a face of another (corresponding to $4 = 3+1$).

(iii) Two faces of different polyhedra and an edge of a third one intersect at a point (corresponding to $4 = 2+1+1$).

(iv)   Four faces, each of a different polyhedron, intersect at a point (corresponding to $4 = 1+1+1+1$).

This convenient dimensional classification of critical configurations will reappear below.

3.   CONTINUOUS VARIATIONS OF COMPONENTS AT NONCRITICAL ORIENTATIONS AND CROSSING RULES AT CRITICAL ORIENTATIONS

Next suppose that the purely polyhedral problem studied above arises by projecting from a space of higher dimension, as in the case of a rigid rotating rod discussed in Section 1. In this context, our polyhedra and their extreme points will depend continuously on additional parameters, which for notational convenience we denote simply by θ. In conformity with the case of a rotating rod, we will sometimes refer to θ as an 'orientation.' The procedure described above assigns a discrete labeling for each connected component of $P(\theta)$ for orientations θ at which condition (*) holds. Let us agree to call such θ *noncritical orientations*, and to call orientations θ at which condition (*) is violated *critical orientations*. Our next aim is to study the way in which the components of $P(\theta)$ and their labels change as θ varies in a small neighborhood U of a critical orientation θ_0. We prepare for this by a few additional definitions and lemmas.

In what follows we will say that a convex polyhedron $K(\theta)$ depending on one or more parameters θ *varies continuously with* θ if the set of half-spaces bounding $K(\theta)$ can be written as $\{H_1(\theta), \ldots, H_n(\theta)\}$, where all the coefficients of each of the H_j depend continuously on θ.

LEMMA 6:   Let $K_i(\theta)$ be a set of convex polyhedra varying continuously with one or more orientation parameters θ, and suppose that for $\theta = \theta_0$ the polyhedra $K_i(\theta)$ are in general position. Then they remain in general position for all θ sufficiently near θ_0.

Proof: Suppose that $\theta_j \to \theta_0$ and that for all j the polyhedra $K_i(\theta_j)$ are not in general position. Then for each j there exist $n+1$ distinct half-spaces bounding certain of the polyhedra $K_i(\theta_j)$ which intersect. As $\theta_j \to \theta_0$, these half-spaces converge to half-spaces $H_1, \ldots, H_{n+1}$ bounding the various $K_i(\theta_0)$. Then plainly $H_1 \cap \cdots \cap H_{n+1} \neq \varnothing$. Since all these limit faces are distinct, this is a contradiction. Q.E.D.

LEMMA 7:   As in Lemma 6, let the convex polyhedra $K_i(\theta)$ vary continuously with θ, and be in general position for $\theta = \theta_0$. Let $K_i(\theta_0) \cap K_j(\theta_0)$ be non-null. Then $K_i(\theta) \cap K_j(\theta)$ is non-null for all θ sufficiently near θ_0, and the vertices of $K_i(\theta) \cap K_j(\theta)$ vary continuously and converge to the vertices of $K_i(\theta_0) \cap K_j(\theta_0)$. Similarly, if $F(\theta_0)$ is a face of $K_i(\theta_0)$ and for θ near θ_0 $F(\theta)$ is the corresponding face of $K_i(\theta)$ (i.e., the vertices of $F(\theta)$ converge to the vertices of $F(\theta_0)$), and if $F(\theta_0) \cap K_j(\theta_0)$ is non-null, then $F(\theta) \cap K_j(\theta)$ is non-null for all θ sufficiently near θ_0, while its vertices vary continuously and converge to those of $F(\theta_0) \cap K_j(\theta_0)$.

Proof: By Lemma 2, the set of vertices of $K_i(\theta) \cap K_j(\theta)$ is the set of all intersections of the interiors of m-faces F of $K_i(\theta)$ with $(n-m)$-faces F' of $K_j(\theta)$, for all $m=1, \ldots, n-1$. At each such intersection, the planes containing the two intersecting faces are linearly independent, so that they remain independent in a sufficiently small neighborhood of θ, and hence their point p of intersection is unique and moves continuously with θ. Our assertion follows obviously from this remark. Q.E.D.

LEMMA 8:   As in Lemma 7, let the orientation θ_0 be noncritical. Then for θ in a sufficiently small neighborhood of θ_0, the vertices and k-dimensional faces, $k=1, \ldots, n-1$, of $P(\theta)$ remain disjoint from each other and vary continuously with θ.

Proof: We proceed by induction on the number m of polyhedra $K_j(\theta)$ involved, for which purpose it is convenient to assume that the polyhedron K_0 containing L is also allowed to vary continuously with θ. Put $K^{(0)}(\theta) = K_0(\theta)$, and successively put $K^{(j+1)}(\theta) = K^{(j)}(\theta) - K_{j+1}(\theta)$. Then we can assume inductively that the assertion of the present lemma is true for some j, and must simply prove that it then remains true for $j+1$. We consider only the step from $j=0$ to $j=1$, which is typical. By Lemma 7, the vertices of $K^{(1)}(\theta)$ are those vertices of $K^{(0)}(\theta)$ which do not lie interior to $K_1(\theta)$, plus those vertices of $K_1(\theta)$ which do not lie exterior to $K^{(0)}(\theta)$, plus all intersections of a-faces of $K_1(\theta)$ with $(n-a)$-faces of $K^{(0)}(\theta)$, $a=1, \ldots, n-1$. That these vertices remain distinct and vary continuously as θ varies follows from Lemma 7.

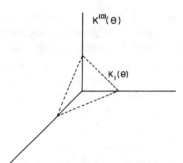

Figure 1. The vertices of the portion of a 2-face of $K_1(\theta)$ lying interior to a region $K^{(0)}(\theta)$.

It can be shown in similar fashion that for each dimension c, each face of $K^{(1)}(\theta)$ having codimension c is either the intersection of a c_1-codimensional face of $K^{(0)}(\theta)$ with a c_2-codimensional face of $K_1(\theta)$, where $c = c_1 + c_2$; or is that portion of a c-codimensional face of $K^{(0)}(\theta)$ which lies interior to $K_1(\theta)$; or is that portion of a c-codimensional face of $K_1(\theta)$ which lies exterior to $K^{(0)}(\theta)$. It is plain that all these sets are disjoint from each other. Moreover, we can easily locate their vertices; for example, to locate the vertices of that portion $G(\theta)$ of a c-dimensional face F of $K_1(\theta)$ which lies interior to $K^{(0)}(\theta)$, we take F and all its c'-dimensional subfaces, for all $c' \le c$, and form all their intersections with $(n-c')$-dimensional faces of $K^{(0)}(\theta)$ (see Figure 1). Lemma 7 plainly implies that these vertices vary continuously with θ; hence so does $G(\theta)$. Q.E.D.

LEMMA 9: Suppose that the polyhedra $K_j(\theta)$ are as above, let the orientation θ_0 be noncritical, and for θ is a sufficiently small neighborhood U of θ_0 let $C(\theta)$ be the connected component of $P(\theta)$ bearing some fixed combinatorial labeling, which is to say the connected component in whose boundary a particular continuously moving vertex $v(\theta)$ appears. Then the set of continuously moving vertices, edges, etc. lying on the boundary of $C(\theta)$ is invariant for $\theta \in U$, and for $\theta \in U$, all these elements move continuously, as do their interiors and the interior of $C(\theta)$.

Proof: Almost everything asserted here has already been proved in Lemma 8; it only remains to show that, for θ in a small enough neighborhood of each noncritical θ_0, the set of faces, vertices, etc. of each fixed component (identified, say, by identifying some one of its vertices) remains fixed. For this, suppose for the moment that we identify the individual components of $P(\theta)$, for θ in a sufficiently small neighborhood N of θ_0, by giving some fixed point Q independent of θ which belongs to each component for all these θ. Now take a vertex $v(\theta)$ and suppose that it is identified by specifying which n planes meet at v. Take a spherical neighborhood U of $v(\theta_0)$ small enough so that for $\theta \in N$ no boundary face of $P(\theta)$, other than the three meeting at $v(\theta)$, enter U. Take a point Q' interior to $P(\theta_0) \cap U$, and a path in $P(\theta_0)$ from Q to Q'. For $\theta \in N$ sufficiently near θ_0, this path lies wholly within the component containing Q; and since the faces intersecting at $v(\theta)$ move smoothly and remain linearly independent, their intersection can always be connected to Q' by a straight line segment lying within this same component. This shows that the component to which a given vertex belongs does not change as θ varies through noncritical positions. Since much the same argument can be given for subfaces of any number of dimensions, our lemma is established. Q.E.D.

Next we consider the phenomena that can occur at critical orientations θ_0, at least for the simplest kind of criticalities. By definition, $P(\theta)$ is the complement of the union of various convex polyhedra $K_1(\theta), \ldots, K_r(\theta)$ in some Euclidean space E^n. We assume that these polyhedra vary continuously with θ. Since θ_0 is assumed to be critical, there must exist $n+1$ hyperplanes $H_1, \ldots, H_{n+1}$ bounding the polyhedra $K_1(\theta_0), \ldots, K_r(\theta_0)$ which all meet at a common point X. Let N denote a sufficiently small convex neighborhood of X. Assuming for the moment that X is the only point at which $n+1$ such hyperplanes meet when $\theta = \theta_0$, our problem reduces to that of studying the behavior of the connected components of $V(\theta) = P(\theta) \cap N$ as θ varies over U. (In the following discussion, we also assume that no other hyperplane bounding any of the polyhedra passes through X; orientations θ_0 at which more than $n+1$ such hyperplanes meet at a point will generally lie on manifolds of codimension at least 2 in θ-space, in which case removal of these points will not affect the connectivity of G; see Lemma 1.9 of Schwartz and Sharir (1983a). Hence, if U and N are sufficiently small, it follows that for each $\theta \in U$, N is disjoint from any hyperplane bounding the $K_i(\theta)$ other than the prescribed hyperplanes $H_1, \ldots, H_{n+1}$.)

As before, we group the $(n-1)$-dimensional hyperplanes $H_1, \ldots, H_{n+1}$ into collections of hyperplanes, each consisting of hyperplanes bounding the same polyhedron. As previously, we can describe this partitioning by a decomposition of the set $\{1, \ldots, n+1\}$ of integers, and accordingly can write

$$\{1, \ldots, n+1\} = \bigcup_{j=1,\ldots,m} C_j$$

where, for each j, all hyperplanes H_i, $i \in C_j$, bound the same polyhedron K_j. Without loss of generality we can assume that $X = 0$, and can characterize each H_i by the normal unit vector α_i drawn from H_i in the outward direction of the polyhedron K which H_i bounds. Let H_i^+ (resp. H_i^-) denote the open half-space $\{X : X \cdot \alpha_i > 0\}$ (resp. $\{X : X \cdot \alpha_i < 0\}$); then we can write $V(\theta)$ as

$$V(\theta) = N \cap \left(\bigcup_{j=1,\ldots,m} \bigcap_{i \in C_j} H_i^+ \right) \tag{1}$$

It follows by DeMorgan Laws that $V(\theta)$ can be written as a union of intersections of the form

$$H_{i_1}^+ \cap H_{i_2}^+ \cap \cdots \cap H_{i_m}^+ \cap N,$$

where $i_p \in C_p$, $p=1, \ldots, m$.

To simplify analysis of the situation before us, we assume that the hyperplanes $H_1, \ldots, H_{n+1}$ are such that any n of them are linearly independent at θ_0. By continuity, this property will also hold for all $\theta \in U$ if U is sufficiently small. In this case the behavior of $V(\theta)$ for $\theta \in U$ is essentially determined by the value of m. To prove this we first state the following simple lemma:

LEMMA 10: Let $H_1, \ldots, H_k$ be linearly independent hyperplanes of E^n (with $k \leq n$), and let $G_1, \ldots, G_k$ be the closed half-spaces bounded by the respective H_i's. Then the intersection of all the G_i's is a convex cone of E^n having nonempty interior.

Proof: The configuration assumed above is linearly isomorphic to that in which the hyperplanes $H_1, \ldots, H_k$ are standard coordinate planes, and since the assertion is plainly true in this transformed configuration, it is also true for the original configuration. Q.E.D.

The fact that any m linearly independent planes, $m \leq n$, are linearly isomorphic to standard coordinate planes is used again, implicitly, in the next few paragraphs.

The parameter m appearing in formula (1) can take on a value in any one of the following ranges, for each of which different behavior is observed.

I. Suppose first that $m < n$. Then $V(\theta)$ is connected for each $\theta \in U$. Indeed, consider any two of the intersection sets

$$A = H_{i_1}^+ \cap H_{i_2}^+ \cap \cdots \cap H_{i_m}^+ \cap N$$

and

$$B = H_{j_1}^+ \cap \cdots \cap H_{j_m}^+ \cap N$$

of which $V(\theta)$ is composed. These two sets are contained in the same connected component of $V(\theta)$, because they are linked by the following chain of sets

$$A = W_1 = H_{i_1}^+ \cap H_{i_2}^+ \cap \cdots \cap H_{i_m}^+ \cap N,$$

$$W_2 = H_{j_1}^+ \cap H_{i_2}^+ \cap \cdots \cap H_{i_m}^+ \cap N,$$

$$\cdots\cdots\cdots\cdots\cdots\cdots\cdots\cdots\cdots\cdots\cdots$$

$$W_m = H_{j_1}^+ \cap H_{j_2}^+ \cap \cdots \cap H_{j_{m-1}}^+ \cap H_{i_m}^+ \cap N,$$

$$B = W_{m+1} = H_{j_1}^+ \cap \cdots \cap H_{j_m}^+ \cap N,$$

and, moreover, each W_i is connected and the intersection of any two successive sets W_p and W_{p+1} is nonempty, by Lemma 1. This shows that in this case $V(\theta)$ is connected for each $\theta \in U$.

II. Next suppose $m = n+1$. Then all the C_j's are singletons, and $V(\theta)$ is the intersection of $n+1$ open half-spaces with N. We can take the first n boundary planes of these half spaces, and since they are linearly independent we can establish coordinates in which they have the form $x_i = 0$, $i = 1, \ldots, n$, and the half spaces they bound have the form $x_i \geq 0$, $i = 1, \ldots, n$. In these coordinates, the $(n+1)$-st plane will have the form $a(\theta) \cdot x \leq b(\theta)$, where without loss of generality we can assume that the coefficient vector has norm $|a(\theta)| = 1$ everywhere in U. For $\theta = \theta_0$ all the $n+1$ planes are concurrent at the origin of our coordinates, i.e., $b(\theta_0) = 0$. If any of the components of the vector $a(\theta_0)$ are zero, (say for definiteness that the first component of $a(\theta_0)$ is zero) then n of our $n+1$ planes (specifically $x_2 = 0, \ldots x_n = 0$, and the $(n+1)$-st plane are linearly dependent for $\theta = \theta_0$, contrary to assumption. This we can assume without loss of generality that all these coefficients are nonzero everywhere in U. Hence none of these coefficients change sign in U, and we can assume that the first k of them are negative and the rest positive. However, $b(\theta_0) = 0$, so $b(\theta)$ can change sign as it crosses the various sheets of the (algebraic) surface $b(\theta) = 0$ passing through θ_0, which we will call Σ. If $b(\theta)$ does not change sign when Σ is crossed, the geometric configuration within N does not change in any significant way. If $b(\theta)$ changes sign, the geometric change which occurs as Σ is crossed depends on the value of k. For $k > 0$, all that happens as $b(\theta)$ goes from positive to negative is that the $(n-k)$-face of $V(\theta)$ which lies in the plane of the last $n-k$ x_j's disappears, and that other minor details of $V(\theta)$ change in corresponding ways. However, the connectivity of $V(\theta)$ does not change. But if $k = 0$, then as $b(\theta)$ goes to zero $V(\theta)$ will always shrink to a point (as Σ is approached) and then disappear entirely when $b(\theta)$ becomes negative. This is shown in the Figure 2.

III. Finally, suppose that $m = n$. Then all the sets C_j are singletons, with the exception of one doubleton set, say C_1. It follows from the preceding discussion and from (1) that $V(\theta)$ is the union of two intersections having the forms

$$G^+ \cap H_2^+ \cap \cdots \cap H_n^+ \cap N.$$

and

$$H^+ \cap H_2^+ \cap \cdots \cap H_n^+ \cap N.$$

It follows from Lemma 1 that each of these sets is nonempty, and will remain nonempty if θ is sufficiently near θ_0. Hence, without loss of generality, we can assume that for $\theta \in U$, $V(\theta)$ is nonempty, and has at most two connected components. In this case one connected component of $P(\theta)$ may split into two components, or vice versa, as θ passes through $\theta_0 \in U$, though, as the Figure 3 shows, it is also possible that no topological change should occur.

These observations show that only three types of *crossing rules* are to be expected in the purely polyhedral case, provided that the critical point being crossed is, so to speak, of a regular criticality. However, the simple analysis which we have given rests upon the assumption that the critical intersecting hyperplanes H_1, $\ldots, H_{n+1}$ are such that any n of them are linearly independent. If this assumption fails to hold, the structure of $V(\theta)$ may become more complicated. As an example illustrating this remark, consider the situation shown in Fig. 4, in which the polyhedral space is assumed to be 5-dimensional. Assume that $F_1, \ldots, F_6$ are faces of the polyhedra in question whose normals $\alpha_1, \ldots, \alpha_6$ span only a two-dimensional subspace of E^5. Figure 4 displays a structure that $V(\theta)$ might have at θ_0, and at two

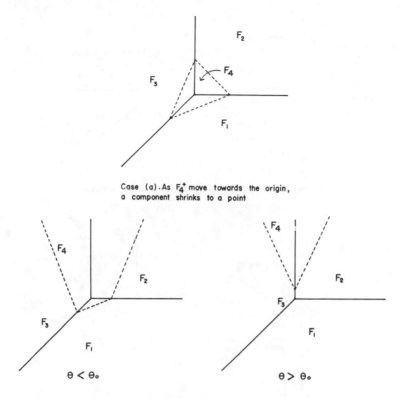

Case (a).As F_4^+ move towards the origin,
a component shrinks to a point

Case (b). Component boundary changes but component does not vanish

Figure 2.   Two possible crossings of type II.

nearby orientations, all drawn in a planar cross-section of E^5 spanned by $\alpha_1, \ldots,$ α_6. As we see, $V(\theta)$ may have more than two components, which can split and merge in various ways as θ crosses through θ_0.

However, the following observation implies that the crossing rules in this degenerate case are still similar to those established above.

LEMMA 11: Let θ_0, N, U and $H_1, \ldots, H_{n+1}$ be as above. Let $\theta_t \rightarrow \theta_0 \in U$, and let B_t be a connected component of $V(\theta_t)$, such that B_t converges (in the Hausdorff metric) to some set B_0. Then the interior of B_0 is a union of connected components of $V(\theta_0)$.

Proof: It is easy to see by continuity that $int(B_0)$ is a subset of $V(\theta_0)$. On the other hand, each point Z on the boundary of B_0 (but lying inside N) must be a limit point of a sequence of points Y_t which lie on one of the hyperplanes $H_1(\theta_t), \ldots, H_{n+1}$ (θ_t) for each t, so that by continuity Z must also lie on one of our hyperplanes H_1 $(\theta_0), \ldots, H_{n+1}(\theta_0)$. These simple observations plainly imply our assertion. Q.E.D.

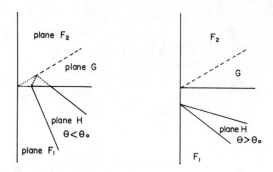

Case (a). Separation of upper octant from lower polyhedron

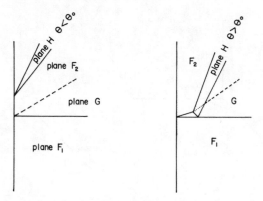

Case (b). Modification of boundary without component separation

Figure 3.   Two possible crossings of type 3.

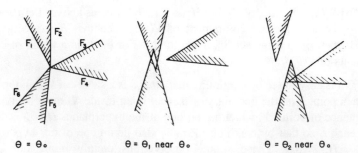

Figure 4.   Crossing a degenerate critical point

Lemma 11 implies that, if θ approaches θ_0 along a continuous arc consisting of noncritical orientations only, then the limit behavior of the components of $V(\theta)$ at θ_0 will always be a mixture of the crossing rules I–III stated above, that is, some components of $V(\theta)$ may split into several subcomponents; other components of $V(\theta)$ may shrink to a point, while the rest of the components will remain intact and disjoint from each other.

We note that degenerate configurations of the sort that we have forbidden in the main part of our discussion generally correspond to degenerate layouts of some of the walls or of some of the convex subparts of the body B. Degeneracies of this kind were analyzed more carefully in Schwartz and Sharir (1983a), where a way of treating such configurations systematically is suggested. Specifically, critical surfaces involving these more complex crossing rules can be considered to consist of several layers, infinitesimally close to each other, such that across each of these layers only one standard crossing rule (of the form I–III) applies, so that the final crossing rule is obtained as an appropriate composition of standard rules. A very similar procedure can be devised to handle the case of degenerate purely polyhedral configurations, but we omit the (purely technical) details.

4. THE CASE OF A ROTATING 3-D ROD

Having now completed our preparatory work, we return to study the movers' problem for the case of a rigid rod moving in 3-space amidst polyhedral obstacles. As in Section 1, we will assume that the wall region V^c has been cut into (finite) collection $W_1, \ldots, W_n$ of convex polyhedra of 'wall blocks' having disjoint interiors, and intersecting only in 2-dimensional faces common to an intersecting pair W_i, W_j of wall blocks (thus we exclude cases in which two sets W_i, W_j meet only at an edge or at a vertex). As before, let $K_i = K_i(\theta)$ denote the convex polyhedron $conv(ext(W_i) - ext(B_0))$. Let $w_1, \ldots, w_p$ denote all the wall vertices (these are fixed and independent of θ). Note that since some of the W_i's will have coincident faces, there will generally exist polyhedra K_i having coincident faces, i.e., if two wall polyhedra W_1, W_2 have a common face F, then for each extremity $e \in ext(B_0)$ the polyhedra K_1 and K_2 can have $F - e$ as a common face. Let us agree to call such common faces *a priori coincident* faces of K_1 and K_2. Any such a priori coincidence of faces exists independently of θ, and so is an invariant feature of our problem. In the presence of such faces condition (*) will fail *a priori* to hold for the collection $\{K_i\}$ of polyhedra. However, the a priori coincident faces can be ignored in the analysis of Section 2, since they can never form any part of the boundary of the complement of the polyhedra K_i. The presence of a priori coincidences requires only obvious changes in the arguments used in the preceding section and with obvious rewordings they remain valid, so that connected components of $P(\theta)$ can still be found and labeled using the recursive procedure given in Section 2.

Note also that redundancies of hyperplanes bounding the polyhedra $K_i(\theta)$ occur when a face of such a polyhedron shrinks to a face of lower dimension. This however occurs only when two distinct vertices of the displaced polyhedra come together, a situation which occurs only at orientations at which two corners of the rod touch two wall corners. Hence if we rule out degenerate wall configurations at which this can happen, we conclude that at a critical orientation either four faces meet at a point, or two faces simultaneously intersect and become parallel. For this latter kind of criticality, we will also assume that no two wall faces are parallel to one another.

The following revised definition indicates what reformulation of the concept 'critical orientation' is necessary.

DEFINITION: (a) An orientation θ is called *noncritical* if the following modified version of (*) holds:
(**) No four faces of the polyhedra $\{K_i(\theta)\}$, no two of which are a priori overlapping faces, have a non-empty intersection, and no two faces which are not a priori coincident simultaneously intersect and become parallel.
(b) An orientation θ for which condition (**) is violated is called a *critical* orientation.

More on the Geometry of Critical Orientations
The crude but useful classification of critical orientations θ introduced at the end of the preceding section can not be repeated, and separates such orientations into the five following categories:

Type I critical orientation: These are orientations θ for which a vertex of one polyhedron $K_i(\theta)$ lies on a face of another such polyhedron.

Type II critical orientation: These are orientations θ for which an edge of one polyhedron $K_i(\theta)$ meets an edge of another such polyhedron.

Type III critical orientation: These are orientations θ for which a face of one polyhedron K_1, a face of another K_2, and an edge of a third polyhedron K_3 meet at a point.

Type IV critical orientation: These are orientations θ for which four faces, each of a different polyhedron, meet at a point.

Type V critical orientation: These are orientations θ for which two faces simultaneously intersect and become parallel.

The following lemma gives a more direct characterization of faces of the polyhedra $K_i(\theta)$.

LEMMA 12: Let K and K' be two convex 3-dimensional polyhedra. Then the boundary of $K - K'$ is a union of 2-dimensional polygons, which we call *subfaces* of $K - K'$, each of which has one of the three following forms
(a) (a face of K) $-$ (a vertex of K');
(b) (an edge of K) $-$ (an edge of K');
(c) (a vertex of K) $-$ (a face of K').

Similarly, each edge of $K - K'$ is either an edge of K minus a vertex of K', or a vertex of K minus an edge of K', and finally each vertex of $K - K'$ is the difference of a vertex of K and a vertex of K'.

Proof: A straightforward exercise in convex analysis. Note that some of the polygons (a)–(c) may be coplanar, and that they may overlap one another. Q.E.D.

Remarks: (1) In the rod case which concerns us, the second polyhedron K' in the lemma is B_0 and is therefore a degenerate polyhedron having no (2-dimensional) faces. Thus case (c) cannot arise in this context. Moreover, when K' is a rod, subfaces S_1 of $K - K'$ having type (a) can be written as $S_1 = F_1 - V_1$, where F_1 is a face of some wall polyhedron W_1, and where V_1 is either $P = 0$ or $Q = L\theta$. Then $X \in S_1$ means that if B is placed with the point P at X and is given orientation θ, then the vertex V_1 will lie on the face F_1 of the walls. The reader will find it easy to give a similar geometric interpretation for subfaces of type (b).

(2) As already noted in the proof of Lemma 12, for some orientations θ two subfaces of some displaced polyhedron may be coplanar. For example, if the direction θ lies in the plane of some face wall F, then the old face F, the face F displaced by $L\theta$, and some boundary edge e of F displaced by B_0 will be all coplanar subfaces of the displaced polyhedron, and they might also overlap one another. This situation occurs precisely at critical orientations of type V.

It is not hard to show that the converse of the condition (**) appearing in the definition of a critical orientation θ can be formulated as follows:

(***) There exist four subfaces $S_1, \ldots, S_4$ of the polyhedra $K_i(\theta)$, none of which lies on an a priori coincident face of the K_i's, which meet at a common point X.

By considering all possible combinations of four subfaces, each being of one of the types (a) or (b) appearing in Lemma 12, we can show that only the following 8 critical configurations enter into our rod-motion problem, provided that the geometry of the walls is not 'exceptional' relative to the length of the rod, in the manner described at the end of the preceding section.

Type L1: orientations θ for which one endpoint of the ladder B can touch a wall corner, while the other endpoint lies on a wall face.

Type L2: orientations θ for which both endpoints of B can lie on wall edges.

Type L3: orientations θ for which one endpoint of B can touch a wall corner, while some point interior to or at the end of B meets a wall edge.

Type L4: orientations θ for which one endpoint of B can lie on a wall edge, while the other point lies on a wall face, and some point interior to B touches a wall edge.

Type L5: orientations θ for which B can simultaneously touch three wall edges. (Or, equivalently, one endpoint lies on an edge while two other points touch other edges.)

Type L6: orientations θ for which B can simultaneously touch two wall edges, while both of its endpoints lie on wall faces.

Type L7: orientations θ for which B can touch a wall corner, while both of its endpoints lie on wall faces.

Type L8: orientations θ which are parallel to some wall face.

That these are the only critical orientations that need concern us can be deduced as follows. Faces of one or more polyhedra $K(\theta)$ can become parallel to one another when the rod touches two parallel edges (which we ignore by assuming that no two edges of our polyhedra are parallel) or when the rod becomes parallel to a wall face (case (L8)). By Lemma 12, every face, edge, and corner of a polyhedron $K(\theta)$ is either an 'old' face, edge or corner (that is, a face, edge or corner of one of the given polyhedral walls) which we write as 'of', 'oe', 'oc' respectively; or is 'new.' A point on a new face either lies on an old face displaced through the length L of the rod (which has orientation θ), which we write as 'df', or at a position of the endpoint of the rod at which the rod would touch an old edge at some point interior to, or at the end of the rod, which we write as 'tc'. The possible new edges, written in a similar notation, are 'de' and 'te'; the possible new corners are just 'dc.'

ﾠTo form a critical intersection, these various elements must be combined in one of the four possible codimensional patterns $1+1+1+1$, $2+1+1$, $2+2$, and $3+1$ noted above. No intersection can contain two elements bearing the prefix 'o', since this would correspond to a nonexistent singularity of the walls (e..g, two wall faces belonging to different wall polyhedra would have to intersect); similarly, no two elements bearing the prefix 'd' can occur.

Using these observations, and noting symmetries where possible, we see at once that the possible patterns of critical intersection for the codimensional pattern $1+1+1+1$ are:

(a)   *of, df, te, te*
(b)   *of, te, te, te*
(c)   *te, te, te, te*

(Here we have, e.g., used the fact that the two endpoints of the rod are symmetric, so that there is no need to distinguish between pattern (b) and the ostensibly different pattern *df, te, te, te*.) Pattern (a) describes a situation in which the two ends of the rod touch two wall faces, while two of its interior points touch edges; this is our critical configuration L6. On the other hand, pattern (b) is a subcase of L5, which (as will be seen just below) is already critical. Pattern (c) can arise only in degenerate wall configurations, in which four distinct edges intersect the same line.

For the codimensional pattern $2+1+1$ we have the possibilities

(d)   *eo, df, te*
(e)   *oe, te, te*
(f)   *tc, of, df*
(g)   *tc, of, te*
(h)   *tc, te, te*

(Here again, we have used the 'o' and 'd' symmetry to limit our enumeration.) Case (d) is L4. Case (e) is equivalent to L5, since given any configuration in which the rod touches three edges we can slide it, without change of orientation, along its own length until one of its endpoint comes to lie along an edge. Case (f) is our critical intersection L7; case (g) is just the subcase of L3 in which the rod has been pushed along its length, without change of orientation, till one of its points touches a wall face. Case (h) will occur only for isolated directions, and hence can be ignored; it is in effect an exceptional case of L5.

Next we consider the codimensional pattern 2+2. Here the possibilities are

(i)   *oe, de*
(j)   *oe, tc*
(k)   *tc, tc*

Possibility (i) is just case L2, and case (j) can be converted into L3 by sliding the rod along its own length till one of its endpoints lies along the edge that it originally touches. Case (k) can clearly occur only for isolated orientations, and thus can safely be ignored.

Finally we must consider the codimensional pattern 3+1. The possibilities here are

(l)   *oc, df*
(m)   *oc, de*

Case (l) is L1, and as already explained case (m) is equivalent to L3.

Thus the list L1–L8 appearing above exhausts all the critical configurations that need concern us.

As we shall see below, some crucial geometric details of $P(\theta)$ at a critical orientation θ will depend on its type. Nevertheless we can make a few general remarks which apply to all eight types of critical curves. To do this, let θ be a critical orientation, which we can regard as a point $(\theta_1, \theta_2, \theta_3)$ on the unit 3-sphere. Critical orientations of type L8 lie on great circles on the sphere. To handle the other types of critical curves, we write the equations of the planes containing the four subfaces S_i whose nonempty intersection defines the criticality of θ, and assume for simplicity that the length L of the rod is 1. Each of the subfaces S can have one of the three following forms:

(i) S is a subface of type (a) having the form $F - P = F$, where F is a face of one of the wall polyhedra. In this case the coefficients of the plane containing S are constants independent of θ.

(ii) S is a subface of type (a) having the form $F - Q = F - \theta$, where F is as in (1). In this case, the coefficients of the plane containing S are a, b, c, $d - a\theta_1 - b\theta_2 - c\theta_3$, where a, b, c, d are the coefficients of the plane containing the face F.

(iii) S is a subface of type (b) having the form $e - B$, where e is one of the wall edges. Let (a,b,c) be a point on e and let (k,l,m) be a vector in the direction of e. Then the coefficients of the plane containing S are easily seen to be the modified cross product

$$\alpha = l\theta_3 - m\theta_2, \ \beta = m\theta_1 - k\theta_3, \ \gamma = k\theta_2 - l\theta_1, \ -a\alpha - b\beta - c\gamma$$

Write the equations of the four planes containing S_j, $j=1, \ldots ,4$ as

$$a_1x + b_1y + c_1z + d_1 = 0$$

$$a_2x + b_2y + c_2z + d_2 = 0$$

$$a_3x + b_3y + c_3z + d_3 = 0$$

$$a_4x + b_4y + c_4z + d_4 = 0$$

so that these planes will meet at a point (which may lie at infinity) if and only if we have

$$\begin{vmatrix} a_1(\theta) & b_1(\theta) & c_1(\theta) & d_1(\theta) \\ a_2(\theta) & b_2(\theta) & c_2(\theta) & d_2(\theta) \\ a_3(\theta) & b_3(\theta) & c_3(\theta) & d_3(\theta) \\ a_4(\theta) & b_4(\theta) & c_4(\theta) & d_4(\theta) \end{vmatrix} = 0 \tag{1}$$

To obtain the algebraic form of the critical curves corresponding to each of the eight geometric types of criticality, we have only to specialize condition (1) to each of these types, as follows.

Type L1: Here three of the intersecting planes are of type (i) and one of type (ii) (or vice versa). Equation (1) then clearly defines a plane in θ, whose intersection with the unit sphere is a circle.

Type L2: Here two of the intersecting planes are of type (i) and two are of type (ii). Again, Equation (1) defines a plane in θ, so that the corresponding critical curve is also a circle.

Type L3: Here three of the intersecting planes are of type (i) or of type (ii), and one is of type (iii). It is easy to see geometrically that equation (1) must define a plane in θ which passes through the origin, so that the corresponding critical curve is a great circle.

Type L4: Here two of the intersecting planes are of type (i), one is of type (ii) and one of type (iii). In this case Equation (1) defines a quadric surface in θ, whose intersection with the unit sphere gives a curve which assumes one of the forms discussed in Ocken, Schwartz, and Sharir (1983).

Type L5: Here two of the intersecting planes are of type (i) and two are of type (iii). Again, it is easily seen that Equation (1) defines a quadric surface in θ, leading to critical curves as in type L4 above.

Type L6: Here one of the intersecting planes is of type (i), one of type (ii) and two of type (iii). In this case Equation (1) defines a cubic surface in θ, and the corresponding critical curve is the intersection of this surface with the unit sphere.

Type L7: Critical orientations of this type can be regarded as degenerate variants of L6, in which the two edges of L6 have come together at a point. Hence type L7 critical curves are also intersections of the unit sphere with a cubic surface.

Finally, as already noted, type L8 critical curves are simply great circles.

It follows from the foregoing discussion that the critical orientations lie on finitely many algebraic curves which decompose the set of noncritical orientations into finitely many *noncritical regions*, i.e., connected open regions of noncritical orientations. Let R be such a region. Lemma 1 implies that for each $\theta \in R$ the set $P(\theta)$ has a fixed number of connected components, each of which varies continuously with θ, and our analysis of the polyhedral case assigns each such component a label which does not depend on θ. By $\sigma(R)$ we shall therefore denote the set of all labels of connected components of $P(\theta)$ for any, hence every, $\theta \in R$, and as in our preceding papers on the movers' problem we call this set the *characteristic* of R. Before continuing with the analysis of the structure of FP, we first give a few significant details relevant to the specialization of the labeling procedure described in Section 3 to the case of a rod. First, we need to modify the labeling scheme because, since our displaced polyhedra depend continuously on θ, the O-faces of the complement $P(\theta)$ of the union of the polyhedra $K_i(\theta)$ will generally depend on θ, so that the labeling of section 2, which represents each such vertex by its coordinates, will not be discrete, but depend on the continuous parameter θ. To avoid this difficulty, we can change the labeling scheme simply by changing the representation of vertices of $P(\theta)$ as follows. Each such vertex X is the unique intersection of three faces $F_j(\theta)$, $j=1,2,3$, and of no other face of $K_i(\theta)$. Moreover, each face (or rather each subface) of the $K_i(\theta)$, is, by Lemma 12, the difference of a wall face and a rod vertex or of a wall edge and the rod itself. Thus, each such subface can be discretely labeled in one of the forms $[W,P]$, $[W,Q]$ (designating a difference of a wall face W and an appropriate endpoint of B), or $[E,B]$ (designating a difference of a wall edge E and this same B). These labels for faces induce obvious discrete labeling for the corners of $P(\theta)$, namely we can label each such corner by the triple of labels of the faces which pass through it. These labels can then be extended to label edges, faces, and finally components of $P(\theta)$ in the manner described in Section 2 (see also the proof of Lemma 9). To bound the complexity of this labeling procedure, recall that it proceeds recursively downward in dimension, and decomposes each face of the polyhedra that it processes into cells, and classifies these cells as being either subsets of one of the given polyhedra, or as bounding the "free" complemented space. Consider the recursive step of the procedure in which 2-dimensional faces are analyzed. Assume that the polyhedra constituting the boundary of V have altogether $O(n)$ vertices, edges and faces. Each face of each of the displaced polyhedra is either a face of a wall polyhedron, or a face of a wall polyhedron displaced by the vector $L\theta$, or the area swept as a wall edge is displaced

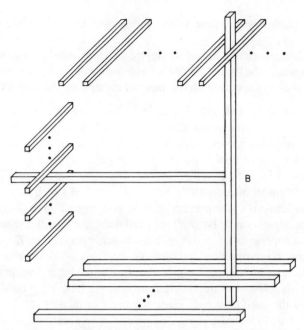

Figure 5. A wall configuration and an orientation θ for which $P(\theta)$ has $\Omega(n^3)$ components in the case of a moving polyhedral body.

by the vector $L\theta$. Thus the displaced polyhedra have altogether at most $O(n)$ vertices, edges and faces. Each 2-dimensional face F of a displaced polyhedron can intersect other displaced polyhedra in a convex polygon, and the portion F_0 of F bounding the free space is the complement of the union of these polygons. We claim that the number of corners and edges in this portion of F, summed over all faces F of the displaced polyhedra, is at most $O(n^3)$, which implies that the maximal number of connected components of $P(\theta)$ for any orientation θ, and also the size of the label for each connected component of $P(\theta)$, is at most $O(n^3)$. Indeed, our claim is obvious, because a corner X of F_0 is either a displaced corner, or the intersection of an edge of one of the $K_i(\theta)$ with a face of some other $K_j(\theta)$, or a point at which three faces F, F', F'' of the displaced polyhedra meet each other, etc. so that there are $O(n^3)$ such corners in all. It seems rather likely that the actual number of components of $P(\theta)$ in the case of a rod is smaller, possibly $O(n^2)$. However, if instead of a rod we consider a general polyedron moving amidst polyhedral walls in 3-space, the example given in Figure 5 shows that there exists a layout of n wall faces for which there exist orientations θ such that $P(\theta)$ contains $\Omega(n^3)$ connectec components.

Note also that, if θ is not a critical orientation, then each corner C of $P(\theta)$ belongs to the boundary of exactly one connected component of $P(\theta)$. Indeed, if C

is the intersection point of three faces $F_j(\theta)$, $j = 1,2,3$, then the intersection of $P(\theta)$ with a sufficiently small neighborhood of C is contained in the intersection of three open half spaces each bounded by the plane passing through one of the faces F_j, and this intersection is contained in just one component of $P(\theta)$.

The considerations set forth in the preceding paragraphs lead to the following lemmas.

DEFINITION: Let R be a noncritical region, and let $L \in \sigma(R)$. For each $\theta \in R$ we let $\phi(\theta,L)$ denote the component of $P(\theta)$ whose label is L.

LEMMA 13:   Let R be a noncritical region, and let $L \in \sigma(R)$. Then for each $\theta \in R$ the set $\phi(\theta,L)$ can be expressed as a Boolean combination of nonempty convex sets having the form

$$\phi(\theta,L) = \bigcup_{i,j_1,\ldots,i,j_r} \left(\bigcap_{l=1}^{r} F_{i_l j_l}^{+}(\theta) \right)$$

where $i_1, \ldots, i_r$ range over distinct displaced polyhedra, and where for each $l = 1, \ldots, r$, $F_{i_l j_l}(\theta)$ is a face of the i_l-th polyhedron $K_{i_l}(\theta)$, and $F_{i_l j_l}^{+}(\theta)$ is the open half-space bounded by the plane passing through $F_{i_l j_l}(\theta)$ and not containing $K_{i_l}(\theta)$.

Proof: By definition, for each $\theta \in R$ we have

$$P(\theta) = \left[\bigcup_i K_i(\theta) \right]^c$$

$$= \left[\bigcup_i \bigcap_j F_{ij}^{-}(\theta) \right]^c$$

where $F_{ij}(\theta)$ is the j-th face of the i-th polyhedron, and where $F_{ij}^{-}(\theta)$ is the closed half-space bounded by the plane passing through $F_{ij}(\theta)$ anc containing $K_i(\theta)$. Thus,

$$P(\theta) = \bigcap_i \bigcup_j F_{ij}^{+}(\theta)$$

Using distributivity, we see that $P(\theta)$, and hence each of its connected components $\phi(\theta,L)$, is the union of (nonempty and convex) intersections of the form $\bigcap_{l=1}^{r} F_{i_l j_l}^{+}(\theta)$, as asserted. Q.E.D.

LEMMA 14:   Let R be a noncritical region, and let

$$C(R) = \{[X,\theta] : \theta \in R, X \in P(\theta)\}.$$

Then the connected components of $C(R)$ are the sets

$$C(R,L) = \{[X,\theta] : \theta \in R, X \in \psi(\theta,L)\}$$

where $L \in \sigma(R)$ and where $\psi(\theta,L)$ is defined to be the connected component of $P(\theta)$ having the label L.

Proof: We first show that each of the sets $C(R,L)$ is connected. Let $[X_1,\theta_1]$, $[X_2, \theta_2] \in C(R,L)$ so that $X_i \in \psi(\theta_j,L)$, $j=1,2$. Let M_j be a corner of $\psi(\theta_j,L)$, i.e., a point where three faces $F_1(\theta_j)$, $F_2(\theta_j)$, $F_3(\theta_j)$ of the displaced polyhedra meet, such that the two points M_1 and M_2 have the same label (i.e., for each $i=1, \ldots , 3$ the faces $F_i(\theta_1)$ and $F_i(\theta_2)$ have the same label). Let $\alpha(t)$, $t \in [0,1]$ be a continuous path connecting θ_1 and θ_2 in R. Then, since α does not cross any critical curve, it follows that the three faces $F_i(\alpha(t))$, $i=1, \ldots ,3$ vary continuously with t, and that the side of $F_i(\alpha(t))$ contained in $\psi(\alpha(t),L)$ remains constant. Let $M(t)$, $e_1(t)$, $e_2(t)$, and $e_3(t)$ denote respectively the point at which the faces $F_i(\alpha(t))$ intersect, and three unit vectors pointing from $M(t)$ along the three edges at which these faces intersect in pairs. Plainly, all these quantities vary continuously with t. It follows that, if $\delta > 0$ is sufficiently small, then the point

$$y(t) \equiv M(t) + \delta(\varepsilon_1 e_1(t) + \varepsilon_2 e_2(t) + \varepsilon_3 e_3(t))$$

is contained in $\psi(\alpha(t),L)$ for all $t \in [0,1]$, for some *fixed* triple of signs ε_1, ε_2, ε_3 (e.g., $\varepsilon_1 = \varepsilon_2 = \varepsilon_3 = +1$, if M_1, M_2 are convex corners; $\varepsilon_1 = \varepsilon_2 = \varepsilon_3 = -1$, if M_1, M_2 are concave corners, etc.). Hence the continuous path $t \rightarrow (\alpha(t), y(t))$ connects $[X_1,\theta_1]$ to $[X_2,\theta_2]$ in $C(R,L)$, proving that this set is connected. To conclude the proof, since the sets $C(R,L)$ are obviously disjoint from each other, it suffices to show that these sets are open. However, Lemma 13 implies that $C(R,L)$ can be expressed in the form

$$C(R,L) = \bigcup_{i_1 j_1, \ldots, i_r j_r} (\bigcap_{l=1}^{r} \{[X,\theta] : \theta \in R, X \in F_{i_l j_l}^{+}(\theta)\})$$

and since $F_{i^+ ji}(\theta)$ varies continuously with θ, each of the individual sets appearing in the above equation is plainly open in $C(R)$. This completes the proof of the lemma. Q.E.D.

LEMMA 15:   Let β be a smooth critical curve section not intersected by other critical curves, let R be a noncritical region bounded by β, and let $L \in \sigma(R)$. Put $\xi = [R,L]$. Then for each $\theta \in \beta$ and each sequence of orientations $\theta_n \in R$ converging to θ, the sets $\psi(\theta_n,L)$ converge (in the Hausdorff metric of sets) to a unique closed limit set, which we denote by $\phi(\theta,\xi)$, whose interior is contained in $P(\theta)$ and is a union of connected components of $P(\theta)$. Moreover, if $L' \neq L \in \sigma(R)$ and $\eta = [R,L']$, then $int(\phi(\theta,\xi))$ and $\phi(\theta,\eta)$ are disjoint from one another for each $\theta \in \beta$. Thus for each $Z \in V$ the set $\{\theta \in \beta : Z \in int(\phi(\theta,\xi))\}$ is open in β.

Proof: By Lemma 13 we can write

$$\psi(\theta_n,L) = \bigcup_{i_1 j_1, \ldots, i_r j_r} (\bigcap_{l=1}^{r} F_{i_l j_l}^{+}(\theta_n))$$

Each of the half-spaces $F_{iji}^+(\theta)$ depends continuously on θ throughout the whole space of admissible orientations. Thus as $\theta_n \to \theta$, the sets $\psi(\theta_n,L)$ converge to a similar union of intersections of closed half-spaces of the form $\bigcap_{l=1}^{r} closure(F_{iji}^+(\theta))$, and this is the set $\phi(\theta,\xi)$. The interior of this set is equal to the union of the intersections of the corresponding *open* half spaces $F_{iji}^+(\theta)$, which is plainly a subset of $P(\theta)$. Moreover, any point Z in the boundary of $\phi(\theta,\xi)$ is plainly a limit of points Z_n on the boundaries of the sets $\psi(\theta_n,L)$. Since each position $[Z_n,\theta_n]$ is semi-free but not free, so is $[Z,\theta]$. Thus the boundary of $\phi(\theta,\xi)$ is contained in $SFP - FP$, where SFP is the set of all *semi-free configurations* of B, that is configurations at which B is either free or touches some wall, but does not penetrate any wall. The set $\phi(\theta,\xi)$ need not be connected, but our last observations imply that $\phi(\theta,\xi)$ contains every connected component of $P(\theta)$ which it intersects.

This proves the first part of the lemma. As to the second part, note that since each of the intersections appearing in the formula displayed above is open, it follows that two such intersections can overlap at θ if they overlap for all $\theta' \in R$ sufficiently near θ. Hence two of the terms appearing in the definitions of $int(\phi(\theta,\xi))$ and $\phi(\theta,\eta)$ can overlap only if the connected components to which they belong would be identical for all θ' near θ. But this would imply $\xi = \eta$, proving that $int(\phi(0,\xi))$ and $\phi(\theta,\eta)$ are disjoint if $\xi \neq \eta$. Finally, we show that for each $Z \in V$ the set $\beta(Z) \equiv \{\theta \in \beta : Z \in int(\phi(\theta,\xi))\}$ is open in β. For this, let θ belong to $\beta(Z)$. By what precedes, $[Z,\theta]$ is a free position of the rod B, and hence for all $\theta' \in R$ lying in a sufficiently small neighborhood U of θ, the point Z belongs to some set $\psi(\theta,L')$ with L' *fixed*. (Indeed, choose U to be an arcwise connected neighborhood of θ in R sufficiently small so that $[Z,\theta'] \subset FP$ for all $\theta' \in U$. Then $\{Z\}\times U$ is a connected subset of $C(R)$, and so must be contained in one of its connected components.) Therefore $Z \in \phi(\theta,[R,L'])$, and since we have just shown that $\phi(\theta,[R,L'])$ and $int(\phi(\theta,[R,L]))$ are disjoint if $L' \neq L$, we must have $L' = L$. It therefore follows that $Z \in int(\phi(\theta',\xi))$ for all $\theta' \in \beta$ sufficiently near θ, so that the set $\beta(Z)$ must be open, as asserted. Q.E.D.

LEMMA 16: Suppose that (a portion of) a smooth critical curve β separates two connected noncritical regions R_1 and R_2 and that $R_1\cup R_2\cup\beta$ is open. Let L_1 (resp. L_2) be an element of $\sigma(R_1)$ (resp. $\sigma(R_2)$). Put $\xi_1 = [R_1,L_1]$ and $\xi_2 = [R_2,L_2]$, and let $C_1 = C(\xi_1)$, $C_2 = C(\xi_2)$. Then the following conditions are equivalent:
Condition A: There exists a point $\theta \in \beta$ such that the open sets $int(\phi(\theta,\xi_1))$, $int(\phi(\theta,\xi_2))$ have a non-null intersection.
Condition B: There exists a smooth path $c(t) = [z(t),\theta(t)]$ in FP which has the following properties:
(i) $c(0) \in C_1$ and $c(1) \in C_2$;
(ii) $\theta(t) \in R_1\cup R_2\cup\beta$ for all $0\leq t\leq 1$;
(iii) $\theta(t)$ crosses β just once, transversally, when $t=t_0$, $0<t_0<1$, and $Z(t)$ is constant for t in the vicinity of t_0.

Proof: The proof is completely analogous to that of Lemma 1.8 of Schwartz and Sharir (1983c), and moreover is general and purely topological. Hence we omit it here.

LEMMA 17: Let the smooth critical curve β separate the two noncritical regions R_1 and R_2. Let β' be a connected open segment of β not intersecting any other critical curve, and suppose that $R_1 \cup R_2 \cup \beta'$ is open. Let L_1, L_2, ξ_1, and ξ_2 be defined as in Lemma 5. Then the set

$$M \equiv \{\theta \in \beta' : int(\phi(\theta, \xi_1)) \cap int(\phi(\theta, \xi_2)) \neq \varnothing\}$$

is either all of β' or is empty.

Proof: Since M is the union of all sets of the form

$$\{\theta \in \beta' : Z \in int(\phi(\theta, \xi_1))\} \cap \{\theta \in \beta' : Z \in int(\phi(\theta, \xi_2))\},$$

for $Z \in V$, and since by Lemma 15 each of these sets is open, it follows that M is open. Hence we have only to show that M is closed. Suppose the contrary; then there exists $\theta \in \beta'$ such that $int(\phi(\theta, \xi_1))$ and $int(\phi(\theta, \xi_2))$ are disjoint, but for which there also exists a sequence θ_n of points on β' converging to θ such that for all n the sets $int(\phi(\theta, \xi_1))$ and $int(\phi(\theta, \xi_2))$ intersect each other.

By Lemma 15, for each n the set

$$D_n \equiv int(\phi(\theta_n, \xi_1)) \cap int(\phi(\theta_n, \xi_2))$$

is a union of components of $P(\theta_n)$. Passing to a subsequence if necessary, we can assume that each D_n contains a connected component C_n of $P(\theta_n)$ such that all the C_n's have the same label.

Now since it does not meet any other critical curve other than β, the subsection β' of β is characterized by the property that for each $\theta \in \beta'$ four fixed faces $H_j(\theta), j=1, \ldots, 4$ of the displaced polyhedra $K_i(\theta)$ meet at a common point $Q(\theta)$ while no four other faces meet. Since each of the connected components C_n must contain more than one corner, we can assume, passing to a subsequence if necessary, that for each n there exists a corner U_n of C_n which does not belong to the intersection of the faces $H_j(\theta_n)$, such that all these corners have the same label, i.e., U_n is the unique point of intersection of three and only three faces $F_1(\theta_n)$, $F_2(\theta n)$, and $F_3(\theta_n)$ of the polyhedra $K_i(\theta n)$, which bear fixed designations independent of $\theta = \theta_n$. It is then clear that the sequence U_n converges to some point U as $n \to \infty$.

At least one of the three faces $F_j(\theta_n)$, for specificity say $F_1(\theta_n)$, must be different from any of the four faces H_j, and for each n the intersection of $P(\theta_n)$ with a sufficiently small neighborhood of U_n is contained in C_n (i.e., U_n is not a corner common to more than one connected component of $P(\theta_n)$). By continuity,

$$U \in \bigcap_{j=1}^{3} F_j(\theta);$$ moreover, the three faces $F_j(\theta), j = 1,2,3$ must still meet at a single point through which no other plane passes, for otherwise at θ either $F_1(\theta)$ would be

a fifth face passing through $Q(\theta)$ or θ would lie on the critical curve defined by some other set of four intersecting faces. Thus in either case θ would have to lie at the intersection of two different critical curves, contrary to our assumption concerning β'. For large n, the three faces F_1, F_2, F_3 will retain their independence and move continuously to the corresponding faces of the polyhedra $K_i(\theta)$ and the interior of the intersection of a small sphere about U with the interior of the region bounded by these faces remains connected. Hence the intersection of $P(\theta_n)$ with any sufficiently small sphere about U is connected for n large and therefore is contained in some fixed component C_n of $P(\theta_n)$.

Hence every open neighborhood of U has a nonempty intersection with $P(\theta)$. Let N be a sufficiently small neighborhood of the origin, and take a point $W \in P(\theta) \cap (U + N)$. Since $[W, \theta] \in FP$ it follows easily that there exist $L'_j \in \sigma(R_j)$, $j=1,2$ such that W belongs to the intersection of the two sets $\phi(\theta, \eta_j)$, where $\eta_j = [R_j, L'_j]$. Thus, by the last assertion of Lemma 15, $W \in int(\phi(\theta_n, \eta_j))$ for all sufficiently large n and for both $j=1$ and $j=2$. But for all sufficiently large n the point W also belongs to $P(\theta_n) \cap (U_n + N)$, and since we have shown that the intersection of $P(\theta_n)$ with every sufficiently small neighborhood of U belongs to C_n, W must belong to C_n, and thus to $int(\phi(\theta_n, \xi_j))$ for both $j=1$ and $j=2$. Lemma 15 now implies that $\eta_j = \xi_j$, $j=1,2$, so that the sets $int(\phi(\theta, \xi_j))$, $j=1,2$ have a nonempty intersection, contrary to assumption.

This proves that M is also closed, which completes the proof of the lemma. Q.E.D.

Using the lemmas that have now been established, we can proceed with our analysis in a manner completely analogous to that used in our preceding papers (e.g., Schwartz and Sharir (1983a,c)). That is, we can define a *connectivity graph* CG whose nodes are the cells $[R,L]$ of FP, where R is a noncritical region and where $L \in \sigma(R)$, and whose edges connect adjacent cells in FP, i.e., cells $[R_1,L_1]$ and $[R_2,L_2]$ which satisfy the conditions of Lemma 5 at any (hence every) point θ on the critical curve section β separating the adjacent noncritical regions R_1 and R_2.

The connectivity graph is a finite combinatorial object which models the connectivity of FP exactly. The following main theorem, which is completely analogous to similar theorems of Schwartz and Sharir (1983a,c), makes this point:

THEOREM 1:   There exists a continuous motion c of B through the space FP of free configurations from an initial configuration $[X_1,\theta_1]$ to a final configuration $[X_2, \theta_2]$ if and only if the nodes $[R_1,L_1]$ and $[R_2,L_2]$ of the connectivity graph CG introduced above can be connected by a path in CG, where R_1, R_2 are the noncritical regions containing θ_1, θ_2 respectively, and where L_1 (resp. L_2) is the label of the connected component of $P(\theta_1)$ (resp. $P(\theta_2)$) containing X_1 (resp. X_2).

Remark: As in Schwartz and Sharir (1983a,c), we assume that θ_1 and θ_2 are not critical orientations. If either θ_1 or θ_2 lies on a critical curve, we first move θ_1 (or θ_2) into a noncritical region, and then apply the above theorem.

Proof: Completely analogous to the proof of Theorem 1.1 of Schwartz and Sharir (1983a) and Theorem 1.1 of Schwartz and Sharir (1983c), and hence omitted.

The algebraic form of the critical curves and the crossing rules associated with them have been described above. Moreover, the algorithms required for the construction of noncritical regions and the connectivity graph are quite similar in nature to those sketched in Schwartz and Sharir (1983a,c). For this reason we refrain from giving here additional geometric and algorithmic details concerning the motion-planning technique just outlined.

A crude upper bound on the size of the connectivity graph for an instance of the problem involving n wall faces, edges and vertices, is easily derived, as follows: The number of critical curves is at most $O(n^4)$ (each curve of type L6 is determined by four distinct wall objects chosen out of n, which is the worst case). Since all these curves have a bounded degree, the number of possible intersections between them, and thus also the number of noncritical regions, is $O(n^8)$. For each noncritical region R, the number of connected components of $P(\theta)$ for any $\theta \in R$ has been shown earlier to be at most $O(n^3)$. Hence the size of the connectivity graph CG is at most $O(n^{11})$, and it is easily seen that CG can be constructed within the time bound.

The motion planning algorithm we have derived is obtained by specializing the polyhedral theory developed in preceding sections to the case of a moving rod in 3-space. The next generalization derivable along these lines is to the case of a general polyhedral body B moving in 3-space amidst polyhedral barriers. Here we must replace the angular orientation θ by a point parametrizing the 3-dimensional group of rotations in E^3. As is well known (see Hamilton, 1969), we can parametrize this group by unit quaternions, and hence can think of θ as belonging to the unit sphere S^3 in 4-space. The moving body B can be decomposed into convex polyhedral parts, and then for each θ the set of free positions of the body becomes the complement of a collection of convex polyhedra of the form $K_i(\theta) = conv(ext(W_i) - ext(B_j(\theta)))$ where W_i designates some one of the convex parts into which the walls have been cut, and $B_j(\theta)$ designates some convex part of the body, rotated by the matrix corresponding to θ. By Lemma 12, the faces, edges, and corners of these $K_i(\theta)$ are as follows:

(a) faces, edges, and corners of the W_i, translated by some corner c of $B_j(\theta)$. For purposes of subsequent enumeration, we can designate these geometric elements as f_c, e_c, and c_c respectively.

(b) edges and corners of the W_i, swept respectively into faces and edges of $K_i(\theta)$ by differencing with some edge e of $B_j(\theta)$. We designate these elements as e_E and c_E respectively.

(c) corners of W_i swept to faces by differencing with some face f of $B_j(\theta)$, and designated as c_F.

Here the possible critical configurations of four face planes classify as follows. In the codimensional pattern $1+1+1+1$, of the four faces entering into the critical intersection, i can have the designation f_c (corresponding to positions in which a body corner lies along a wall face), j have the desingation e_E (corresponding to positions in which a body edge touches a wall edge), and k the designation c_F (wall corner on a body face). We must have $i+j+k = 4$, giving $5+4+3+2+1 = 15$ types of singularities. In the codimensional pattern $2+1+1$, an edge of one of the forms e_C or c_E combines with two faces. Then e_C designates situations in which a corner of the body lies along a wall edge, and c_E describes situations in which a body edge touches a wall corner. This gives $2*(3+2+1) = 12$ more singularity types. Three more types of singularity have the codimensional pattern $2+2$, and three have codimensional pattern $3+1$. To these we must also add orientations at which a wall edge becomes parallel to a body face, or at which a body edge becomes parallel to a wall face. Counting these as two more types of singular orientations, we obtain altogether $15+12+3+3+2$ or 35 possible types of critical orientations.

If rotations are parametrized by unit quaternions in the manner suggested above, all of these 2-dimensional critical subsurfaces of S^3 appear as algebraic surfaces of degree at most eight. Indeed, to apply the rotation θ to a vector x, we simply regard x as the quaternion with zero first component, and form the quaternion product $\theta x \bar{\theta}$, where $\bar{\theta}$ designates the conjugate of θ. This makes it plain that the various geometric coefficients of the rotated $B(\theta)$, and thus also of the displaced polyhedra $K_i(\theta)$, depend only quadratically on θ. Hence concurrency of four of the faces of these polyhedra occurs along the intersection of S^3 with an algebraic surface of degree at most eight. Hence, in this case, as in the simpler rod case, we have $O(n^4)$ critical surfaces, but since three of these surfaces can intersect to form corners of the noncritical regions into which they divide S^3, there can exist $O(n^{12})$ noncritical regions. In much the same way as in the case of a rod, for any fixed orientation θ the number of connected components of $P(\theta)$, can be shown to be at most $O(n^3)$. Hence there exists at most $O(n^{15})$ nodes in the connectivity graph.

Lemmas 13, 14, 15, 16, and 17 extend verbatim from the rod case to the more general case presently under consideration. Hence Theorem 1 extends to the case of a moving polyhedron as well. The crossing rules for critical surfaces can be derived from the general principles outlined at Section 3.

However, since in the case of a moving polyhedron the orientation space is three dimensional, its decomposition by critical surfaces into noncritical regions is more complicated than that of the 2-dimensional θ-space in the case of a rod. This can be done using techniques like those described in Ocken, Schwartz, and Sharir (1983) for the topological analysis of 3-dimensional bodies defined as general semi-algebraic sets. Roughly speaking, one proceeds by computing the intersection curves of all pairs of critical surfaces. Then, for each critical surface Σ, one considers the collection of all curves of intersection of Σ with every other critical surface. These curves partitions the 2-dimensional set Σ into connected regions, which can be computed in a way similar to the computation of noncritical regions in the case of a rod. After collecting all these regions on all critical surfaces Σ one can compute adjacency relationships between these regions. Two regions are said to be adjacent if they lie on two critical surfaces Σ_1, Σ_2, and if their boundaries overlap along a portion of the curve of intersection of Σ_1 with Σ_2. From these adjacency relationships one can easily construct connected components of the boundary of each noncritical region, in much the same way as the equivalence relation Ξ_1 was constructed in Section 2. To find what boundary components bound the same noncritical region, one can again use the technique described in Section 2, this time extending a great circle on S^3 connecting a point on each boundary component with some fixed point on S^3, and noting all its pairs of successive intersections with other components, and then equivalencing these pairs of components which must bound the same noncritical region (see Ocken, Schwartz, & Sharir, 1983 for more detail).

This completes our brief outline of the way in which the case of a moving polyhedron can be treated. We note that a restricted version of this problem has already been studied in Lozano-Perez and Wesley (1979). There, however, the moving polyhedron is assumed either not to rotate at all, or rotate only a few times at some discrete positions.

Remark: Other generalizations of this 'polyhedral' scheme to more complex systems, such as hinged polyhedral bodies, are also possible. Although such problems can be considered within the general scheme presented in this paper, their analysis becomes substantially more complicated due to the increasing number of nontranslational degrees of freedom involved.

CHAPTER 6

A Retraction Method for Planning the Motion of a Disc

COLM Ó'DÚNLAING
CHEE K. YAP

Courant Institute of Mathematical Sciences
New York University

A new approach to certain motion-planning problems in robotics is introduced. This approach is based on the use of a generalized Voronoi diagram, and reduces the search for a collision-free continuous motion to a search for a connected path along the edges of such a diagram. This approach yields an $O(n \log n)$ algorithm for planning an obstacle-avoiding motion of a single circular disc amid polygonal obstacles. Later papers will show that extensions of the approach can solve other motion-planning problems, including those of moving a straight line-segment or two coordinated discs in the plane amid polygonal obstacles.

0. INTRODUCTION

The Piano-Movers' Problem has been the subject of a number of recent papers (Hopcroft, Joseph & Whitesides, 1982; Reif, 1979; Schwartz & Sharir, 1983a,b). The 2-dimensional version of the problem which will concern us here is to determine whether there exists a continuous motion of a given rigid body B free to move within a 2-dimensional region Ω bounded by a set S of finitely many polygonal obstacles, between two specified placements, during which B continually avoids collision with these obstacles, and to produce such a motion when one exists. The method described in the previous papers is roughly as follows: for a given set S of obstacles and a body B, the set FP, of all pairs [position, orientation] which represent placements of B in which B is not incident on any obstacle in S, is an open manifold, and a continuous obstacle-avoiding path between two given placements of B exists if and only if these placements belong to the same connected component of FP. Thus the problem is reduced to that of finding the connected components of FP, which is solved by partitioning FP into a collection of smaller cells having relatively simple structure, and by determining the adjacency relationships between these cells. These relationships yield a finite 'connectivity graph,' which can then be searched to determine the connected components of FP. This approach can be applied in general situations, and has been shown in Schwartz and Sharir (1983a,b)

The encouragement and generous help which Jack Schwartz and Micha Sharir gave us are much appreciated and have greatly improved the quality of this paper.

to yield polynomial-time algorithms for planning the motion of various bodies or groups of bodies B.

In this paper we suggest a different approach to the movers' problem and apply it to the simple case in which the body B is a circular disc. A motion-planning algorithm of complexity $O(n \log n)$ is shown for this case. As suggested in Lozano-Perez and Wesley (1979) and used in Moravec (1980), the idea of solving the motion-planning problem for a disc circumscribing a 2-dimensional body of more complex shape can be used as an initial heuristic in motion-planning for the body. Only if no continuous motion exists for this disc should more refined algorithms (taking more detail of the body shape into account) be brought to bear.

Throughout this paper we assume that the collection S of obstacles is finite, and that each obstacle consists of a simple polygon together with its interior (or exterior).

The basic idea is to take the set S of obstacles and construct the associated *Voronoi diagram Vor(S)* (see Kirkpatrick, 1979). This diagram is a planar network of straight and parabolic arcs, and is characterized as the set of points which are equidistant from at least two distinct obstacles; it can be computed in time $O(n \log n)$ by an algorithm due to Kirkpatrick (1979) (a somewhat simpler $O(n \log^2 n)$ algorithm due to Lee and Drysdale, 1981, and an alternative $O(n \log n)$ algorithm of Yap's (1984), are also available). Let B be a circular disc of radius r. Our technique is based on the fact (proved below) that there exists a continuous obstacle-avoiding motion π of the disc B between two specified positions x_0 and x_1 of its center C if and only if there exists another continuous obstacle-avoiding motion π' of B from x_0 to x_1 during which, except for its initial and final portions (in which C is moved to and from the Voronoi diagram), the center C moves entirely along the Voronoi diagram $Vor(S)$. This enables us to reduce the problem to searching the Voronoi diagram.

1.   MOTION-PLANNING FOR A DISC

We first introduce some terminology. 'Distance' will always mean Euclidean distance in the plane. Let S be the set of obstacles, let r be the radius of the moving disc, and let x_0 and x_1 denote the two given positions of its center C, between which a continuous obstacle-avoiding motion of B is sought. Write Ω for the set of points not lying in any obstacle. An instance of our movers' problem is a quadruple $\langle S, x_0, x_1, r \rangle$, where S, x_0, x_1 and r are as defined above. A *solution path* π to $\langle S, x_0, x_1, r \rangle$ is a continuous path from x_0 to x_1 such that if a circle B of radius r is moved with its center along π it will not collide with (or touch) any of the objects in S.

Let P be any nonempty set of points in the plane. Given a point x, define the *Hausdorff distance* of x from P as

$$d(x,P) = inf\,\{d(x,p): p \; \varepsilon \; P\}.$$

It is fairly easy to show that for some p in the closure $\bar{P}$ of P, $d(x,P) = d(x,p) = d(x,\bar{P})$. It is also well-known that the Hausdorff distance $d(x,P)$ is continuous in x.

For the time being, let s be a nonempty compact convex set in the plane. Again, it is well-known that for every point x in the plane there is a unique point $\text{Nearest}_s(x)$ in s closest to x, and that the map $\text{Nearest}_s(\cdot)$ is a retraction of the plane onto s. (A retraction onto s is a continuous map which fixes every point in s.)

Now consider a set of obstacles S (which, by abuse of notation, we will identify with their union). Let Ω denote the (open) complement of S. Clearly the Hausdorff distance from x to S is attained on the boundary of S. Let $\textit{Clearance}(x)$ denote this Hausdorff distance. As noted above, $\text{Clearance}(\cdot)$ is continuous. Note that the boundary of S can be represented as a finite union of closed line-segments: thus $\text{Clearance}(\cdot)$ can be represented as the minimum of a finite set of continuous functions which are easily calculated.

For each point x in Ω define $\textit{Near}(x)$ to be the set of points y in (the boundary of) S such that $d(x,y) = \text{Clearance}(x)$. Since S is closed, $\text{Near}(x)$ is always nonempty. The $\textit{Voronoi diagram}\ Vor(S)$ of S is the set of points x in Ω such that $\text{Near}(x)$ contains more than one point.

LEMMA 1.   The diagram $Vor(S)$ is closed in Ω (though not in the plane).

Proof. It is enough to show that the set $W = \Omega \setminus Vor(S)$ is open. Let x be any point in W, so the closed disc D_x of radius clearance (x) intersects the complement Ω' of Ω at a unique point p. Let D be a slighly larger open disc centered at x such that $D \cap \Omega'$ is convex. Clearly for all x' in a sufficiently small neighborhood U of x in Ω, $D_{x'} \subseteq D$, and hence $D_x \cap \Omega'$ is convex: i.e., a unique point. Thus $U \subseteq W$, so W is open. Q.E.D.

Define the mapping Im from Ω onto $Vor(S)$ as follows: given a point x in Ω, choose any point p in $\text{Near}(x)$, and let H be the half-line from p through x. Then $Im(x)$ is the point closest to p in the intersection of H with $Vor(S)$. If H does not intersect $Vor(S)$ then $Im(x)$ is undefined.

LEMMA 2.   If Ω is bounded then $Im(x)$ is defined uniquely for all x in Ω, and Im is a retraction of Ω onto $Vor(S)$, i.e., a continuous map onto $Vor(S)$ which fixes every point in $Vor(S)$. Moreover, for any x in Ω, $\text{Clearance}(\cdot)$ is strictly increasing along the line-segment joining x to $Im(x)$.

Proof. First consider any point x in Ω; let p belong to $\text{Near}(x)$. Claim that for any point z on the open line-segment joining p to x, $\text{Near}(z) = \{p\}$. This is because, given q in $\text{Near}(z)$, $d(x,q) \geq d(x,p) = d(x,z) + d(z,p) \geq d(x,z) + d(z,q)$; therefore xzq is a straight angle, so $p = q$, as asserted.

In particular, if x belongs to $Vor(S)$ then $Im(x) = x$. This resolves the apparent ambiguity when $\text{Near}(x)$ is not a singleton. Therefore, to show that Im is well-defined, it is enough to show that given x in $\Omega \setminus Vor(S)$, so $\text{Near}(x)$ is a singleton $\{p\}$, the half-line H from p through x intersects $Vor(S)$.

Consider the points p_1 and p_2, defined to be either the endpoints of the wall-segment containing p (if p is not a corner), or the other endpoints of the two wall-segments touching p (if p is a corner). The angles xpp_1 and xpp_2 cannot be acute.

This implies that there exists a neighborhood U of p (in S, not in the plane) such that U and x lie on opposite sides of the line through p perpendicular to H. Thus, p is the unique point in U nearest to every point in H.

Since Ω is bounded, the set D of points x' in H such that Near(x') $\neq \{p\}$ is nonempty. Let y be the point nearest to p in the closure of D. Note that y does not lie in the open line-segment px. By continuity, $d(y,p) =$ Clearance(y) $\geqslant d(x,p)$, so y lies in Ω. Again by continuity, $d(y,p) = d(y,S \setminus U)$, so y lies in $Vor(S)$: indeed, $y = Im(x)$. This concludes the proof that Im is well-defined and fixes every point in $Vor(S)$.

It follows from the above argument that, if Near(x) $= \{p\}$, then Near(x') $= \{p\}$ for every point x' in the open line-segment joining x to $Im(x)$: thus Clearance($\cdot$) is strictly increasing along this segment, and in general, Clearance($Im(x)$) $\geqslant$ Clearance(x).

It remains to show continuity of Im. If x_n in $Vor(S)$ converges to x in Ω, then by Lemma 1 x is in $Vor(S)$, so $Im(x_n)$ also converges to x. Thus we need only consider a sequence x_n in $\Omega \setminus Vor(S)$ converging to some point x in Ω. By passing to a subsequence if necessary, we can assume that Near(x_n) $= \{p_n\}$ and $Im(x_n) = y_n$ where the sequences p_n and y_n are convergent, to p and y respectively. By continuity, Clearance(y) $\geqslant$ Clearance(x), so y is in Ω, so by Lemma 1 y is in $Vor(S)$. Again by continuity, x lies on the closed line-segment joining p to y, and $d(p,y) =$ Clearance (y) (so $p \; \varepsilon$ Near(y)). Therefore Near(z) $= \{p\}$ for every point z in the open line-segment joining p to y. It follows immediately that $y = Im(x)$. Q.E.D.

The retraction map Im is simpler than might appear from the proof of Lemma 2. Given that the boundary of S consists of straight line-segments, $Vor(S)$ is a planar graph whose edges are composed of finitely many straight and parabolic line-segments: the wall-edges and corners are respectively the directrices and foci for these parabolic segments, and the retraction Im moves each point x in a perpendicular direction away from the edge nearest x or in a radial direction outward from the corner nearest x. (Note that the Voronoi diagram as defined here is a subgraph of the generalized Voronoi diagram defined in Kirkpatrick, 1979, which includes straight edges perpendicular to the open wall-segments through their endpoints.)

We can now state and prove the main result of this paper:

THEOREM. Suppose that Ω is bounded, and x_0 and x_1 are free placements for the disc B in S. Then $\langle S,x_0,x_1,r \rangle$ admits a solution path π if and only if $\langle S,Im(x_0),im(x_1),r \rangle$ admits a solution path π' which lies entirely in $Vor(S)$.

Proof. Note that, since Ω is bounded, the map Im is well-defined and continuous.

First suppose that $\langle S,x_0,x_1,r \rangle$ admits a solution path π. Let $\pi' = Im(\pi)$. Clearly, π' lies entirely in $Vor(S)$ and joins $Im(x_0)$ to $Im(x_1)$. Furthermore, by Lemma 2, Clearance($Im(x')$) $\geqslant$ Clearance(x') $> r$ for every point x' in π, so π' is a solution path.

Conversely, let π' be a solution path joining $Im(x_0)$ to $Im(x_1)$ and lying entirely in

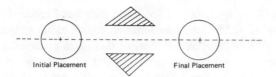

Figure 1. where the disc must leave the Voronoi diagram.

$Vor(S)$. Define π as follows: first move (the center C of) B directly from x_0 to $Im(x_0)$, then move C along π', and finally move C directly from $Im(x_1)$ to x_1. Since x_0 and x_1 are free placements, it again follows from Lemma 2 that every placement between x_0 and $Im(x_0)$ and between $Im(x_1)$ and x_1 is free, so π is a solution path for $\langle S, x_0, x_1, r \rangle$. Q.E.D.

Our method will not work if the space FP of free placements is not bounded. For example, in Figure 1, there is a motion from x_0 to x_1, but it must leave the Voronoi diagram. This problem can be fixed by 'framing' the region in which the disc is allowed to move; within this frame, all Voronoi cells are bounded, and the problem disappears.

To ensure that FP is bounded, choose a large rectangle M and define a further obstacle F (a frame) consisting of all points on or outside M. Choose M large enough so that all the corners in S (of which there are only finitely many), and also x_0 and x_1, are at a distance greater than $2r$ from F. Let the new set S' of obstacles be the union of S with $\{F\}$. We claim that there exists a solution path for $\langle S, x_0, x_1, r \rangle$ if and only if there exists a solution path for $\langle S', x_0, x_1, r \rangle$. To see this, note first that any solution path for $\langle S', x_0, x_1, r \rangle$ is also a solution path for $\langle S, x_0, x_1, r \rangle$. Conversely, any solution path π for $\langle S, x_0, x_1, r \rangle$ gives rise to a path π' for $\langle S', x_0, x_1, r \rangle$, as follows: Let M' be the set of points inside M which are at distance $(3r/2)$ from F. Note that M' is itself a closed rectangle. Then π' is obtained by 'truncating' those portions of π which lie outside M' and replacing them with appropriate segments of M'.

Remark. The above framing transformation is apparently parameterized by r, but it can be handled in a symbolic manner independent of the value of r. Alternatively, we can work in the projective plane and then $Im(p)$ is always defined (possibly as a point at infinity). This observation will be pertinent to the discussion in section 3.

The preceding discussion justifies the following algorithm:

1. (Preprocessing) If necessary, modify S by framing it along the lines described above. Decompose the boundary of S into a set W of corners and open wall-segments. Let n be the number of objects in W.
2. Construct the Voronoi diagram $Vor(S)$, using, for instance, Kirkpatrick's $O(n \log n)$ algorithm (1979). The diagram $Vor(S)$ is a union of $O(n)$ straight and parabolic arcs. Kirkpatrick's algorithm can specify it as a (combinatorial) graph N, in which each edge is labelled with an algebraic equation defining it as a curve in the plane, and each vertex is labelled with its co-ordinates. Finally,

from the equation defining an edge e and the co-ordinates of its endpoints, one can compute its *width*—the minimum clearance along e; this information can also be linked to e. The overall cost of this extra processing is $O(n)$.

3. (Motion planning) Given two points x_0 and x_1 in the plane, it is required to move the center of the disc B from x_0 to x_1 while not touching any obstacle. By considering in turn every object in W, one can compute the set Near(x_0) and the quantity Clearance(x_0). At this point, if Clearance(x_0) $\leq r$, the initial placement is not free: stop with an error message. If not, one can then compute $Im(x_0)$, as follows. If Near(x_0) contains more than one point, then $Im(x_0) = x_0$. Otherwise, say Near(x_0) = $\{p\}$. Again by considering all objects in W, one can locate the first point y_0 on the semi-infinite line H from p through x_0 such that y_0 is equidistant from p and some other point in S. Then $y_0 = Im(x_0)$. By searching the graph N, one can determine the edge e_0 containing y_0. Similarly, one can determine the edge e_1 of N containing $y_1 = Im(x_1)$, and stop if x_1 is not free. All this can be done in time $O(n)$.

4. If $e_0 = e_1$, then one can determine explicitly whether the clearance exceeds r between y_0 and y_1 along this edge. This takes constant time.

5. If $e_0 \neq e_1$, then one can determine which, if any, of the endpoints of e_0 can be reached from y_0 while maintaining sufficient clearance; similarly for y_1. Assuming that an endpoint z_0 of e_0 and z_1 of e_1 can be so reached, the motion-planning problem now reduces to finding a path from z_0 to z_1 in N along edges of width greater than r. This can be done in time $O(n)$.

3. FINAL REMARKS

It is instructive to compare our *Voronoi diagram* approach to that used in previous papers (Reif, 1979; Schwartz and Sharir, 1983a,b) which can be called the *critical region* approach. (We discuss only the case of a disc.)

When applied to a single moving disc, the critical region approach also gives rise to an $O(n \log n)$ algorithm. Indeed, this particularly simple path problem reduces to computing the "r-fringe" of the given set of obstacles S, where the r-fringe consists of those points in Ω which are at distance r from the nearest obstacle. It is easy to give an $O(n)$ algorithm for computing the r-fringe given the Voronoi diagram. Once this 'fringe' is available, the connected components of its complement are easily obtained. (However, we do not know how to obtain the r-fringe in $O(n \log n)$ time without using an $O(n \log n)$ algorithm to compute the Voronoi diagram).

In practical terms, the solution path obtained by the Voronoi diagram approach should be a particularly reliable way to avoid collisions, since it keeps the disc as far as possible from the obstacles. Again, note that for both approaches the preprocessing time is $O(n \log n)$ and the path-finding time is $O(n)$. However, in the critical region approach the data-structure established during preprocessing can only be used to plan motions for a disc of fixed size; a significant advantage of our approach is that the data-structure established in the preprocessing phase can be used to solve motion-planning problems for discs of any size.

CHAPTER 7

Retraction: A New Approach to Motion-Planning

COLM Ó'DÚNLAING
MICHA SHARIR
CHEE K. YAP
Courant Institute of Mathematical Sciences
New York University

The two-dimensional *Movers' Problem* may be stated as follows: Given a set of polygonal obstacles in the plane, and a two-dimensional robot system B, determine whether one can move B from a given placement to another without touching any obstacle, and plan such a motion when one exists. Efficient algorithms are presented for the two specific cases in which B is either a disc or a straight-line segment, running respectively in time $O(n \log n)$ and $O(n^2 \log n \log^* n)$. To solve the problem for a disc one uses the planar Voronoi diagram determined by the obstacles; in the case of a line-segment one generalizes the notion of Voronoi diagram to the 3-dimensional configuration space of the moving segment.

1. INTRODUCTION

This paper studies the following version of the motion-planning problem, known as the "Piano Movers" problem, which arises in robotics: Consider a rigid two-dimensional body B which is free to move within a two-dimensional space Ω bounded by polygonal obstacles, along with two specified placements Z_1 and Z_2 of B. Determine whether there exists a continuous motion that will take B from placement Z_1 to placement Z_2 during which B avoids collisions with the boundary of Ω, and plan such a motion if it exists. In this paper we shall consider two special cases of this general problem. First we consider the case where B is a disc possessing two (translational) degrees of freedom. Then we consider the case where B is an oriented line-segment ("ladder") possessing three degrees of freedom: translational and rotational.

Various versions of this problem have been studied previously by Lozano-Perez and Wesley (1979), Reif (1979), Schwartz and Sharir (1983a,b,c), and Hopcroft, Joseph, and Whitesides (1982). The general problem, in which B can be an arbitrary robot system, moving in 2-D or 3-D space and consisting of an arbitrary collection of jointed subparts, is shown in Reif (1979) and Hopcroft, Joseph, and Whitesides (1982) to be PSPACE-hard if one allows the number k of degrees of freedom of the system to grow arbitrarily. However, if we focus our attention on a particular robot system (so that k is fixed), then the general motion-planning problem is solvable in time polynomial (but with a large exponent) in the number of geometric constraints imposed on B, provided that all these constraints are algebraic (Schwartz & Sharir, 1983b). Motion-planning algorithms for severl specific robot systems have also been developed. Schwartz and Sharir describe one general approach to the problem, which might be called a "projection" approach, and which is based on the observation that essentially the problem is to decompose the k-dimensional configuration space FP of *free placements* of the system B into its (pathwise) connected components, and then to determine whether the two given placements of B lie in the same connected component of FP. To achieve this decomposition, one projects FP onto a subspace A of lower dimension and then partitions A into connected regions R, each one of which has the property that, as X varies in R, the collection of connected components of the "fiber space" $P(X)$ of points in FP projecting to X varies continuously and remains qualitatively unchanged. This enables one to decompose the space of all placements in FP which project into the same region R into finitely many cells (as many as there are components of $P(X)$ for $X \in R$). Then the adjacency relationships between cells of this sort are determined, which lead to the construction of a discrete *connectivity graph* CG whose nodes are these cells and whose edges connect adjacent cells. In this way, determining connectivity in FP reduces to determining connectivity in CG, so that the originally continuous movers' problem is reduced to searching a graph.

This "projection" technique has been applied to various special robot systems, yielding motion-planning algorithms for a ladder and for a rigid polygonal body moving amidst polygonal barriers (Schwartz & Sharir, 1983a), and for two or three separate circular bodies moving within the same environment (Schwartz & Sharir, 1983c). However, this technique leads to relatively inefficient algorithms. For example, the motion-planning algorithm for a ladder given in Schwartz and Sharir (1983a) runs in time $O(n^5)$, where n is the number of wall segments.

In this paper we describe another approach to the motion-planning problem, which we call the "retraction" approach. Intuitively speaking, the safest way to move a body B amidst obstacles is not to let it get too close to any one of the obstacles; roughly speaking, by keeping B equidistant from at least two obstacles at all times during its motion. Imposing this condition reduces the number of degrees of freedom available to B by 1, and thus simplifies the analysis of possible motions of B.

To carry this approach out in detail, we introduce a generalized notion of *Voronoi diagram*, defined as the subset of the configuration space FP of B consisting of placements of B simultaneously nearest to two or more obstacles. We then define an "image map" Im of FP onto the Voronoi diagram. When B is a disc, the map is a (continuous) "retraction" (in the standard topological terminology), but when B is a ladder the map may be discontinuous. For any initial placement Z, this map "pushes" B away from the obstacle closest to B at Z until B reaches a placement $Im(Z)$ on the Voronoi diagram. We prove that there exists a continuous wall-avoiding motion of B between two placements Z_1 and Z_2 if and only if there exists a similar motion between $Im(Z_1)$ and $Im(Z_2)$ which is constrained to move along the Voronoi diagram. It then remains to study the existence of motions within the diagram.

This paper applies this general approach to motion-planning for a disc and for a ladder. In the case of a disc, our notion of a Voronoi diagram is a 1-dimensional subspace of *FP*, and is a proper subset of the standard Voronoi diagram as described in Kirkpatrick (1979). This observation leads to an $O(n \log n)$ motion-planning algorithm for a disc, using the fact that the Voronoi diagram of a set of n points and line-segments can be constructed in $O(n \log n)$ time (Kirkpatrick, 1979).

In the case of a ladder, the generalized Voronoi diagram is a 2-dimensional subspace of *FP*, consisting of two-dimensional *Voronoi surfaces* meeting at one-dimensional *Voronoi edges*, which are bounded by zero-dimensional *Voronoi vertices*. Since the diagram is 2-dimensional, planning motions along the diagram is still a complicated task. To accomplish this task we define a second retraction from the diagram onto a network *N* of curves lying in it, and prove that the existence of a continuous motion along the diagram between two placements Z_1 and Z_2 is equivalent to the existence of a motion between the two corresponding retracted placements which is constrained to move only along edges of *N*. Since the network *N* is 1-dimensional, motion in *N* can be planned easily using graph search.

Efficient construction of the network *N* turns out to require many new techniques of computational geometry. Specifically, we prove that the size of the required network is $O(n^2 \log^* n)$, and provide algorithms for its construction which run in total time $O(n^2 \log n \log^* n)$, thereby obtaining an $O(n^2 \log n \log^* n)$ motion-planning algorithm for a ladder, which is substantially faster than the $O(n^5)$ algorithm of Schwartz and Sharir (1983a).

2. MOTION-PLANNING FOR A DISC

In order to introduce the idea of retraction, the first part of this paper is devoted to the relatively easy problem of moving a disc. Any placement of a disc is specified by giving the co-ordinates of its center, so this problem involves only two degrees of freedom. The algorithm presented below involves $O(n \log n)$ prepocessing, after which paths between any two given placements of the center of the disc *B* can be planned in time $O(n)$ for discs of arbitrary size.

Let *S* be the set of polygonal obstacles; without risk of confusion we also use *S* to denote the union of all the obstacles. Given two points x and y in the plane, $d(x,y)$ denotes the Euclidean distance between them. If *Y* is a nonempty set of points, the *Hausdorff distance* $d(x,Y)$ is defined as $\inf \{d(x,y) : y \in Y\}$, and if *X* is another nonempty set, $d(X,Y) = \inf \{d(x,Y) : x \in X\}$. (Note that $d(X,Y)$ is not a distance function.) For each point x define *Clearance*(x) to be the Hausdorff distance $d(x,S)$ of x from *S*. We assume that the union of all obstacles forms a closed set, and that its complement, denoted Ω, is *bounded*. Note that this implies that for every point x in Ω there exists a point p in (the boundary of) some obstacles such that $d(x,p) =$

Clearance (x). Hence, if we define $Near(x)$ to be the set of points p in S attaining the minimum distance Clearance (x) from x, then $Near(x)$ is nonempty.

DEFINITION. The (standard) *Voronoi diagram* Vor_0 (S) is the set $\{x$ in Ω: Near (x) contains more than one point$\}$.

$Vor_0(S)$ can be constructed in time $O(n \log n)$ using Kirkpatrick's algorithm using Kirkpatrick's algorithm (Kirkpatrick, 1979) or Yap's (Yap, 1984), where n is the number of open straight segments ("wall edges") and their end-points ("corners") constituting the boundary of S (an alternative $O(n \log^2 n)$ algorithm of Lee and Drysdale (1981) is also available). Note that although Kirkpatrick's algorithm constructs a planar graph G strictly containing the standard Voronoi diagram, the latter may easily be constructed from G.

Any placement of B may be identified with the location of its center. Thus if r is the radius of the disc, the space FP of free placements is identical to the set of points in Ω whose clearance exceeds r. We now show how to "retract" Ω into the Voronoi diagram.

DEFINITION. For each x in $\Omega \setminus Vor_0(S)$, let y be the unique element in $Near(x)$. The *image $Im(x)$* of x is the first point where the half-line from y through x intersects the Voronoi diagram. For x in $Vor_0(S)$ we define $Im(x) = x$.

THEOREM 2.1. The map Im is well-defined and is a continuous retraction of Ω onto the Voronoi diagram $Vor_0(S)$. Furthermore, Clearance$(\cdot)$ increases strictly along the line joining x to $Im(x)$.

Let p and q be two points in FP. If there is a path from p to q in FP, then by Theorem 2.1 this path can be projected into the Voronoi diagram using the map Im, yielding a path from $Im(p)$ to $Im(q)$ lying along the Voronoi diagram and contained in FP. Furthermore, all points in the line-segments joining p and q to their respective images are in FP. This gives the following:

THEOREM 2.2. There exists a path from p to q within FP if and only if there exists a path from $Im(p)$ to $Im(q)$ within the intersection of $Vor_0(S)$ with FP.

These observations justify the following algorithm, which runs in time $O(n \log n)$:

1. (Preprocessing). Using Kirkpatrick's algorithm, construct $Vor_0(S)$. In this diagram, each edge is either a straight or parabolic segment, with $O(n)$ edges overall. The regions in Ω bounded by the walls and these edges are the *Voronoi cells,* and each cell consists of points in Ω closest to a specific wall edge or corner. Thus it is possible to compute in overall time $O(n)$ the *width $w(e)$* of e— the minimum clearance along e—for every edge e.
2. Let x and y be two points in Ω. It is required to determine whether they represent free placements and whether there exists an obstacle-avoiding path between them. First search the Voronoi diagram to find the cell containing x, if

x is not already on the diagram. Then the point z in S closest to x can be determined immediately; another search along the half-line from z through x will determine $x' = Im(x)$. We can obtain $y' = Im(y)$ similarly.

3. It remains to determine whether x' and y' can be joined by a path of free placements along the Voronoi diagram. If x' and y' are on the same Voronoi edge, it is enough to verify that the minimum clearance between them on the edge is greater than r. Otherwise, one must establish that at least one of the two endpoints of the edge containing x' can be reached while maintaining sufficient clearance, and similarly for y'. This reduces the path-finding problem to that of finding a path between two Voronoi vertices.

4. There exists a path between two Voronoi vertices in the diagram if and only if there exists a path in the Voronoi graph consisting of all those edges whose width is greater than r.

Complexity: The necessary preprocessing can be done in time $O(n \log n)$ (Kirkpatrick, 1979). Locating points in Voronoi cells and computing their clearances can be done naïvely in linear time, by computing the minimal distance between the point and all walls and corners. The points x' and y' can be obtained in linear time by checking the points of intersection of the two corresponding semi-infinite lines with all Voronoi edges. Computing minimal clearance between two points on an edge takes constant time. Searching the graph takes linear time. Thus the overall cost of the preceding algorithm is the sum of O(n log n) time for preprocessing with an additional cost of $O(n)$ time for each subsequent path-planning task.

3. MOTION-PLANNING FOR A LADDER

3.1. Overview of the Approach

Having described the motion-planning algorithm for a disc, we now consider the much harder problem of efficient motion-planning for a ladder B. Here again, we use an initial preprocessing phase given the set of obstacles and the length d of the ladder, followed by a path-planning phase for each given pair of initial and final placements of B. Here the strategy is as follows: (a) generalize the Voronoi diagram to the 3-dimensional configuration space FP of B; (b) construct a 'retraction' Im of FP onto the Voronoi diagram; (c) construct a network N of 1-dimensional curves in the Voronoi diagram; and (d) define a retraction Im_2 of the Voronoi diagram into N. Once N has been constructed and the appropriate retractions defined, to plan a motion reduces essentially to searching the graph N.

The actual path-planning phase consists of applying the two retractions Im and Im_2 to each of the two given placements Z and Z' of B, thus obtaining the two placements $W = Im_2(Im(Z))$ and $W' = Im_2(Im(Z'))$ in N. Let e (resp. e') be the edge of N containing W (resp. W'). We then only need to check whether these two edges are connected to each other along N. If not, then Z and Z' are not connected to each other in FP. On the other hand, if such a path exists then it can be extended to a path from Z to Z' by combining it with canonical paths from Z to W and from W' to Z'.

3.2. Generalized Voronoi Diagrams for a Ladder

We go on to fill in details concerning the various steps of the procedure described above. We may regard S as a finite collection of *objects* which are either open line segments ("wall edges") or their end-points ("corners"). Two objects s and s' are *separated* if $d(s,s') > 0$. Each placement of the ladder B can be denoted as (x,y,θ), where (x,y) is the position of a specified endpoint P of the ladder, and θ is its orientation. For each such triple $Z = (x,y,\theta)$ we let $B(Z)$ denote the set of points in the plane occupied by the ladder in placement Z. The manifold FP of *free placements* of B consists of those placements Z for which $B(Z)$ is wholly contained in Ω (i.e., in which the ladder does not touch nor penetrate the complement of Ω.

For each Z in FP, define Clearance (Z) to be:

$$inf\{d(p,B(Z)): p \notin \Omega\}$$

and let Near(Z) be the set of all points in the boundary of Ω at distance Clearance(Z) from $B(Z)$.

DEFINITION. The generalized Voronoi diagram $Vor_d(S)$ for a ladder of length d is the set of free placements Z for which Near(Z) is *disconnected*.

Note that, if Z does not belong to $Vor_d(S)$, then Near (Z) must be either a point or a closed line-segment, and if $Z \in Vor_d(S)$, then Near (Z) must intersect at least two separated objects of S. For each object s in S we define the *Voronoi cell* $V(s)$ of s to be the set of all placements Z such that Near(Z) intersects s.

Let s_1 and s_2 be separated objects. Then the *Voronoi surface* $\Sigma(s_1,s_2)$ is the set of placements Z in $Vor_d(S)$ such that

$$d(s_1,B(Z)) = d(s_2,B(Z))$$

$$= Clearance(Z) < d(p,B(Z))$$

for any point p in $S\setminus\{s_1,s_2\}$. Clearly, Voronoi surfaces are disjoint. Two surfaces can share a common boundary consisting of placements where the ladder is equidistant from three closest objects, and three surfaces can share common points, "vertices," where the ladder is equidistant from four closest objects. We assume that the walls are in "general position" to avoid certain complications due to degeneracy: this requirement will not be discussed here. Intuitively, degeneracies may be avoided by an arbitrarily small perturbation of the objects. One consequence of the requirement is that the Voronoi diagram is contained in the union of the closures of its surfaces. Voronoi surfaces can be open, closed or neither.

Note that the closure of $Vor_d(S)$ need not be contained in FP, but may also include placements Z in which $B(Z)$ touches objects in S. For each pair s_1,s_2 of separated objects in S, we will call the locus of placements Z of B for which $B(Z)$

simultaneously touches s_1 and s_2 (but such that $B(Z)$ does not penetrate any other wall) the *Voronoi pseudo-edge* associated with s_1 and s_2. These pseudo edges can meet free Voronoi edges at *Voronoi pseudo-vertices,* which are placements Z for which $B(Z)$ simultaneously touches three distinct objects s_1, s_2, s_3 in S (but does not penetrate any other wall).

3.3. Retracting onto $Vor_d(S)$

We go on to show that, using an appropriate retraction of FP onto $Vor_d(S)$, it suffices to study only motions in which B is constrained to move within the Voronoi diagram. The required retraction *Im* from FP onto $Vor_d(S)$ is defined as follows.

DEFINITION.   Given a free placement Z not in $Vor_d(S)$, choose a point p on an obstacle and q on the ladder such that $d(p,q) = $ Clearance (Z). Keeping the orientation fixed, move the ladder away from p so that q remains on the same straight line from p, until the ladder becomes equidistant from another obstacle: this placement is denoted $Im(Z)$. If Z is already on $Vor_d(S)$ let $Im(Z) = Z$.

It is easily seen that *Im* does not depend on the choice of p and q. Moreover, since we have assumed that Ω is bounded, and since, when B is moved in the manner described above, Clearance (Z) increases as long as B does not reach $Vor_d(S)$, it follows that B will eventually reach the Voronoi diagram and $Im(Z)$ is well-defined.

LEMMA 3.1.   *Im* is continuous on $FP \setminus Vor_d(S)$. Any discontinuity of *Im* at Z in $Vor_d(S)$ is of the following sort: If Z_n converges to Z but $Im(Z_n)$ converges to some placement $W \neq Im(Z)$, then W and Z both have the same orientation, and the ladder can be translated from W to Z (as in the definition of *Im*) without leaving the Voronoi diagram.

This lemma leads to the following theorem:

THEOREM 3.2.   Let Z, $Z' \in FP$ and let $W = Im(Z)$, $W' = Im(Z')$. Then there exists a continuous path in FP between Z and Z' if and only if there exists a continuous path in $Vor_d(S)$ between W and W'.

Sketch of Proof. If there exists a path π in $Vor_d(S)$ between W and W', then the concatenation of π with the two translations $Z \rightarrow W$ and $W' \rightarrow Z'$ yields a continuous path from Z to Z' in FP. Conversely, let π be a continuous path between Z and Z' in FP. By perturbing π slightly we can assume that π intersects $Vor_d(S)$ at only finitely many points $Z_1, \ldots, Z_n$ which break it up into n+1 open subpaths π_0, $\ldots, \pi_n$. Let $\pi_i' = Im(\pi_i)$ for $i = 0, \ldots, n$. By Lemma 3.1, each of the subpaths π_i' is connected and contained in $Vor_d(S)$, and the segments τ_i connecting the corresponding endpoints of π_i' and π_{i+1}' (in the sense of Lemma 3.1) are also contained in $Vor_d(S)$. Concatenation of all the paths π_i', τ_i yields a continuous path in $Vor_d(S)$ between W and W'.

3.4. The Network N and the Second Retraction

Having thus reduced our problem to that of motion-planning within the Voronoi diagram, we reduce the problem still further to a task of planning paths along a certain network N of 1-dimensional curves in the Voronoi diagram. In the following discussion, a Voronoi *sheet* is the closure of a connected component of some Voronoi surface $\Sigma(s_1,s_2)$.

THEOREM 3.3.   The boundary of any sheet K has one or two components. In the latter case all orientations are possible in K.

The proof of Theorem 3.3 is quite involved, and consists of the following steps.

Let Σ be a sheet of $\Sigma(s_1,s_2)$. By changing the coordinate system, we can assume that Σ has one of the following canonical forms. We will also define 'virtual sheet' Σ' which contains Σ and which has a very simple structure:

    I.   If the two objects s_1,s_2 defining Σ are wall corners, then s_2 is at the origin O and s_1 is on the positive y-axis. Σ' is defined to be the set of placements Z that $d(s_1,B(Z)) = d(s_2,B(Z))$. Note that Σ consists of those Z in Σ' such that for all points p in objects *except s_1 and s_2* $d(p,B(Z)) > d(s_1,B(Z))$.

    II.   If s_1 is a wall corner and s_2 is a wall edge, then s_2 is on the x-axis and s_1 on the positive y-axis. Σ' is the set of placements Z such that $d(s_1,B(Z)) = d(X,B(Z))$ where X is the x-axis.

    III.   If both s_1 and s_2 are wall edges, then they are contained in two rays R_1,R_2 (resp.) from the origin where R_1 and R_2 are both in the $x \geq 0$ half-plane, and the x-axis bisects the angle between the rays. Σ' is the set of placements Z such that $d(R_1,B(Z)) = d(R_2,B(Z))$.

In these modified coordinates the placements (x,y,θ) in Σ' (and hence in Σ) are uniquely parameterized by their (x,θ) coordinates, except for the special case of sheets of type I above where the two corners are closest to the same side of B. In this exceptional case, the (x,y) coordinates can serve as the parameterization. In what follows we shall only discuss the nonexceptional cases, but the technique easily generalizes.

Let K be a sheet (which is in one of the above canonical forms). Let Φ be the set of all possible orientations of the ladder in K. For each θ_0 in Φ define the *streamline* $L(\theta_0)$ to be the set of placements in K with orientation θ_0. We remark that the set $\{(x,y) : (x,y,\theta_0) \in K\}$ is very simple: namely, a finite union of straight and parabolic arcs. Define the "left" endpoint $l(\theta_0)$ {resp. the "right" endpoint $r(\theta_0)$} to be the placement on $L(\theta_0)$ having smallest (resp. largest) x-coordinate.

LEMMA 3.4:   Let $Z_0 = (p_0,\theta_0)$ be an interior point of K. Then $\theta_{min} < \theta_0 < \theta_{max}$. Moreover, any neighborhood of Z_0 contains placements with θ greater than and less than θ_0.

Sketch of Proof: By considering the various types of sheet, one shows that it is always possible to perturb Z_0 slightly so as to obtain placements $Z = (p,\theta)$ in K for which $\theta > \theta_0$ and for which $\theta < \theta_0$.

In what follows it will be convenient to introduce the concept of a "racetrack" (so called because of its shape). For placement Z and distance r, let $Racetrack(Z,r)$ denote the set of points at distance $\leq r$ from $B(Z)$. Also for $Z \in FP$ let $Racetrack\ (Z)$ abbreviate Racetrack (Z, Clearance (Z)).

LEMMA 3.5. For each θ in Φ, the streamline $L(\theta)$ is connected. Moreover, if (p,θ), (q,θ) both lie on $L(\theta)$ and are *interior* points of K, then the whole portion of $L(\theta)$ between these two placements is contained in the interior of K.

Sketch of Proof. For a fixed streamline $L(\theta)$, parameterize placements on $L(\theta)$ by their x-coordinates and write $L(\theta) = \{Z(t) : t \in M\}$ for some set M of real numbers. We will show that $L(\theta)$ is connected by proving that M must be an interval of the real line. Given t in M, let a_t (resp. b_t) be the point on s_1 (resp. s_2) which is in contact with the boundary of Racetrack($Z(t)$). (a_t, b_t are uniquely defined except when $Z(t)$ is on the boundary of the sheet.) The boundary of Racetrack($Z(t)$) is naturally broken up into two portions by a_t and b_t, and we may call them the "left-boundary" λ_t and the "right-boundary" ϱ_t of the racetrack. (Figure 1 illustrates this for one type of sheet.) Every point in the plane can be classified so that it is either (i) "left" of (or on) λ_t, or (ii) "right" of λ_t, but "left" of ϱ_t, or (iii) "right" of (or on) ϱ_t. Furthermore, a point is in case (ii) if it is in the interior of Racetrack($Z(t)$). It is also seen that the λ- and ϱ- boundaries vary continuously with θ and t (in the Hausdorff metric of sets). To show our lemma, suppose that $t_1, t_2 \in M$ represent placements in the streamline, where $t_1 < t_2$, but there exists t_3, where $t_1 < t_3 < t_2$, not in M. This means there exists a point p in some obstacle $s \neq s_1, s_2$ such that $p \in$ Racetrack ($Z(t_3)$). From elementary geometry we can show that p is to the "right" of ϱ_{t_1} but to the "left" of λ_{t_2}. Since $Z(t_1)$ and $Z(t_2)$ are in the same sheet, we can connect them by the path Π which (except for its endpoints) lies entirely in the interior of the sheet K. By the continuity of the λ- and ϱ-boundaries, we conclude that for some $Z \in \Pi$, p lies on the boundary of Racetrack (Z). This contradicts the fact that Π is in the interior of K.

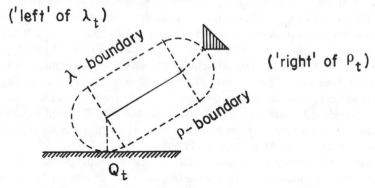

Figure 1. The λ- and ϱ- boundaries of a racetrack

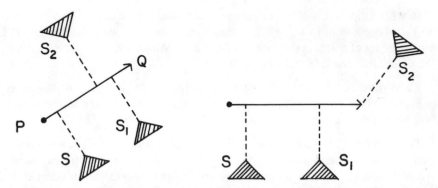

Figure 2.   The two types of sheet on which $l(\theta)$ may be discontinuous

LEMMA 3.6.   The function $l(\theta)$ is piecewise continuous on Φ, and has only a finite number of discontinuities. At each such discontinuity θ_0 with $\theta_{min} < \theta_0 < \theta_{max}$, $l(\theta_0^-)$ and $l(\theta_0^+)$ exist, and the whole "linear segment" of $L(\theta_0)$ between $l(\theta_0^-)$ and $l(\theta_0^+)$ is contained in the boundary of K. A similar statement holds also for the function $r(\theta)$.

Sketch of Proof. By perturbation analysis, we can show that $l(\theta)$ is continuous in θ except for the two situations illustrated in Figure 2. For these two types of sheet, there is a discontinuity at θ only if another corner s is aligned with either s_1 or s_2 at an angle θ. The property stated in the lemma is seen to hold in both cases. Finally, the number of discontinuities is $O(n)$ by virtue of the fact that there are $O(n)$ such directions of alignment.

COROLLARY.   Let the *left boundary* $bd_L(K)$ of K be the union of $l(\Phi)$ with all the "linear segments" connecting $l(\theta_0^-)$ to $l(\theta_0^+)$ at each point θ_0 of discontinuity of l. Then $bd_L(K)$ is connected. The *right boundary* $bd_R(K)$ of K, defined in a symmetric fashion, is also connected.

Finally, if not all orientations are possible in K, consider the two extreme streamlines $L(\theta_{min})$ and $L(\theta_{max})$; these could degenerate to single points. By Lemma 3.5 they must be contained in $bd(K)$, and they connect $bd_L(K)$ to $bd_R(K)$. It follows that $bd(K)$ consists of the union of the four portions $bd_R(K)$, $bd_L(K)$, $L(\theta_{min})$ and $L(\theta_{max})$, and is connected. This concludes the proof of Theorem 3.3. Following Theorem 3.3 we define the *skeleton* of a sheet K to be the union of $bd(K)$ with its core streamlines, those at orientations $\theta = 0$ or π relative to the standard co-ordinate systems introduced above for the three kinds of sheet. By Theorem 3.3 the skeleton of a sheet is a connected 1-dimensional network within the sheet. Finally the network N is the union of all such skeletons.

Theorem 3.3 allows us to reduce the motion-planning problem still further. Specifically we define a second mapping Im_2 from $Vor_d(S)$ onto N as follows. Let Z

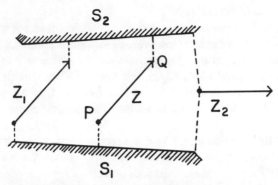

Figure 3. The boundary of the surface $\Sigma(s_1,s_2)$ has two connected components

$= (p,\theta) \in Vor_d(S)$. If $Z \in N$ we put $Im_2(Z) = Z$. Otherwise, let K be the sheet containing Z in its interior, and let l be the "left side" map for streamlines on K. Then put $Im_2(Z) = l(\theta)$.

THEOREM 3.8: Let $Z, Z' \in Vor_d(S)$, and let $W = Im_2(Z)$, $W' = Im_2(Z')$. Then there exists a continuous path in $Vor_d(S)$ between Z and Z' if and only if there exists a continuous path in N between W and W'.

Proof: The "If" part is clear. Conversely, the existence of such a path between Z and Z' implies that there exists a chain $K_1, \ldots , K_t$ of adjacent sheets where $Z \in K_1$ and $Z' \in K_t$. By Theorem 3.3, the union U of the boundaries of K_i $(i = 1, \ldots , t)$ is a connected subgraph of N; but W and W' both belong to U. Q.E.D.

Theorem 3.8 thus implies that, to plan a path in FP between two given placements Z and Z', it suffices to map these placements into the two placements $W = Im_2(Im(Z))$, $W' = Im_2(Im(Z'))$ in N, and then to plan a continuous path in N between W and W'. This latter task is accomplished by a graph search through the network N, which can be easily implemented to run in time $O(|N|)$ once the network N has been constructed.

Remarks:

1. Notice that a connected component of a Voronoi surface $\Sigma(s_1,s_2)$ need not have a connected boundary, as seen in Figure 3. It is therefore not sufficient to construct N using Voronoi edges and vertices only, because then N would not have the same connectivity as $Vor_d(S)$.

2. Some of the edges of N may be pseudo-edges, and so correspond to motions of B in which it touches some obstacle in S. Strictly speaking, this type of motion is unacceptable, but such a motion can be infinitesimally perturbed into a free motion. A careful analysis shows that pseudo-edges can be omitted from N without affecting its connectivity [OSY2].

3.5. Efficient Construction of N

In the preceding paragraphs we have described and justified a retraction approach for planning motions of a ladder. However, to make this approach into an efficient algorithm, we still need to develop techniques for efficient construction of the network N of pseudo- and quasi-edges and vertices. We will first show how to construct all quasi-vertices, and then explain how to find all pairs of vertices adjacent to each other along some quasi-edge. This will determine the topology of the network N.

Note first that the introduction of Voronoi pseudo- and quasi-vertices will lead to a large number of distinct types of vertex. Roughly speaking, a Voronoi vertex is a placement Z in FP for which there exist four objects s_j, $j = 1, \ldots, 4$, where $s_j \in S$, such that

$$d(s_1,B(Z)) = d(s_2,B(Z)) = d(s_3,B(Z)) = d(s_4,B(Z)) = Clearance(Z)$$

It is possible that some of the objects s_j, $j = 1, \ldots, 4$, be contiguous. Figures 4 and 5 show a few examples of these vertices. The notion of a "racetrack" introduced in Section 3.4 will again prove useful. The boundary of a racetrack is naturally divided into four parts, two of which are straight line segments ("sides" of the racetrack) and the other two are semicircular arcs ("bends" of the racetrack). The racetrack is *empty* if its interior lies entirely in Ω. We also call Racetrack(Z,r) the *r-racetrack* of Z.

For purposes of classifying the Voronoi vertices it is convenient to label these four parts P, Q, PQ, and QP, denoting, respectively, the bends nearest P and Q and the left and right sides: here the sides are open line-segments and the bends are closed semicircles. Then a Voronoi vertex Z at which Racetrack(Z) intersects four separated objects $s_i \in S$ may be labelled with its *characteristic* $\{(s_i,p_i): 1 \leq i \leq 4\}$ where $p_i \in \{P,Q,PQ,QP\}$, indicating that the part p_i of the Z-racetrack intersects s_i ($1 \leq i \leq 4$).

For brevity, we shall consider only vertices involving four distinct and separated

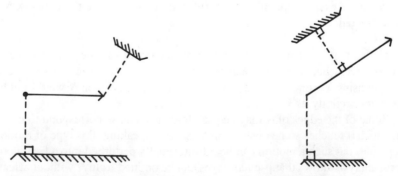

Figure 4.   Quasi-Voronoi vertices determined by two separated objects

Figure 5.   Quasi-Voronoi vertices determined by three separated objects

objects: these are the hardest to find. We can classify such vertices into the following four types:

I.   Vertices Z for which $p_1 = p_2 =$ an endpoint of B.
II.   For all other vertices at least two of the p_j belong to $\{PQ, QP\}$. These are further classified as follows:
   a.   Vertices Z for which $p_1 = p_2 = PQ$, and p_3 and p_4 are opposite endpoints of B.
   b.   Vertices Z for which $p_1 = p_2 = PQ$, $p_3 = QP$ and p_4 is an endpoint of B.
   c.   Vertices Z for which $p_1 = PQ$, $p_2 = QP$ and p_3 and p_4 are opposite endpoints of B.

Figure 6 gives examples of such vertices.

Calculating Voronoi Vertices of Type I. Calculation of such vertices proceeds as follows. Let Z be a Voronoi vertex of Type I having characteristic $\{[s_1,P],[s_2,P],[s_3,\alpha],[s_4,\beta]\}$ (we will refer to such a Z as being associated with the pair s_1,s_2 of objects). Then, at placement Z, the P-end of the ladder must lie on the standard (i.e., planar) Voronoi diagram $Vor_0(S)$ (as introduced in Section 2); in fact P lies on an edge of $Vor_0(S)$ whose points are simultaneously nearest to s_1 and s_2. But $Vor_0(S)$ contains only $O(n)$ edges, so that there are at most $O(n)$ pairs s_1,s_2 in S with which Voronoi vertices of Type I can be associated.

Our aim is to show that, for each pair s_1,s_2 in S of obstacles which share an edge of $Vor_0(S)$, there are at most $O(n \log^* n)$ associated Type I vertices, and to construct them all in time $O(n \log n \log^* n)$. We split each edge determined by $s_1,s_2 \in S$ in $Vor_0(S)$ into at most two *segments* such that the distance $\Delta(\cdot)$ from s_1 and s_2 along the segment is monotonic. We then parametrize such segment e^r by a "moving point" X so that $\Delta(X)$ is strictly increasing. For each $s \in S - \{s_1,s_2\}$ and each $\theta \in (-\pi/2,\pi/2)$ we denote by $X_s(\theta)$ the position $X \in e^r$ for which

$$d(B(X,\theta),s) = d(B(X,\theta),s_j), \; j=1,2,$$

and moreover the distances from $B(X,\theta)$ to s_1 and s_2 are attained at the endpoint P of B, whereas the distance from $B(X,\theta)$ to s is attained at another point of B. (See Figure 7.) We show that, for each $s \in S$ and each $\theta \in (-\pi/2,\pi/2)$, there exists at

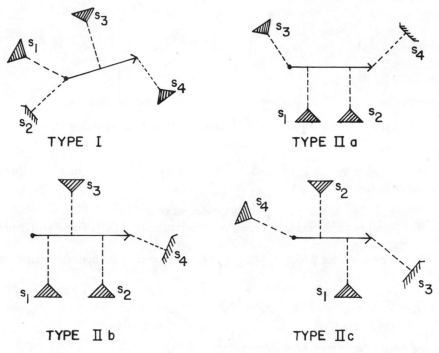

Figure 6.   The four types of Voronoi vertex

most one point $X_s(\theta) \in e^r$ having the above property, and that if $s \in S$ and $\theta \in (-\pi/2,\pi/2)$ are such that $X_s(\theta)$ is defined, then for points $X \in e^r$ left of $X_s(\theta)$ we have $d(B(X,\theta),s) > d(B(X,\theta),s_1)$, and conversely for points $X \in e^r$ right of $X_s(\theta)$ we have $d(B(X,\theta),s) < d(B(X,\theta),s_1)$.

(Note that $X_s(\theta)$ need not always be defined. This might happen either when for all points X on e^r the distance from $B(X,\theta)$ to s is less than the distance from X to s_1 and s_2, or when for all such points X the distance from $B(X,\theta)$ to s is larger than that to s_1

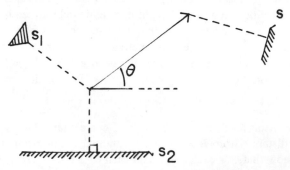

Figure 7.   The critical position $X_s(\theta)$.

and s_2. In the first case we put $X_s(\theta) = -\infty$ and in the second case we put $X_s(\theta) = +\infty$.)

The implication of the above results is that, for a given orientation θ, the only positions $X \in e^r$ for the endpoint P of B at which (X,θ) can be a Voronoi vertex of type I generated by s_1 and s_2 are those lying left of $X_s(\theta)$ for every $s \in S' = S - \{s_1,s_2\}$. In other words, such points X must be less than (i.e., left of) $M(\theta)$, where

$$M(\theta) = \min_{s \in S'} X_s(\theta)$$

Moreover, the type I vertices (X,θ) generated by s_1 and s_2 are precisely the placements $(M(\theta), \theta)$ for which there exist two objects s, $s' \in S'$ such that $X_s(\theta) = X_{s'}(\theta) = M(\theta)$.

Hence, to find the vertices that we seek, we only have to compute the function $M(\theta)$, or more precisely to find all those θ's at which the function $X_s(\theta)$ realizing M is replaced by another such function. Let K denote the number of such θ's.

In this setting, we can handle our problem by reducing it to a combinatorial problem studied by Davenport (1971) and by Davenport and Schinzel (1965), by Szemeredi (1974) and recently by Atallah (1983), as follows. Let $s \in S'$ and let I be the set of $\theta \in (-\pi/2,\pi/2)$ at which $X_s(\theta) < +\infty$ (i.e., is either defined or is $-\infty$). We can assume without loss of generality that I is a connected subinterval of $(-\pi/2,\pi/2)$, and extend the graph of $X_s(\theta)$ as follows: Let θ be the left (resp. right) endpoint of I. We extend continuously the graph of $X_s(\theta)$ left (resp. right) of θ as a decreasing (resp. increasing) straight line with a very steep slope. Similarly, if J is a subinterval of I at which $X_s(\theta) = -\infty$, then we extend continuously this graph to J as a very steeply decreasing line segment followed by a very steeply increasing segment. It is plain that any pair of these extended $X_s(\theta)$ functions intersect in at most d points, where d is a constant which depends only on the degrees of these functions, but not on n. It then follows (cf. Davenport & Schinzel, 1965; Atallah, 1983) that the number K of changes in M that we seek is at most the length $\lambda(n,d)$ of the longest sequence σ, each of which components belongs to $\{1, \ldots ,n\}$, which satisfies the following two conditions:

(i)   For each i, $\sigma_i \neq \sigma_{i+1}$.
(ii) There do not exist indices $i_1 < i_2 < \cdots < i_{d+2}$ such that $\sigma_{i_1} = \sigma_{i_3} = \sigma_{i_5} = \cdots = a$, $\sigma_{i_2} = \sigma_{i_4} = \sigma_{i_6} = \cdots = b$, and $a \neq b$.

It is known that $\lambda(n,2) = 2n - 1$ (Davenport & Schinzel, 1965; Atallah, 1983) and that $\lambda(n,d) = O(n \log^* n)$ for each $d > 2$ (with the constant of proportionality depending on d) (Szemeredi, 1974). It follows that $K = O(n \log^* n)$, and since there are only $O(n)$ standard Voronoi edges e^r to be considered, we conclude that there are $O(n^2 \log^* n)$ Voronoi vertices of type I.

To calculate all these vertices in time $O(n^2 \log n \log^* n)$, we use the following straightforward divide and conquer technique. We first compute the standard Voronoi diagram of S, using e.g., the $O(n \log n)$ algorithm of Kirkpatrick (1979).

For each of the $O(n)$ portions $e^r = e^r(s_1, s_2)$ of this diagram we compute the lower envelope $M(\theta)$ of the functions $X_s(\theta)$ as defined above, as follows.

We first partition the remaining objects in S into two equal sized sets S_1 and S_2, and compute recursively the lower envelope M_i of the functions $X_s(\theta)$ with $s \in S_i$, for $i = 1, 2$. Each such computation produces a partitioning of $(-\pi/2, \pi/2)$ into subintervals such that M_i coincides with the same function $X_s(\theta)$ throughout each such subinterval, and assigns to each such subinterval the corresponding function $X_s(\theta)$. We then compute $M(\theta) = \min (M_1(\theta), M_2(\theta))$ by merging the two partitions of $(-\pi/2, \pi/2)$ corresponding to M_1 and M_2, and then by scanning $(-\pi/2, \pi/2)$ from left to right, computing in constant time the minimum M within each of the refined subintervals.

It is easily seen from the above results that this algorithm will run in time $O(n \log n \log^* n)$, and we therefore conclude that The collection of all Voronoi vertices of type I can be found in total time $O(n^2 \log n \log^* n)$.

Remark: The bounds established above depend on the maximal number d of intersections of any pair of the functions $X_s(\theta)$. As noted above, if $d = 2$ the number $K(n)$ of changes in the lower envelope of the $X_s(\theta)$'s is $O(n)$, which implies that the total number of type I vertices is only $O(n^2)$ and that they can all be calculated in time $O(n^2 \log n)$. If on the other hand, $d \geq 3$, we obtain the slightly worse bounds given above. We have not carried through the (messy) calculations needed for the actual determination of d, and we leave it as an open problem. However, initial investigations into this problem lead us to conjecture that $d = 2$.

Calculation of Voronoi Vertices of Types IIa and IIb. Let s_1 and s_2 be two corners in S. Without loss of generality they lie on the x-axis; we want to locate those placements Z where the ladder is horizontal at distance y above the x-axis, and the y-racetrack around it is empty but touches s_1 and s_2 on its lower side and two other obstacles s and s' at a bend and an edge or on opposite bends. (See Figure 8.)

Given $y \geq 0$, let $F(y)$ be the set of all x such that in placement $Z = (x, y, 0)$ the y-racetrack is empty and touches s_1 and s_2 on its lower side. Each such set $F(y)$ is either empty or a closed interval $[f(y), g(y)]$; furthermore, $f(y)$ is nondecreasing and $g(y)$ nonincreasing. Let y_0 be the maximal y for which $F(y)$ is nonempty. Let $x_0 = f(y_0)$ and $x_1 = g(y_0)$. If $x_0 = x_1$ then it is easy to verify that $(x_0, y_0, 0)$ is a vertex of Type IIa or IIb. Otherwise, $(x_0, y_0, 0)$ (resp. $(x_1, y_0, 0)$) is either a Type IIb vertex or a quasi-vertex (where one end of the ladder is directly above s_1 or s_2). To estimate the number of these vertices, and to find them all in an efficient manner, we can proceed as follows. For each $0 \leq y \leq y_0$, let $\alpha(y)$ (resp. $\beta(y)$) denote the set of objects in S which lies on the left (resp. right) bend of the y-racetrack at $(f(y), y, 0)$ [resp. $(g(y), y, 0)$]. For initial values of y (possibly all y) $\beta(y)$ may be empty: Here $g(y)$ coincides with s_1 and the ladder is on a quasi-edge; similarly for $\alpha(y)$. Note that $f(y)$ is continuous and that $\alpha(y)$ is constant except for at "critical" heights y where one of the following conditions hold (see Figure 8):

(a.1)   Two objects s_3, s_4 in S touch the left bend of the y-racetrack R.

(a.2)   *One object s_3 in S touches the upper side of R to the left of s_1 and another*

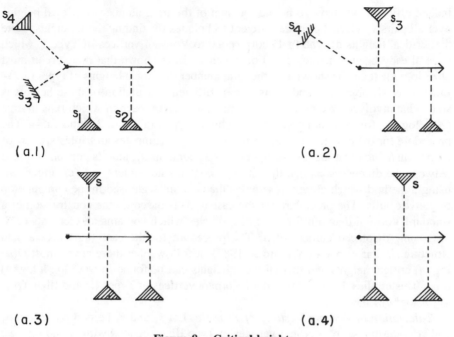

(a.1)   (a.2)

(a.3)   (a.4)

Figure 8. Critical heights

object s_4 in S touches the left bend of R. (In this case $Z = (f(y),y,0)$ is a Voronoi vertex of Type IIb.)

(a.3)   The right hand end of the ladder lies directly above s_2 and another object s in S touches the upper side of R to the left of s_1, or the left bend of R.

Similar and symmetric conditions govern the occurrence of discontinuous changes in $g(y)$ and $\beta(y)$.

If $F(y_0)$ is not a singleton, the following condition must hold:

(a.4)   There exists a corner s in S such that for every x in $[f(y_0),g(y_0)]$, Racetrack $((x,y_0,0))$ is empty and touches s on its upper side. This implies that the rectangular part of the racetrack contained between its sides is empty and touches only s_1, s_2, and s.

The preceding observations justify the following procedure. We find all the "critical" heights defined by conditions (a.1)–(a.4) and process them in increasing order, updating α and β each time y crosses a critical height, and checking whether $F(y)$ can shrink to a point within each interval of noncritical heights (knowing α and β for this interval, this check can be accomplished in constant time). Processing continues until either $F(y)$ shrinks to a point or the critical height y_0 satisfying (a.4) is encountered.

To see that this procedure, which must be applied to each pair of corners in S, is indeed efficient, we have to bound the sum of the total number of critical heights over all pairs s_1,s_2, and provide efficient techniques for finding these heights. Note that critical heights of type (a.1) correspond to Voronoi vertices of Type I, which have already been constructed and of which we have shown that there are at most $O(n^2 \log^* n)$. It can be shown that the total number of critical heights of types (a.2)– (a.4), over all objects s_1 and s_2, is $O(n^2)$. Efficient calculation of these heights is somewhat involved. We give three separate procedures, each of which runs in time $O(n^2 \log n)$, for computing critical heights of types (a.2), (a.3), and (a.4). The procedure for the case (a.2) and part of (a.4) simply computes all triples s_1,s_2,s_3 of corners in S such that the disc of diameter s_2s_3 contains s_1, and its portion bounded between the diameter s_2s_3 and the line s_1s_2 contains no corners of S (cf. Figure 8), using a method which closely resembles the Graham scan used in the computation of convex hulls. The procedure for the case (a.3) is based on the construction of a standard Voronoi diagram for a set of obstacles which contains all elements of S, plus some displaced corners of S. The procedure for the case (a.4) uses a data structure similar to one of Willard's (1982). It follows that there are at most $O(n^2 \log^* n)$ critical heights and that all these heights can be found in $O(n^2 \log n \log^* n)$ time. This implies that we can find all Voronoi vertices of Types IIa and IIb in $O(n^2 \log n \log^* n)$ time.

Calculation of Voronoi Vertices of Type IIc. Let s_1 and s_2 be two corners in S, and consider the set of Voronoi vertices of Type IIc associated with s_1 and s_2, i.e., placements in which the racetrack is empty but s_1 and s_2 touch opposite sides while two other objects touch opposite bends.

Let D be the midpoint of the segment s_1s_2. By shifting the coordinate system, we can assume that D is the origin and that s_1,s_2 lie on the x-axis at distance $2a$ apart, with s_1 to the left of s_2. Note that, for any Voronoi vertex Z of Type IIc associated with s_1,s_2, D must lie on $B(Z)$, and Z has an orientation which lies strictly between 0 and π.

For each orientation $0 < \theta < \pi$ let $F(\theta)$ denote the set of all $r < 0$ such that Racetrack $(Z,a \sin \theta)$, where $Z = (re^{i\theta},\theta)$, is empty and touches s_1 on its left side and s_2 on its right side. Where $F(\theta)$ is nonempty, write it as $[f(\theta),g(\theta)]$, and let $\alpha(\theta)$ [resp. $\beta(\theta)$] be the set of lower (resp. upper) objects touching Racetrack $(f(\theta))$ [resp. Racetrack $(g(\theta))$].

In contrast to the preceding analysis, in which $F(y)$ had been monotone, here $F(\theta)$ need not be monotone. Nevertheless we can track the way in which it varies in much the same way as before. Specifically, we note that $g(\theta)$ is continuous and that $\beta(\theta)$ is constant except at "critical" orientations θ at which one of the following conditions hold (see Figure 9):

(c.1)   Two objects s_3,s_4 in S meet the upper bend of $R \equiv$ Racetrack $(g(\theta))$.

(c.2)   A third corner s_3 in S touches one of the sides of R, so that the projection of s_3 onto the ladder lies above the projections of the other two corners; in addition, another object s_4 in S may touch the upper bend of the racetrack.

(c.3)   A third corner s_3 in S touches one of the sides of R, so that the projection of

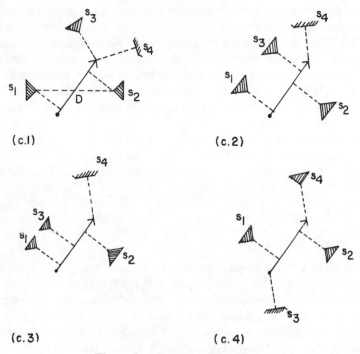

(c.1)   (c.2)

(c.3)   (c.4)

Figure 9.   Critical orientations

s_3 onto the ladder lies between the projections of the other two corners; in addition, another object s_4 in S may touch the upper bend of the racetrack. This implies the following condition which is easier to check: s_3 lies in the upper semicircle C whose diameter is $s_1 s_2$, and furthermore the section of C between $s_1 s_2$ and the ray $s_1 s_3$ (or $s_2 s_3$) is empty—i.e., contained in Ω.

(c.4)   Finally, the placement $Z = (g(\theta)e^{i\theta}, \theta)$ may be a Voronoi vertex of Type IIc that we seek.

Similar and symmetric conditions governing the discontinuous changes in $f(\theta)$ and $\alpha(\theta)$ can be given. Let $N(s_1, s_2)$ be the number of critical orientations satisfying one of the conditions (c.1)–(c.3) (or one of their symmetric counterparts) and suppose that all these orientations are already available. Then in each of the noncritical intervals of orientations bounded by two adjacent critical orientations the functions $f(\theta)$ and $g(\theta)$ vary continuously and the sets $\alpha(\theta)$ and $\beta(\theta)$ remain constant. This implies that we can determine in constant time whether such an interval contains orientations θ for which $f(\theta) = g(\theta)$, each of which corresponds to a Voronoi vertex of the type we seek, i.e., to a critical orientation of type (c.4). (Note that there can be at most $O(1)$ such vertices in each noncritical interval.) Thus we can find all such vertices in time $O(N(s_1, s_2) \log N(s_1, s_2))$ once the critical orientations are available.

To find the set of all critical orientations θ satisfying one of the conditions (c.1)–(c.3) for some pair of corners $s_1, s_2 \in S$, we note that orientations θ satisfying condition (c.1) correspond to Voronoi vertices of Type I, of which there are at most $O(n^2 \log^* n)$, and which have already been found. Placements at which (c.2) is satisfied are essentially the same as placements satisfying condition (a.4) of the preceding analysis, so that all $O(n^2)$ such placements are also already available. Similarly, placements at which (c.3) is satisfied are the same as those satisfying the preceding condition (a.2).

All this implies that there are altogether at most $O(n^2 \log^* n)$ critical orientations (summed over all pairs s_1, s_2 of corners in S), all of which are already available. Hence the total time required to calculate Voronoi vertices of Type IIc is

$$\Sigma \, N(s_1, s_2) \log N(s_1, s_2) = 0(n^2 \log n)$$

All this gives the following:

THEOREM 3.9. The total number of Voronoi vertices in $Vor_d(S)$ is $O(n^2 \log^* n)$ and they can all be found in time $O(n^2 \log n \log^* n)$.

Construction of the Network N. The next and final algorithmic step required in the preprocessing phase of our overall procedure is to build the network N by connecting pairs of Voronoi vertices which are adjacent along some quasi-edge or pseudo-edge. Since the number of vertices has been shown to be $O(n^2 \log^* n)$, it follows that the number of edges on which they lie is also $O(n^2 \log^* n)$. It is easy to parametrize each such edge (in fact we show that each edge on which θ is not constant can be parametrized by θ, whereas edges of constant orientation can be parametrized by either x or y when cast into an appropriate canonical form). This allows us to distribute the vertices among the $O(n^2 \log^* n)$ possible edges, sort all vertices lying on the same edge in increasing order of values of the edge parameterization, and connect adjacent pairs Z, Z' of such vertices, provided that the segment of e between Z and Z' does indeed belong to $Vor_d(S)$ (this latter condition can be easily tested by applying appropriate infinitesimal perturbations at Z and Z'). All this can be done in $O(n^2 \log n \log^* n)$ time.

4. DISCUSSION

We have presented a new approach to motion-planning, based on "retracting" the configuration space of the problem to one of lower dimension. This involved extending the notion of "Voronoi diagram" to non-Euclidean metrics. When applied to the problem of moving a ladder, we sketched a satisfactory solution, which employs many efficient techniques of computational geometry. The previous upper bound of $O(n^5)$ has now been reduced to $O(n^2 \log n \log^* n)$. We conjecture that this is near-optimal, since the set of free placements of a ladder B amid $O(n)$ obstacles can have $\Omega(n^2)$ connected components; see Figure 10.

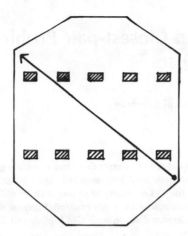

Figure 10. FP has $\Omega(n^2)$ connected components

It is likely that the methods described here may be generalized to the problem of motion-planning for a rigid polygonal body in 2-space, of a ladder in 3-space, of a hinged ladder in 2-space, etc. In a recent paper (Yap, 1983), the third author has used a "retraction" principle to plan the motion of two independent discs in the plane in time $O(n^2)$ thus improving an $O(n^3)$ algorithm in Schwartz and Sharir (1983c).

CHAPTER 8

Intersection and Closest-pair Problems for a Set of Planar Discs

MICHA SHARIR

Department of Mathematical Sciences
Tel Aviv University

Efficient algorithms for detecting intersections and computing closest neighbors in a set of circular discs, are presented and analyzed. They adapt known techniques for solving these problems for sets of points or of line segments. The main portion of the paper contains the construction of a generalized Voronoi diagram for a set of n (possibly intersecting) circular discs in time $O(n \log^2 n)$, and its applications.

1. INTRODUCTION

Let S be a set of n closed convex two-dimensional bodies of relatively simple structure (e.g., circular discs, straight segments, polygons of few sides, or expansions by some amount of such objects). In this paper we consider a variety of problems associated with such a set S. Typical problems are:

I. Do any two objects in S intersect?

II. More generally, suppose that we assign a 'color' to each object in S. Does there exist a pair of objects in S having different colors and intersecting each other?

III. If no two objects in S intersect, what is the smallest distance between any two objects in S (or, in the 'colored' version, what is the smallest distance between two objects in S having different colors)? More generally, for each object B in S find the object in S (of different color) nearest to B.

IV. Preprocess S so that, given an arbitrary 'query point' X, the object in S nearest to X can be found quickly.

Problems of this sort arise in robotics and are related to the problem of detecting and avoiding collisions between a moving subpart of a robot system and stationary objects, or between two or more moving subparts of such a system. In this note we will simplify the problem by assuming that each of the robot subparts and the stationary obstacles is either a closed convex object of a simple form, or else is

This research has been supported in part by ONR Grant N00014-82-K-0381 and by a grant from the U.S.-Israeli Binational Science Foundation.

The author wishes to thank Jacob Schwartz for suggesting the problems discussed in this paper, and for carefully reviewing the manuscript. His comments on the manuscript have led to substantial improvements in its presentation.

covered by finitely many such objects. Note that, in the 'colored' setting, some of the objects involved in our problem may be known a priori to intersect (e.g., circles covering two robot subparts hinged together, or two objects covering some robot subpart which overlap each other may always intersect). The 'colored' version of our problems allows for this situation by looking only for intersection of subparts which would not intersect under normal conditions (e.g., subparts belonging to two distinct robot 'arms,' or a robot subpart and an obstacle, etc.).

Efficient solution of these problems in the three-dimensional case would facilitate construction of an 'off-line' debugging system for robot control programs to check whether collision occurs along a planned path of motion, and would also make it more feasible to check in real time whether a moving subpart of the system is getting dangerously close to another (moving or stationary) object. This paper addresses the much simpler 2-dimensional case and uses generalized Voronoi diagrams for solving some of the problems just noted. In the 3-dimensional case, efficient algorithms for these problems would have to use other techniques, since Voronoi diagrams in 3 dimensions are generally quadratic in the number of objects involved. A 3-dimensional version of the simplest problem mentioned above, namely that of detecting intersection in a set of n arbitrary spheres, has been recently solved in Hopcroft, Schwartz, and Sharir (1983) by an $O(n \log^2 n)$ algorithm which uses sweeping methods.

This paper is organized as follows. In section 2, we describe a simple sweeping technique for detecting intersection in a set of n planar "monotone" objects. This technique has already been noted by various other researchers, but was not published. It is in fact a straightforward generalization of Shamos's algorithm for detecting intersection of line segments. We include it here for the sake of completeness, to indicate that the simplest intersection detection problem mentioned above can be solved efficiently for rather general planar objects. In section 3, we define the generalized Voronoi diagram associated with a set of n arbitrary and possibly intersecting planar discs, and analyze some of its properties. Section 4 presents an efficient algorithm for the construction of this diagram. Section 5 contains some applications of the diagram for the problems noted above, and Section 6 presents another unrelated application, namely that of efficient calculation of the area of a union of many discs.

The study of the two-dimensional case carried out in this paper may find applications in collision detection and avoidance for robot systems whose underlying motion is 2-dimensional, such as roving robots moving on a floor, etc. We hope that some of the ideas suggested here would also be useful in attacking the three-dimensional case.

2. DETECTING INTERSECTION OF MONOTONE OBJECTS

In this section, we consider the first problem posed above and show that it can be solved in time $O(n \log n)$ by a straightforward modification of an algorithm due to

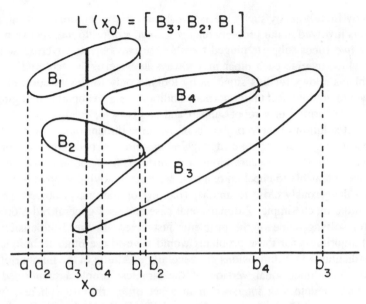

$$L(x_0) = [B_3, B_2, B_1]$$

Figure 1. An instance of the intersection problem.

Shamos (1975) which tests for intersection of straight line segments. As noted in the introduction, the material presented in this section has already been noted by several researchers. We assume that the bodies in S are monotone in the x-direction (i.e., the boundary of each object $B \in S$ consists of an upper portion and a lower portion, and both portions extend monotonously from left to right). Thus, for each such B there exist exactly two vertical lines tangent to it, and any vertical line between these lines intersects B in a closed segment. The structure of the objects in S is assumed to be simple enough so that each of the following operations takes constant time:

(i)   Check whether two specific bodies in S intersect each other.
(ii) For each $B \in S$, find the smallest and largest abscissae of points in B.
(iii) For each $B \in S$, find, for a given abscissa x, a point $(x,y) \in B$.

Let $B_1, \ldots, B_n$ be the objects in S. For each $j=1, \ldots, n$ let a_j (resp. b_j) be the smallest (resp. largest) abscissa of a point in B_j. For simplicity, we assume that the $2n$ numbers $a_j, b_j, j=1, \ldots, n$, are all distinct (see Figure 1).

Note that, if we draw a vertical line $x=x_0$ it will cut (some of) the objects in S in straight segments which, if the objects do not intersect each other along that line, are disjoint from each other. Hence, the objects in S which intersect the line $x=x_0$ can be linearly ordered in a list $L(x_0)$ in which an object B_i precedes another B_j if the segment cut off B_i by $x=x_0$ lies below the corresponding segment cut off B_j. Note that since the objects in S are monotone, the list $L(x_0)$ remains unchanged as x_0 increases until either the line $x=x_0$ meets a new object B_k (this will happen when

$x_0 = a_k$), or when it stops making contact with some object B_k (just after $x_0 = b_k$), or when two of the objects in $L(x_0)$ intersect at $x = x_0$. Moreover, the leftmost intersection (if any exists) of any two objects $B, B' \in S$ will occur at some $x = x_0$ such that, for x slightly less than x_0, the list $L(X)$ contains B and B' as adjacent elements. (Here we ignore the special case in which B and B' intersect at a point which is leftmost in one of these objects, which requires a slightly different argument.)

In view of these observations, to detect the presence or absence of an intersection one simply has to deck repeatedly whether any two adjacent elements in the list $L(x_0)$ intersect each other. Plainly, these checks have to be performed only at points where $L(x_0)$ changes (i.e., at the points $x_0 = a_j, b_j, j = 1, \ldots, n$), and one only needs to test newly adjacent pairs in $L(x_0)$. To facilitate execution of these operations, the list L can be maintained as a 2–3 tree, allowing all the required list-maintainance operations to be performed in time $O(\log n)$.

Details are as follows. The algorithm begins by sorting the $2n$ numbers $a_j, b_j, j = 1, \ldots, n$, in increasing order, and then processes them from left to right. Initially, the list L is empty. Suppose that the abscissa currently being processed is one of the a_j. Then L is updated by inserting the object B_j into L in its proper place, using a standard 2–3 tree search during which comparison of two objects B, B' is accomplished by comparing two representative points in the intersection of $x = a_j$ with B, B' respectively (we have assumed that this can be done in constant time). After insertion, the algorithm finds the two objects B, B' immediately preceding and succeeding B_j in L, and checks whether either B or B' intersects B_j.

Similarly, if the abscissa currently being processed is b_j, then the object B_j is deleted from L, using essentially the technique just outlined. After deletion, the algorithm finds the two objects in L which immediately preceded and followed B_j prior to its deletion (these will have become newly adjacent in L after deletion of B_j), and determines whether they intersect each other.

The algorithm halts whenever an intersection is detected, or, if no intersection has been detected, when all the abscissae a_j and b_j have been processed, in which case the algorithm reports that there is no intersection between the objects in S.

The correctness of the algorithm follows from the preceding observations. The time complexity of the algorithm is $O(n \log n)$ since processing of each of the $2n$ abscissae a_j, b_j can be accomplished in $O(\log n)$ time, using a 2–3 tree representation for the list L.

3. VORONOI DIAGRAMS FOR CIRCULAR OBJECTS

The algorithm presented in the preceding section does not solve the more complicated 'colored object' intersection problem posed in the introduction. Indeed, the argument justifying the correctness of the algorithm breaks down as soon as an intersection is detected, so that if the first intersection detected is between two objects having the same color we can no longer use the procedure described to find additional intersection points. To handle the colored intersections problem, we therefore propose a different approach based on generalized Voronoi diagrams. We will show such an approach which can be used to handle the special case where all the objects in the set S are circular discs, not necessarily of equal radii.

Let each of the objects $B_j \in S$ be a disc of radius r_j about the center x_j, for $j=1, \ldots ,n$. These circular discs need not be disjoint from each other, and may intersect, or even contain, one another. We define a generalized Voronoi diagram $Vor_0(S)$ associated with the set S as follows. For each $i \neq j$ define

$$H(i,j)=\{y \in E^2 : d(x_i,y)-r_i \leq d(x_j,y)-r_j\},$$

i.e., the set of all points whose distance from B_i is no greater than their distance from B_j (note that the distance of y from B_i is taken as the distance of y from the boundary of B_i with a positive sign if y lies outside B_i and with a negative sign if y lies inside B_i). Then define the (closed) *Voronoi cell* $V(i)$ associated with B_i to be

$$V(i) = \bigcap_{j \neq i} H(i,j),$$

i.e., the set of all points y whose distance from B_i is no greater than y's distance from any other element of S. (We will sometimes refer to the point x_i as the *center* of this Voronoi cell; hence the center of $V(i)$ is the same as the center of the disc B_i.) Finally, the *Voronoi diagram* $Vor_0(S)$ is defined to be the set of points which belong to more than one Voronoi cell. For simplicity we assume that no point in $Vor_0(S)$ lies in more than three Voronoi cells. This assumption generalizes the familiar assumption concerning Voronoi diagrams associated with a set of points, in which one requires that no more than three of these points be cocircular. As in the case of points, this assumption is not essential, and is made just to simplify the description of the algorithm. Figure 2 shows an example of such a Voronoi diagram.

Generalized Voronoi diagrams have been previously introduced and analyzed by various authors; see, e.g., Kirkpatrick (1979), who considers Voronoi diagrams for a set of line segments and points, and Lee and Drysdale (1981), who consider Voronoi diagrams for sets of line segments and for sets of circular objects. Our setup is somewhat different from that of Lee and Drysdale (1981), in that we allow the objects in S to intersect each other, or even to contain one another, whereas in Lee and Drysdale the circles are required to be disjoint. As we shall see below, there are some difficulties in adapting the technique of Lee and Drysdale to the case of intersecting circles. The algorithm of Lee and Drysdale for constructing the Voronoi diagram of n *disjoint* circles runs in time $O(n \log^2 n)$, but is given in a very sketchy form without any analysis of the structure of the diagram, and without any proof of its correctness. Another algorithm for constructing generalized Voronoi diagrams for circles, which runs in time $O(n \, c^{(\log n)^{1/2}})$, has been presented by Drysdale and Lee (Drysdale, 1979; Drysdale & Lee, 1978).

The generalized Voronoi diagram just defined has the following properties.

(1) The collection of Voronoi cells covers the whole plane.

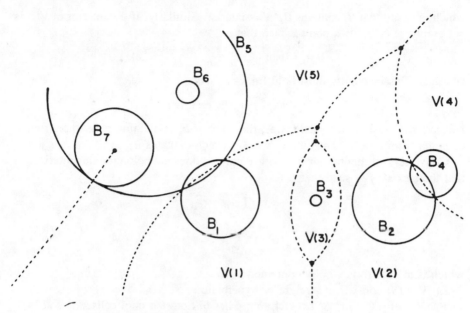

Figure 2.   The Voronoi diagram of a set of circular discs.

Proof: Immediate. Indeed, given any $y \in E^2$, it will belong to the cell $V(i)$ for which

$$d(x_i, y) - r_i = min\ \{d(x_j, y) - r_j: j = 1, \ldots, n\}.$$

(2) $V(i)$ is empty iff B_i is wholly contained in the interior of another disc B_j; $V(i)$ has an empty interior iff B_i is wholly contained in another disc B_j.

Proof: For the first assertion, note that, if $V(i)$ is not empty, it contains a point x such that

$$d(x, x_i) - r_i \leq min\ \{d(x, x_j) - r_j : j = 1, \ldots, n\}.$$

The triangle inequality then shows that this same inequality holds for $x = x_i$, i.e.,

$$-r_i = min\ \{d(x_i, x_j) - r_j : j = 1, \ldots, n\},$$

which implies that, for any of the other circles B_j of S, the point of B_i at maximum distance from x_j is not interior to B_j. Thus, there is no B_j whose interior contains B_i. Conversely, if $V(i)$ is empty, then $x_i \notin V(i)$, so that there exists $j \neq i$ such that

$$d(x_i, x_j) - r_j < -r_i$$

i.e.,

$$d(x_i, x_j) + r_i < r_j$$

which is to say that B_j contains B_i in its interior. Similarly, if the interior of $V(i)$ is not empty, it contains a point x such that

$$d(x,x_i)-r_i < min \{d(x,x_j)-r_j : j=1, \ldots ,n\},$$

and again this inequality must hold for $x=x_i$. Thus

$$-r_i<d(x_i,x_j)-r_j, \text{ for each } j\neq i,$$

Hence $r_j<d(x_i,x_j)+r_i$ for each j, i.e., the point of B_i at maximum distance from x_j does not belong to B_j. Thus there is no B_j which contains B_i.

Conversely, if the interior of $V(i)$ is empty, x_i does not belong to this interior, so that there exists $j\neq i$ such that

$$d(x_i,x_j)-r_j\leq-r_i$$

or

$$d(x_i,x_j)\leq r_j-r_i,$$

which is to say, B_i is wholly contained in B_j.

(3) $Vor_0(S)$ consists of straight or hyperbolic arcs.

Proof: Let $y \in Vor_0(S)$ be such that y lies in both Voronoi cells $V(i)$ and $V(j)$. Then we have

$$d(x_i,y)-d(x_j,y)=r_i-r_j$$

The locus of points satisfying this condition is a hyperbolic arc having x_i and x_j as foci, or, if $r_i=r_j$, the perpendicular bisector to the segment $[x_i,x_j]$. (Note also that this (generally hyperbolic) locus degenerates into a half-line if $d(x_i,x_j)=\pm(r_i-r_j)$, i.e., if the discs B_i, B_j are tangent to one another with one of them wholly containing the other.

(4) Each nonempty Voronoi cell $V(i)$ is star-shaped with respect to the point x_i. Moreover, if $V(i)$ has nonempty interior, then the interior of a segment connecting x_i to a point on the boundary of $V(i)$ does not intersect the interior of any other Voronoi cell, and such a segment can intersect another Voronoi cell $V(j)$ only if the corresponding disc B_j is wholly contained in B_i (so that, by remark (3), $V(j)$ has empty interior).

Proof: Let $y \in V(i)$, and let I be the segment connecting x_i to y. We first claim that each point z in the interior of I is contained in $V(i)$. Indeed, if this were false then there would exist a point z in the interior of I which does not belong to $V(i)$. By (1) there would exist $j\neq i$ such that z belongs to $V(j)$. Hence

$$d(x_j,z)-r_j<d(x_i,z)-r_i.$$

By the triangle inequality (and since z lies between x_i and y on a straight line) we have

$$d(x_j,y)-r_j\leq d(x_j,z)+d(z,y)-r_j<d(x_i,z)+d(z,y)-r_i=d(x_i,y)-r_i.$$

Thus y cannot belong to $V(i)$, a contradiction which proves that $V(i)$ is star-shaped.

Next, suppose that $V(i)$ has nonempty interior. Let $y \in V(i)$ and I be as above, and suppose that I contains an interior point z which also belongs to some other Voronoi cell $V(j)$. Then we have

$$d(x_j,z)-r_j=d(x_i,z)-r_i.$$

Using the triangle inequality as before, we obtain

$$d(x_j,y)-r_j \leq d(x_j,z)+d(z,y)-r_j=d(x_i,z)+d(z,y)-r_i$$

$$=d(x_i,y)-r_i \leq d(x_j,y)-r_j,$$

since $y \in V(i)$. Hence x_j lies on the line containing I, and z lies between x_j and y on this line. However, this implies that

$$d(x_i,x_j)=\pm(r_i-r_j),$$

i.e., that one of the discs B_i,B_j contains wholly the other. But by (2) B_i is not wholly contained in any other disc, so that B_j is wholly contained in B_i, and consequently $V(j)$ has empty interior. This establishes our two final assertions.

DEFINITION: The modified Voronoi diagram $Vor(S)$ is defined to consist of the boundaries of all cells in $Vor_0(S)$ having nonempty interior. All other cells are discarded from the modified diagram.

(5) The intersection I of three Voronoi cells $V(i)$, $V(j)$, $V(k)$ in $Vor(S)$ consists of at most two points.

Proof: Let y be a point in I. By (4) the interior of the segment L_i (resp. L_j,L_k) connecting y with x_i (resp. x_j,x_k) is contained in $V(i)$ (resp. $V(j),V(k)$) and in no other cell. Let y' be another point in I, and let L_i',L_j',L_k' be the segments connecting y' with x_i,x_j,x_k respectively. It is clear that no two of these six segments can intersect one another (except at an endpoint). It follows that I cannot contain a third point z, because at least one of the segments connecting z to the three centers x_i,x_j,x_k would have to intersect one of the preceding six segments, which is impossible.

(6) Let D^* denote the *dual* of $Vor(S)$, defined to be a graph whose vertices are the points $x_i,i=1, \ldots ,n$ for which $V(i)$ has nonempty interior, and which contains an edge $[x_i,x_j]$ if $V(i)$ and $V(j)$ have a nonempty intersection including at least one open arc. If $V(i)$ and $V(j)$ intersect in more than one arc, define the graph D^* to contain multiple edges connecting x_i to x_j, one for each such arc. Then D^* is a planar graph, and its natural planar embedding (described below) has all of its faces (except for the outer one) containing at least three edges.

Proof: We define an embedding of D^* in E^2 as follows. Each vertex x_i is mapped to itself (as a point in E^2). Let $e=[x_i,x_j] \in D^*$ be an edge corresponding to an open arc α in the intersection of $V(i)$ and $V(j)$. It follows from (5) that there must exist at least one point $y \in \alpha$ which does not belong to any other Voronoi cell. We then map the edge e to the path consisting of the two segments $[x_i,y]$ and $[y,x_j]$. To see that the resulting graph G is indeed a planar embedding of D^*, suppose to the contrary that two distinct edges $e_1=[x_i,x_j], e_2=[x_k,x_l]$ of D^* map to paths which intersect each other at a point z which is not a vertex of G. Let z_1 (resp. z_2) be a point on the common boundary of $V(i)$ and $V(j)$ (resp. $V(k)$ and $V(l)$) such that e_1 (resp. e_2) appears in G as the union of the segments $[x_i,z_1]$ and $[z_1,x_j]$ (resp. $[x_k,z_2]$ and $[z_2,x_l]$). Suppose without loss of generality that the segments $s=[x_i,z_1]$ and $s'=[x_k,z_2]$ intersect at a point other than a common endpoint $x_i=x_k$. Since by (4) the interior of the segment s (resp. s') is contained in $V(i)$ (resp. $V(k)$) and in no other Voronoi cell having nonempty interior, we must have either $i=k$ or $z_1=z_2$. In the first case s and s' meet at x_i, and so if they meet at another point they must overlap each other, which is possible only if either $z_1=z_2$ or if one of these points (say z_1) lies in the interior of the other segment s'. The latter assumption would contradict the fact that the interior of s' is wholly contained in $V(i)$ and in no other cell having nonempty interior. Thus we must have $z_1=z_2$, and then by the choice of these points it follows that $j=l$ too. Then it is plain that the two arcs of the common boundary of $V(i)$ and $V(j)$ which define our two paths are identical. Hence the two edges e_1 and e_2 are not distinct, contrary to assumption. All this shows that G is a planar embedding of D^*.

Note that the inner faces of the embedding G of D^* stand in 1-1 correspondence to the Voronoi vertices in $Vor(S)$. Hence the second part of our assertion will follow from the fact that each Voronoi vertex must be incident to three Voronoi edges. This property of Voronoi vertices will be established later in this section (see the Corollary to property (11) below), and will thus imply the property stated above.

(7) Since D^* contains $O(n)$ vertices, and since each of its faces contains at least three edges, it follows by Euler's formula that it has at most $O(n)$ edges, that is, $Vor(S)$ consists of at most $O(n)$ connected straight or hyperbolic arcs.

(8) Let C be the convex hull of the union of all the discs $B_i \in S$. Note that the boundary of C consists of an alternating sequence of straight segments and circular arcs, the circular arcs being boundary portions of some of the discs, whereas the straight segments are tangents to a pair of discs in S. We will say that B_i and B_j are *adjacent* along the boundary of C, if this boundary contains a straight segment tangent to both B_i and B_j. Then the unbounded edges of $Vor(S)$ are those edges that are common to two cells $V(i)$, $V(j)$ for which B_i and B_j are adjacent along the boundary of C. (This property will not be used in the sequel, but is noted as a generalization of similar properties for other types of Voronoi diagrams. See Shamos, 1975).

Proof: Let e be an unbounded edge of $Vor(S)$, common to two Voronoi cells $V(i)$ and $V(j)$. Since e is either a straight or hyperbolic arc, it tends asymptotically to some half-line l. Suppose, without loss of generality, that l is the positive y-axis, and let $a=[0,t]$ be a point on l. For sufficiently large t and for any $k=1, \ldots ,n$, $d(a,B_k)$ behaves asymptotically as $t-\eta_k-r_k$, where η_k is the y-coordinate of x_k, i.e., as $d(a,\bar{B}_k)$ where $\bar{B}_k$, is the image of B_k translated parallel to the x-axis until its center lies on the y-axis. It follows that

$$\eta_i + r_i = \eta_j + r_j \geq \eta_k + r_k,$$

for every $k=1, \ldots, n$. This however is easily seen to imply that B_i and B_j are adjacent to each other along C, since the line $y = \eta_i + r_i$ is tangent to both discs B_i, B_j and all the discs in S lie in the lower half-plane which this line bounds, so that the portion of this line between its points of tangency with B_i and B_j belongs to the boundary of C. (An extreme case that we need consider is that in which the line $y = \eta_i + r_i$ is also tangent to a third disc B_k, at a point lying between its points of contact with B_i and B_j. However, if such a situation arises, it is easy to check that all the points on e sufficiently far away are nearer to B_k than to one of B_i, B_j, contradicting the definition of e. The converse statement, namely that any pair of adjacent discs along C induces an unbounded Voronoi edge, can be proved using the above argument in reverse.

(9) $Vor(S)$ need not be connected. In fact, it can have up to $O(n)$ connected components. However, every connected component of $Vor(S)$ is unbounded.

Proof: Consider the following set S of discs, which consists of the unit disc B_1, and of k additional small discs $B_2, \ldots, B_{k+1}$ all of radius ϱ, such the centers of these discs are placed on the boundary of B_1 at equally spaced positions. If ϱ is chosen to be sufficiently small (e.g., of the order $O(1/k^2)$) then it is easily checked that for each $j=2, \ldots, k+1$ the discs B_1 and B_j are adjacent to each other along the boundary of the convex hull of the B_i's. Moreover, it can also be shown that for ϱ sufficiently small each unbounded edge common to $V(1)$ and some other $V(j)$ is a full branch of the corresponding hyperbola, and that no two such edges intersect each other. This shows that $Vor(S)$ can have as many as $O(n)$ connected components.

Remark: The example just given also shows that the boundary of a single Voronoi cell ($V(1)$ in the example) can have up to $O(n)$ disjoint connected components.

Suppose next that for some set S of discs $Vor(S)$ contains a bounded component K. Then the portion E of a sufficiently small neighborhood of K which lies exterior to K must be contained in a single Voronoi cell $V(i)$, since otherwise some arc of $Vor(S)$ would have to enter any such exterior neighborhood of K, contradicting the assumption that K is a connected component of $Vor(S)$. But if a whole neighborhood of K in the exterior of K lies in $V(i)$, there must exist a point $y \in int(V(i))$ such that the line connecting y to x_i intersects K, contradicting (4). Thus $Vor(S)$ cannot have any bounded connected component.

COROLLARY; $Vor(S)$ does not contain any isolated point. Moreover, by modifying the argument given above one can also show that each Voronoi vertex must belong to at least two edges. (Assertion (11) below will strengthen this claim, by showing that each such vertex must belong to three distinct Voronoi edges.)

(10) No two edges of $Vor(S)$ can be tangent to each other. Also, for each point z on a Voronoi edge e separating two cells $V(i)$ and $V(j)$, the segment connecting z to x_i (or to x_j) is not tangent to e.

Proof: Since Voronoi edges are either straight segments or hyperbolic arcs, it follows that if two Voronoi edges are tangent to each other then any Voronoi edges lying between them at their point of tangency must be tangent to both of them. Hence if there exist any two tangent Voronoi edges, then there exist two such edges which belong to the boundary of the same Voronoi cell. Assume this to be possible, and let $V(i)$ be a Voronoi cell whose boundary contains two edges e, e' which also belong to $V(j)$, $V(k)$ respectively, and which are tangent to each other at some point y. It is plainly impossible for both e and e' to be straight arcs; hence at least one must be part of a hyperbola. Let l be the line which is tangent at y to both hyperbolas (or to the hyperbola and straight line) containing e, e' respectively. As is well known, a line tangent to a hyperbola at a point y bisects the angle $f_1 y f_2$, where f_1 and f_2 are the two foci of the hyperbola; furthermore, this result also holds trivially in case the hyperbola degenerates into the perpendicular bisector of the segment connecting the two points f_1, f_2. Thus, in any case l bisects the angle between the two segments connecting y to x_i and to x_j (resp. to x_i and x_k). It follows that these two angles must be equal, and consequently the three points y, x_j, x_k are colinear, with x_j and x_k lying on the same side of y. Suppose for definiteness that x_j lies between y and x_k. By definition of e and e' we then have

$$d(x_i, y) - d(x_j, y) = r_i - r_j$$

$$d(x_i, y) - d(x_k, y) = r_i - r_k$$

or

$$d(x_k, y) - d(x_j, y) = d(x_j, x_k) = r_k - r_j,$$

which is to say, B_j is wholly contained in B_k, so that by definition $Vor(S)$ does not contain any edge bounding $V(j)$. This contradiction establishes our assertion.

The second assertion follows from the fact that no tangent to a hyperbola can pass through any of its foci. (The only exception is when the hyperbolic arc containing e degenerates into a half-line, but then either B_i or B_j must be wholly contained in some other disc, so that by convention e does not appear in our Voronoi diagram.)

(11) Let e and e' be two adjacent edges along the boundary of a Voronoi cell $V(i)$. Then the interior angle in $V(i)$ between e and e' is less than 180 degrees. (Stated otherwise, in traversing the boundary of $V(i)$ with $V(i)$ to the right, we make a right turn as we pass from one of the edges e, e' to the other.)

Proof: We have already shown that e and e' cannot be tangent to each other. Let y be their point of intersection, and let l be the line containing x_i and y. The two edges e and e' cannot both lie on the same side of l in the vicinity of y, because then the segment connecting x_i to a point on one of them sufficiently near y would have to intersect the other edge, contradicting (4). Suppose then that e lies on the left side of l (oriented from x_i to y), and that e' lies on the right side of l. Extend e along the hyperbolic branch containing it into the right side of l (note that a hyperbola is never tangent to the segment connecting a point lying on it to one of its foci), and denote the extended portion of e by e''. Then e' must lie between x_i and e'', for otherwise the segment connecting x_i to any point on e' lying sufficiently near to y will intersect e'' at a point z, and plainly no point $z \in e''$ can belong to the interior of $V(i)$ (because z is equidistant from B_i and from some other disc in S), again contradicting (4). It therefore follows that the interior angle between e and e' is less that 180 degrees, as asserted.

COROLLARY: This argument shows that each Voronoi vertex must be incident to three distinct Voronoi edges, for if it belonged to just two edges, at least one of the angles between these edges would be interior to some Voronoi cell, and would be greater than 180 degrees, contrary to what we have just shown.

4. EFFICIENT CONSTRUCTION OF GENERALIZED VORONOI DIAGRAMS

Next, we present an algorithm which adapts the divide-and-conquer methods used by Shamos (1975) and by Kirkpatrick (1979) to construct the modified Voronoi diagram of a set of points, and which computes the Voronoi diagram of a set S of n circular bodies in time $O(n \log^2 n)$. The algorithm produces a list of all Voronoi cells having nonempty interior, and for each cell constructs a circular list containing the edges on its boundary, arranged in clockwise order (for unbounded cells this list will also include 'virtual edges' at infinity connecting pairs of unbounded edges e, e' such that the intersection of e' with an arbitrarily large circle lies immediately clockwise to the intersection of this circle with e). Finally, the algorithm produces a table in which each edge points to the two cells containing it.

For simplicity, we will assume in what follows that no two circles in S have distinct leftmost points lying on the same vertical line. If S does not have this property, then we can apply an infinitesimal rotation that will make the abscissae of the leftmost points of all circles in S distinct from each other. Cf. also Schwartz and Sharir (1983a), where a similar technique based on infinitesimal perturbations is used to resolve degenerate configurations arising in other geometric problems. The algorithm begins by dividing S into two subsets R and L of equal size, such that the leftmost point w_i of each $B_i \in L$ lies to the left of the leftmost point w_j of every disc $B_j \in R$. This partitioning of S can easily be done by finding the median of these leftmost points in time $O(n)$. Note that it has the property that no disc $B_i \in L$ is wholly contained in another disc $B_j \in R$, although a reverse containment is possible.

Assume that the Voronoi diagrams $Vor(R)$ and $Vor(L)$ have been computed recursively. The main step of the algorithm is to merge these two diagrams into a single diagram $Vor(S)$. For this, one must compute the set C of points y which are simultaneously nearest to a disc $B_i \in L$ and to a disc $B_j \in R$. Following Kirkpatrick (1979), we call C the *contour* separating R and L. We will see that, once C has been computed, $Vor(S)$ can be obtained by taking the union of $Vor(R)$, $Vor(L)$, and C,

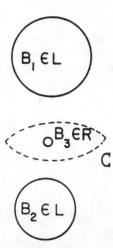

Figure 3. A bounded connected component of C.

and then by discarding (portions of) edges belonging to one of the partial diagrams $Vor(R)$, $Vor(L)$, whose points have become nearer to some object belonging to the other set (these portions will be delimited by the intersections of these edges with C). Note that, during this merging step, some Voronoi cells $V(i)$ with $B_i \in R$ may be wholly deleted from $Vor(S)$ if B_i happens to lie wholly inside some disc $B_j \in L$.

Since C will be a curve consisting of straight and hyperbolic arcs, the complement of C will consist of finitely many open connected planar regions. Each such region M is either a union of cells $V(i)$ (in the final diagram $Vor(S)$) with $B_i \in L$ (in which case we will call M an L-region), or a union of cells $V(i)$ with $B_i \in R$ (in which case M will be called an R-region).

By assertion (9) of Section 3 and the example provided there, the contour C may consist of several disjoint connected components, no matter how S is partitioned into R and L (note that, in the point-based case (Shamos, 1975), appropriate partitioning of S will produce a contour with a single component). Moreover, it is also possible for a connected component of C to be bounded, as shown by the example appearing in Figure 3.

The following lemma restricts this complexity, and begins to develop some of the facts that will be needed to show that tracing all the components of C need not be expensive.

LEMMA 4.1:   (a) C consists of a disjoint union of simple topologically closed curves without endpoints (i.e., each such curve is either closed or stretches to infinity in both directions).
(b) Let $B_i \in L$ and let u be the horizontal half-line whose rightmost endpoint is x_i. Then u does not intersect C (and consequently lies wholly inside an L-region).

(c) There exists precisely one L-region; all other components of the complement of C are R-regions.

(d) Each R-region has a connected boundary, consisting of a single (bounded or unbounded) component of C.

Proof:

(a) It follows from its definition that C is a union of Voronoi edges of $Vor(S)$. It therefore suffices to show that for each Voronoi vertex v of $Vor(S)$ lying on C, there are exactly two Voronoi edges emerging from v which belong to C. Since we have ruled out degenerate configurations, we can assume that v belongs to exactly three Voronoi cells $V(i)$, $V(j)$ and $V(k)$. Moreover, since $v \in C$, one of the discs B_i, B_j, B_k must belong to L, and another of these discs must belong to R. Assume first that B_i, $B_k \in L$ and that $B_j \in R$. Then, in the neighborhood of v, the contour C consists of the two edges separating $V(j)$ from $V(i)$ and $V(k)$ respectively. Much the same argument applies if B_i, $B_k \in R$ and $B_j \in L$. This proves (a).

(b) Suppose the contrary, and let z be a contour point lying on u. Let $w \in u$ be a point at distance r_i from x_i (that is, w is the leftmost point of B_i). It follows that there exist discs $B_j \in R$ (which is not wholly contained in any other disc), and $B_k \in L$ such that

$$d(z,B_j) = d(z,B_k) \leq d(z,B_i) = d(z,w)$$

But then some point on B_j (and in particular its leftmost point) must lie to the left of w, or coincide with w. However, both these possibilities contradict the way in which L and R have been defined, a contradiction which proves (b).

(c) It suffices to show that the centers x_i, x_j of each pair B_i, B_j of discs in L are connected to each other via a path which does not intersect the contour C. Let u_i (resp. u_j) be the horizontal halfline whose rightmost endpoint is x_i (resp. x_j), and let $w_i \in u_i$ (resp. $w_j \in u_j$) be the leftmost point of B_i (resp. B_j). For each t, let y_i (resp. y_j) be a point on u_i (resp. u_j) whose abscissa is t. We claim that if t is negative and has a large enough absolute value, the segment $e = y_i y_j$ does not intersect the contour, which, together with (b), implies that x_i and x_j are connected to each other via the polygonal path $x_i y_i y_j x_j$ which is wholly contained in a single L-region. To see this, suppose to the contrary that there exists a contour point $z \in e$, nearest to some $B_k \in R$ and to some $B_m \in L$. Then

$$d(z,B_k) = d(z,B_m) \leq d(z,B_i) = (d^2(y_i,w_i)+d^2(z,y_i))^{\frac{1}{2}} \approx d(y_i,w_i)$$

and similarly

$$d(z,B_k) \leq d(z,B_j) \approx d(y_j,w_j)$$

as $t \to \infty$. It follows that the leftmost point w_k of B_k does not lie to the right of w_i or of w_j. Since, by definition of L, R, w_k cannot lie to the left of w_i, w_j, it follows that the three points w_i, w_j, w_k all have the same abscissa, again contradicting the way in which L and R are defined.

(d) Suppose that M is an R-region whose boundary consists of at least two disjoint connected components K_1, K_2. By part (a) of the present lemma K_1 (resp. K_2) partitions the plane into two disjoint components, both bounded by K_1 (resp. K_2). Since K_1 and K_2 are disjoint, it follows that they both collectively partition the plane into three components, one of which contains M, whereas the other two contain L-regions. This, however, contradicts (c), thus proving our assertion. Q.E.D.

In general outline, the remainder of our argument is as follows. We first show that, given points z_K on each of the components K of C, the whole of C can be traced in a number of steps bounded by $O(n)$. Next we show how to find such a set of points z_K. This is done by noting that (by definition) every R-region must contain at least one center x_i of a disc $B_i \in R$ such that B_i is not contained in any other disc of R. Hence if we iterate over all such points x_i, and connect each one of them by a straight arc e to a point of an L-region, these e must together intersect all the components of C. We will show that in total time $O(n \log n)$ we can find such arcs e, each intersecting C in just one point z, and in this way can find a z on each component K of C. This leads to an $O(n \log n)$ merging step, and hence to an $O(n \log^2 n)$ overall algorithm. Note that in this generality our algorithm is very similar to that of Lee and Drysdale (1981); they differ in the actual implementation of the various steps just outlined.

The tracing of C during the merge phase of our algorithm is done in a manner quite similar to that described in Kirkpatrick (1979). To facilitate this tracing, we will find it convenient (following Kirkpatrick) to partition each Voronoi cell $V(i)$ (of either $Vor(L)$ or $Vor(R)$) into *subcells* by connecting x_i to each vertex v of $V(i)$ by a straight segment (called a *spoke*, as in Kirkpatrick). Clearly, each subcell is an angular sector bounded by two spokes and one Voronoi edge. Note that, given a directed straight line or hyperbolic arc e, its intersection points with the boundary of any Voronoi subcell can be found in constant time, assuming that we use an appropriate representation of the corresponding diagram $Vor(L)$ or $Vor(R)$.

Suppose that we have somehow found a point $z \in C$ (but such that z is not in either $Vor(L)$ or $Vor(R)$), for which the two discs $B_i \in L$, $B_j \in R$ nearest to z are known, and suppose further that the two subcells of $V(i)$ in $Vor(L)$ and of $V(j)$ in $Vor(R)$ to which z belongs are also known. Then we can trace the component K of C containing z as follows. We first find the Voronoi edge e (in $Vor(S)$) containing z. Note that e is part of the straight line or hyperbolic are equidistant from B_i and B_j, and is an edge lying on K. We begin tracing K by following e from z in some direction, and by computing its intersection points with the boundaries of the two subcells $U(i)$, $U(j)$ of $V(i)$, $V(j)$ respectively, containing z. Suppose for specificity that the nearest of these points along e is the point z' at which e intersects the boundary of $U(i)$. If z' lies on a Voronoi edge, then the contour K crosses this edge to another Voronoi cell $V(k)$ of $Vor(L)$ after z' (by assertion (10) of the previous section, two Voronoi edges are never tangent to one another). In this case, K continues after z' along the Voronoi edge e' containing points equidistant from B_k

and B_j. On the other hand, if z' lies on a spoke, K will continue after z' along the edge e, but will cross into another subcell of $V(i)$ (Note that the contour can never be tangent to a Voronoi spoke, by the second part of assertion (10)).

Tracing the contour in this way, we either come back to z, in which case the component K is a bounded component of the contour, or else we reach an unbounded edge of the contour, in which case we have to repeat the tracing procedure just outlined by starting from z in the other direction of the edge e in order to obtain the entire component K.

Let M be the R-region bounded by K. Each cell $V(i)$ through which K passes is cut by K into several portions; one of which belongs to an R-region (and contains x_i) while the others belong to the L-region. Moreover, all the cell portions belonging to R-regions belong to the same R-region M. Thus, as K is being traced, we can also note that all these cell portions cut by K belong to M. Observe that M may also contain additional *internal* cells of $Vor(R)$ which have not yet been encountered. These will be dealt with during later steps of the algorithm.

Next we show that if a point z_K is available on each of the components K of C, the total cost of constructing C is $O(n)$. The complexity of the tracing procedure just described is plainly $O(n_1^{(K)}+n_2^{(K)})$, where $n_1^{(K)}$ is the number of Voronoi edges in K, and where $n_2^{(K)}$ is the number of intersections of K with Voronoi spokes (in either $Vor(L)$ or $Vor(R)$). As in Kirkpatrick (1979), we can show that the sum over all K of the quantities $n_1^{(K)}+n_2^{(K)}$ is $O(n)$. Namely, we have

LEMMA 4.2:   Each Voronoi spoke (in either $Vor(L)$ or $Vor(R)$) is intersected by the contour C in at most one point. Moreover this remark also holds for any segment one of whose endpoints is the center x_i of some Voronoi cell $V(i)$ (in either diagram) and which is wholly contained in $V(i)$.

Proof: Let e be a Voronoi spoke of a cell $V(i)$ in $Vor(L)$. Then each point z at which e and C intersect each other must lie on the boundary of the cell $V(i)$ in $Vor(S)$, and since this cell is star-shaped with respect to x_i it follows that at most one such intersection point can exist. Q.E.D.

COROLLARY: The total time required to trace the contour, given a point z (and the two subcells in $Vor(L)$, $Vor(R)$ containing it) on each of its connected components, is $O(n)$.

Proof: The total number of edges on C is $O(n)$, because C is a subset of $Vor(S)$. The total number of intersections of C with Voronoi spokes is also $O(n)$, by Lemma 4.2. Hence the total complexity of the tracing procedure applied to each component K of C is

$$\sum_K (n_1^{(K)} + n_2^{(K)}) = O(n)$$

Q.E.D.

The problem that now remains is that of finding a representative point z_K on each component K of the contour C (and also finding the subcells containing z_K). For this, Kirkpatrick (1979) uses a technique which traces edges of minimum spanning trees for R and L. However, this technique, which works nicely for a set S of *points*, is not easily generalizable to sets of more general objects like the circles which now concern us. We will therefore present an alternative approach, which works for sets of circular discs, but whose complexity is $O(n \log n)$, instead of the linear complexity of Kirkpatrick's technique.

We iterate over all the cells $V(i)$ of $Vor(R)$ (and their corresponding circles B_i), proceeding as follows. Let $B_j \in L$ be such that $x_i \in V(j)$ (in $Vor(L)$) so that B_j is the circle of L which is 'closest' to x_i. If B_j contains all of B_i, then, by definition, the final diagram $Vor(S)$ will not contain a cell $V(i)$. In this case we simply do not use the pair x_i, x_j to find a point on the contour, but go on to consider the other discs $B_{i'} \in R$.

Next, suppose that B_j does not contain all of B_i. Then no other disc $B_k \in L$ can have this property, and so it follows that the disc B_l in S for which $d(x_i, x_l) - r_l$ reaches its minimum belongs to R. Similarly, $d(x_j, x_l) - r_l$ attains its minimum when $l = j$, since no disc in R can contain the whole of a disc in L, and since the recursive construction of $Vor(L)$ described in the following paragraphs will eliminate discs that are wholly contained in other discs. Thus the segment $e = x_i x_j$ must contain a point z for which $d(z, x_i) - r_i$ reaches its minimum simultaneously for some $B_i \in L$ and for some other disc $B_{i'} \in R$, which is to say, a point z on C. Moreover, since e emanates from x_j and is wholly contained in $V(j)$, it follows by Lemma 4.2 that the contour C cannot intersect e in more than one point. Hence e intersects C in exactly one point. Note that the entire segment e is contained in a single subcell of $V(j)$ in $Vor(L)$. All that we have to show is that we can either find z, or assure ourselves that a point on the same component K (of C) as z has already been found, in total time $O(n \log n)$.

We can proceed to find the unique intersection z of e with C using a technique quite similar to the contour-tracing procedure described above. That is, we first find the Voronoi subcell (of $V(i)$) in $Vor(R)$ containing points on e near x_i (since e emerges from x_i, this amounts to finding the two Voronoi spokes of $V(i)$ between which e lies). We then find the intersection of e with the Voronoi edge bounding that subcell, beyond which e crosses into another subcell of $Vor(R)$. (In the extreme case in which e coincides with a Voronoi spoke of $V(i)$, e will exit the two subcells of $V(i)$ in which it lies at the Voronoi vertex which is the other endpoint of the spoke; it is then a bit more complicated, but still straightforward, to determine the Voronoi subcell into which e enters past this vertex.) Continuing in this manner, we partition e into subsegments $e_1, \ldots, e_m$, each of which is contained in some Voronoi subcell of $Vor(R)$ or lies along a spoke of $Vor(R)$. As all this is done, we keep track of all the cells $V(k)$ of $Vor(R)$ that have already been encountered. If such a cell is encountered for the second time, tracing of the sequence of edges $e_1, \ldots, e_m$ stops immediately (this rule is justified by Lemma 4.3 below). This guarantees that the total cost of traversing subcells of $Vor(R)$ is bounded by the total number of such subcells, and hence by $O(n)$.

Remark: It is well to compare this tracing technique with the technique of Lee and Drysdale (1981) for locating points on the contour. First of all there is an inaccuracy in the description of their algorithm, even in the case of line segments. Namely (cf. Lee & Drysdale, 1981, top of p. 83) they claim that if t is the nearest point on some segment of L to an endpoint q of a segment in R, then the nearest endpoint of a segment in R to t is q itself. This however is false, as can be easily checked. Lee and Drysdale correct themselves later by considering the midpoint m of the segment qt rather than t itself. In the case of circles, however, they also state an inaccurate statement of this form, but do not correct themselves. In the case of disjoint circles, using the midpoint of the segment uv (in the terminology of Lee & Drysdale, p. 86) will result in a correct algorithm; it is not clear whether this amendment will also work in the case of intersecting circles.

For each $t=1, \ldots, m$ let $V(i_t)$ be the cell in $Vor(R)$ containing e_t. As tracing proceeds through the cell $V(i_t)$, we check whether $V(i_t)$ has been encountered before, and, if not, whether there exists $z \in e_t$ such that $d(z, B_j) = d(z, B_i)$ (this can be done in constant time). As already shown, there will exist a unique point z on e having this property, and this z will be the required intersection point of e with C (the algorithm will reach this z only if tracing is not abandoned earlier, because a previously encountered cell of $Vor(R)$ is encountered again). Let e_s be the subsegment of e containing z. Note that z is found in time $O(p+s)$, where p is the number of subcells of $V(i)$ in $Vor(R)$.

To show that tracing of e can be abandoned as soon as any cell of $Vor(R)$ is encountered for the second time, we will use the following

LEMMA 4.3: Let s be as in the preceding paragraph. Then for each $t \leq s$, $Vor(S)$ contains a cell whose center is x_{i_t}, and the point x_{i_t} lies in the same R-region as x_i.

Proof: Let M be the R-region containing $x_{i_1} = x_i$, and let $1 < t \leq s$. Pick any point $z_t \in e_t$ (but in case $t = s$, z_t must lie between x_i and z). Since $t \leq s$, the segment $x_i z_t$ does not intersect the contour, and therefore is wholly contained in M. As always, let B_{i_t} be the disc in L corresponding to the cell $V(i_t)$ of $Vor(R)$, and let x_{i_t} be its center. Suppose for the moment that we have already shown that a cell $V(i_t)$ (with center x_{i_t}) appears in $Vor(S)$, i.e., that B_{i_t} is not wholly contained in any disc of L. The segment $J = z_t x_{i_t}$ is contained in the cell $V(i_t)$ of $Vor(R)$, which is star-shaped with respect to its center x_{i_t}, by property (4) of Section 3. Moreover, since this segment emanates from x_{i_t}, it can intersect the contour in at most one point, by Lemma 4.2. But such an intersection is impossible, because both the endpoints of J lie in an R-region (x_{i_t} lies in an R-region because by assumption it belongs to $V(i_t)$ in $Vor(S)$). Therefore the polygonal path $x_{i_1} z_t x_{i_t}$ does not intersect the contour. But x_{i_1} and x_{i_t} are connected to each other via this path, and hence lie in the same R-region.

It only remains to show that a cell $V(i_t)$ with center x_{i_t} appears in $Vor(S)$. For this, note that the point z_t lies in an R-region, so that it is nearer to some disc of R than to any disc of L. From this it is plain that z_t is nearer to B_{i_t} than to any other disc of S.

But then z_t must be an interior point of the cell $V(i_t)$ in $Vor_0(S)$, so that, by definition of $Vor(S)$, a cell $V(i_t)$ with center x_{i_t} appears in this diagram, as asserted. Q.E.D.

As we apply the procedure just described to each of the discs $B_i \in R$, one of the following three situations will arise: either

(a) We discard B_i immediately, because the nearest disc B_j in L wholly contains B_i; or

(b) While tracing subcells of $Vor(R)$ crossed by the segment $e = x_i x_j$ (where $B_j \in L$ is the disc for which $x_i \in V(j)$), we encounter a subcell of some cell $V(r)$ whose R-region M in $Vor(S)$ was encountered before. In this case we conclude from Lemma 4.3 that B_i, as well as every other disc of R whose cell in $Vor(R)$ has been crossed by e so far, lies in the R-region M. In this case we can stop tracing e and go on to process other discs of R, since we can be sure that the component of the contour C intersected by e has already been explored. The algorithm will also note that all cells of $Vor(R)$ crossed by e so far belong to M, to avoid repeated processing of these cells later on. (Note that this case will arise only when $V(i)$ is an inner cell in M, i.e., a cell not intersected by the contour); or

(c) The tracing procedure continues until an intersection z of e with the contour is found. In this case only new cells of $Vor(R)$ are being traced, and z will lie on a new component of C (this component must be new, because all the old components of C have already been traced, and all the R-subcells through which they pass have already been encountered and marked; hence before reaching any old component of the contour the scanning will stop by step (b) above). As in (b), we take note of the fact that all cells crossed by e during this tracing belong to the new R-region just found, to avoid repeated processing of these cells later on.

These observations imply that we can find a representative point on each component of the contour in total time $O(n)$, provided that, for each $B_j \in R$, the subcell of the cell $V(i)$ of $Vor(L)$ containing x_j is already known.

To obtain this final item of information, we can use a simple plane-sweeping algorithm, similar to those described by Shamos (1975), Bentley and Ottman (1979), and Nievergelt and Preparata (1982). The algorithm sweeps the plane from left to right and maintains a vertical "front" $T(a)$ consisting of the segments lying along the line $x = a$ and delimited by the points of intersection of this line with the edges and spokes of $Vor(L)$. The structure of $T(a)$ will change only at points a which are either abscissae of Voronoi vertices of $Vor(L)$, or abscissae of centers of discs in L, or points for which the line $x = a$ is tangent to some Voronoi edge of $Vor(L)$. The number of such 'transition points' is plainly $O(n)$, and the total number of segments of $T(a)$ is also $O(n)$ for any real a. To start the algorithm, sort the set A, consisting of all transition points of $Vor(L)$ and all centers x_i of discs $B_i \in R$, by their x-coordinates, and initialize the list $T(a)$ as a 2-3 tree for some large enough negative real a. Both these tasks can be done in time $O(n \log n)$. Then scan A from left to right. For each $a \in A$, if a is a transition point of $Vor(L)$, update the list T by an appropriate combination of deletions, insertions, and merge operations applied to segments in T; this can be done in time $O(k_a \log n)$, where k_a is the number of segments which undergo these changes. Note that if a is the abscissa of a center x_i of some disc $B_i \in L$, then k_a is the number of Voronoi edges on the boundary of the cell $V(i)$ in $Vor(L)$, which may be large. Nevertheless, the total sum of all the k_a's over all transition points a is always $O(n)$. If a is the abscissa of a center c of some disc in R, search T to find the segment in T containing x_i, from which the Voronoi subcell of $Vor(L)$ containing c is readily obtained. Proceeding in this way, we locate all the centers of discs of R in $Vor(L)$ in time $O(n \log n)$.

Together, the details just described yield the following algorithm for constructing $Vor(S)$:

1. Split S into two equal-size subsets L, R such that the leftmost point of each $B_i \in L$ lies to the left of the leftmost point of every $B_j \in R$ (we have assumed that no two leftmost points have the same abscissa).
2. Compute $Vor(L)$ recursively.
3. Apply the plane-sweeping procedure described above to locate the subcell $V(j)$ of $Vor(L)$ containing x_i for all centers x_i of discs $B_i \in R$. Discard the disc $B_i \in R$ if it is wholly contained in B_j.
4. Let R' be the remaining set of discs of R. Compute $Vor(R')$ recursively.
5. Construct the 'contour' C as follows:
   For each disc $B_i \in R$ whose R-region (in $Vor(S)$) has not yet been identified
   a. Connect x_i to the center x_j of the disc $B_j \in L$ whose Voronoi cell $V(j)$ in $Vor(L)$ contains x_i. If B_j wholly contains B_i, then $Vor(S)$ will not contain a cell corresponding to B_i, and we go on to process other discs of R.
   b. Find the unique intersection z of the contour with the segment $e = x_i x_j$ by applying the tracing procedure described above. If that procedure detects an intersection of e with a cell $V(k)$ of $Vor(R)$ whose R-region M has already been found, it assigns M as the R-region of B_i and of all other discs of R whose cells in $Vor(R)$ have been crossed by e before $V(k)$ has been reached, and continues with the main loop of this phase.
   c. Trace the whole contour component K containing z. An R-region indication is thereby assigned to all discs $B_k \in R$ whose cells in $Vor(R)$ are encountered.
6. Obtain the final diagram $Vor(S)$ by taking the union of $Vor(R)$ and of $Vor(L)$ with C, and then by discarding (portions of) edges of $Vor(R)$ or of $Vor(L)$ which are cut off from their cells by C.

The algorithm just sketched runs in time $O(n \log^2 n)$. Its costliest phase is step 3, which locates subcells of $Vor(L)$ containing the centers of discs of R.

Note that it is a simple matter to modify the algorithm so that it also produces a mapping *contain*, which, for each disc B_i deleted from the diagram by the algorithm, gives the disc B_j containing B_i as found in step 5.a of the algorithm.

5. APPLICATIONS OF THE GENERALIZED VORONOI DIAGRAM; POSSIBLE EXTENSIONS

In this section, we show how the generalized Voronoi diagram can be used to solve some of the intersection problems mentioned in the introduction to this paper.

Suppose that the generalized diagram $Vor(S)$ for a set S of circular discs has been constructed, and that it is represented by the data structures described in Section 4.

First consider the problem of detecting the existence of an intersection between any pair of discs in S. This can be tested using the following procedure:

First check whether there exists a Voronoi cell $V(i)$ having empty interior (i.e., a cell which is missing from $Vor(S)$). If so, B_i is wholly contained in some other disc, and an intersection has been found. Otherwise, for each edge e in $Vor(S)$ belonging to the common boundary of two Voronoi cells $V(i)$ and $V(j)$, compute the value

$$\varrho(e)=min \{d(x_i,y)-r_i : y \in e\} = min \{d(x_j,y)-r_j : y \in e\}.$$

If $\varrho(e) \leq 0$ for some $e \in Vor(S)$, then it is clear that the discs in S intersect. On the other hand, if $\varrho(e) > 0$ for each $e \in Vor(S)$, then no two discs in S intersect. Indeed, suppose that two discs B_i and B_j intersect each other. Let I be the segment $[x_i,x_j]$. For each $y \in I$ consider the function

$$f(y)=min (\{d(x_k,y)-r_k : k=1, \ldots ,n\}.$$

Note that $f(y) \leq 0$ for each $y \in I$, because each such y lies either in B_i or in B_j, so that either $d(x_i,y)-r_i$ or $d(x_j,y)-r_j$ is ≤ 0. Moreover, by assumption both $V(i)$ and $V(j)$ have nonempty interiors, which implies by (2) that x_i belongs to $V(i)$ and to no other cell. Hence I must intersect $Vor(S)$ at least once. Let $y \in I$ be a point belonging to some edge e of $Vor(S)$, and let $V(k)$, $V(l)$ be the two Voronoi cells containing e in their boundary. Then we have

$$\varrho(e) \leq d(x_k,y)-r_k=f(y) \leq 0,$$

from which our claim follows immediately.

It is easy to see that a simple modification of the procedure just outlined yields a solution to the more complicated problem in which the discs in S are of several colors and we want to detect intersection between two discs of different colors. The appropriate procedure in this case is:

(a) First check whether there exists a disc B_i which is wholly contained in another disc B_j of a differnt color, i.e., if $V(i)$ has empty interior, and x_i belongs to a cell $V(j)$ of a disc with a different color. Once the Voronoi diagram has been constructed by the method described in the preceding section, and has been supplemented by the mapping *contain*, we can detect such cases in $O(n)$ time. Note that not every containment of a disc B_i of, say, red color in another disc B_j of a different color can be detected from the *contain* map, because B_j might be contained in another red disc B_k, and *contain* may map B_i directly to B_j. Nevertheless, if B_i is the *largest* possible disc contained in a disc of a different color, then *contain* will map B_i to some *differently*-colored disc containing it, so that if any disc is wholly contained in a disc of a different color, the procedure just described will detect at least one such situation.

(b) If step (a) detects no intersection, compute the quantities $\varrho(e)$, as defined above, for all edges $e \in Vor(S)$. Then two discs of different colors intersect each other if and only if there exists an edge e common to two cells $V(i)$ and $V(j)$, for which B_i and B_j have distinct colors, such that $\varrho(e) \leq 0$.

Proof: Plainly if $\varrho(e) \leq 0$ for such an edge e, then e must contain a point y which lies inside both B_i and B_j, so that these two differently-colored discs intersect each other. Conversely, suppose that two differently-colored discs B_i and B_j intersect each other. Define the segment I and the function f on it as in the preceding paragraphs. We can assume without loss of generality that $V(i)$ and $V(j)$ have nonempty interiors, for if B_i (or B_j) had empty interior then it would have been wholly contained in another disc B_k of the same color, so that we could replace B_i (or B_j) by B_k in what follows. Call the colors of B_i, B_j 'red' and 'green,' and call a Voronoi cell $V(p)$ a 'red' (resp. 'green') cell if B_p is colored red (resp. green). Then x_i lies in (the interior of) a red cell, whereas x_j does not. Hence I must intersect $Vor(S)$ at an edge e which separates a red cell $V(k)$ and a cell $V(l)$ of a different color. Arguing as before, it follows that $\varrho(e) \leq 0$ for this edge. Q.E.D.

Next consider the problem of determining the shortest distance between any two discs in S. Suppose that the discs in S do not intersect each other (if they do, the above procedures will detect this fact, and the distance that we seek will be 0), and let B_i, B_j be the two discs closest to each other among all pairs of discs in S. Let y be the point on the segment $I = [x_i, x_j]$ equidistant from B_i and B_j. We claim that $y \in Vor(S)$. For otherwise, there would exist another disc $B_k \in S$ such that $d(x_k, y) - r_k < d(x_i, y) - r_i$. But then, by the triangle inequality,

$$d(B_j, B_k) = d(x_j, x_k) - r_j - r_k \leq d(x_j, y) - r_j + d(x_k, y) - r_k$$

$$< d(x_j, y) - r_j + d(x_i, y) - r_i = d(x_i, x_j) - r_i - r_j = d(B_i, B_j),$$

contrary to assumption. Thus $y \in Vor(S)$. Moreover, the function

$$f(z) = min \{d(x_k, z) - r_k : k = 1, \ldots, n\}$$

attains its minimum value on the whole Voronoi diagram $Vor(S)$ at the point y. This follows since by the triangle inequality we have $2f(z) \geq d(B_k, B_l)$ for each $z \in Vor(S)$, where $V(k)$ and $V(l)$ are the two Voronoi cells containing z. It follows by the definition of f and by the preceding definition of $\varrho(e)$ that $f(y)$ is the smallest of the values $\varrho(e)$, for edges e of $Vor(S)$. Taken together, these arguments show that the shortest distance between two discs in S is equal to

$$min \{2\varrho(e) : e \text{ an edge of } Vor(S)\},$$

and hence this distance can be found in time $O(n \log^2 n)$.

A similar technique can be used to find the nearest neighbor in S of each $B_i \in S$.

Indeed, an easy generalization of the preceding argument implies that if B_j is the nearest neighbor of B_i, then $V(i)$ and $V(j)$ meet at a common Voronoi edge, and the shortest distance between B_i and any other disc in S is equal to

$$min \{2\varrho(e) : e \text{ a boundary edge of } V(i)\},$$

from which the nearest neighbor of B_i is easily found.

These arguments extend easily to the case in which each of the discs of S is assigned a certain color, and we want to find the shortest distance between any two differently-colored discs. For this, let B_i and B_j be two discs in S of different colors such that their distance is the smallest among all distances between two differently-colored discs in S. Let the colors of B_i, B_j be 'red' and 'green,' respectively. Let y be the point on $[x_i, x_j]$ equidistant from B_i and B_j. We claim that $y \in Vor(S)$, for otherwise there would exist another disc B_k such that $d(x_k, y) - r_k < d(x_i, y) - r_i = d(x_j, y) - r_j$. If B_k is colored red, then arguing as above we would obtain $d(B_j, B_k) < d(B_i, B_j)$, i.e., a shorter distance between a red and a green disc. Similarly, if B_k is colored green, then we would have $d(B_i, B_k) < d(B_i, B_j)$, again a contradiction. Finally, if B_k is of another color, then both $d(B_i, B_k)$ and $d(B_j, B_k)$ are smaller than $d(B_i, B_j)$, again a contradiction.

Thus $y \in Vor(S)$, and similar arguments to those used above imply that $f(y)$ is the minimum of all $\varrho(e)$, for edges e of $Vor(S)$ separating cells of different colors. This shows that the shortest distance between two differently-colored discs in S is simply the smallest of the values $2\varrho(e)$, taken over all edges e of $Vor(S)$ separating two differently-colored cells.

We next consider Problem IV listed in the introduction. For this we apply Kirkpatrick's algorithm (1983) for fast searching in planar subdivisions to the subdivision of the plane into Voronoi cells. This algorithm applies in our case since each Voronoi cell is star-shaped with respect to the corresponding disc center. The additional pre-processing cost is only $O(n)$ (after $Vor(S)$ has been constructed), and then one can find the Voronoi cell containing any specified point in time $O(\log n)$.

Using the Voronoi diagram $Vor(S)$ has enabled us to solve some of the problems noted in the introduction, but not all of them. In particular, we would like to use these methods for solving the second part of Problem III (that is, to find the differently-colored disc nearest to any given disc in S), and for solving Problem IV. Concerning Problem III, we note that in some special cases straightforward generalizations of the techniques presented above can be used to solve this problem. This is the case for example if each color class has a constant size. Note also that the following partial solution to the second part of Problem III is available in general: Consider the subgraph H of the dual graph D^* of $Vor(S)$ defined so that x_i and x_j are adjacent in H if and only if (they are adjacent in D^* and) B_i and B_j have the same color. Let P be a connected component of H (i.e., a "clustering" of discs having the same color). Then the shortest distance between some disc whose center appears in P and a differently colored disc can be found by tracing all Voronoi edges which separates a disc in P from a disc with a different color, using the same technique given above. In many applications partial results of this form are sufficient.

6. COMPUTING THE AREA OF A UNION OF DISCS

Let us consider next another interesting application of the generalized Voronoi diagram for discs. Let S be a set of n possibly intersecting discs in the plane, and let K denote the union of all discs in S. The problem at hand is to compute efficiently the area of K. We will show that, once the generalized diagram $Vor(S)$ has been computed, the area of K can be computed in linear time. More specifically, we will show that K can be decomposed in linear time into $O(n)$ disjoint subparts, each being either a circular sector or a quadrangle, so that the area of each such subpart can be readily calculated, and will be assumed to require constant time to calculate. This gives us the stated linear time bound.

Suppose that $Vor(S)$ has already been computed. We partition each cell $V(i)$ of $Vor(S)$ into subcells as in Section 3, by connecting each vertex in $V(i)$ to x_i by a straight "spoke." Evidently there are overall $O(n)$ such subcells. We will make use of the following simple

LEMMA 6.1:   For each $B_i \in S$, we have $V(i) \cap K \subset B_i$.

Proof: Let $y \in V(i) \cap K$, and suppose that $y \in B_j$ for some B_j in S. Since $y \in V(i)$ we have

$$d(x_i, y) - r_i \leq d(x_j, y) - r_j \leq 0$$

so that $y \in B_i$ too. Q.E.D.

Let U be a subcell of some cell $V(i)$. The boundary of U consists of two spokes connecting x_i to two Voronoi vertices v_1, v_2, and of a unique Voronoi edge e separating $V(i)$ from an adjacent cell $V(j)$ and connecting v_1 with v_2 (if e is an unbounded edge, v_1 or v_2 may lie at infinity). Suppose without loss of generality that x_i lies at the origin. By the preceding lemma, $U \cap K$ in polar coordinates is

$$U \cap K = \{(r, \theta) : \theta_1 \leq \theta \leq \theta_2, \, 0 \leq r \leq min \, (r_i, e(\theta))\}$$

where θ_k is the orientation of the spoke connecting x_i to v_k, $k = 1, 2$ and where $e(\theta)$ is the length of the segment at orientation θ connecting x_i to e.

It is now plain that $U \cap K$ can be decomposed into at most three subparts, each of which is either a circular sector (whose area is readily calculable) or a sector of the form

$$A = \{(r, \theta) : \theta' \leq \theta \leq \theta'', \, 0 \leq r \leq e(\theta)\}.$$

Although the area of such a sector can be easily calculated by straightforward integration, the formula giving this area is somewhat complicated and involves trigonometric and logarithmic terms. However we can avoid explicit calculation of the area of such sectors as follows. Let A be such a sector, and let y, y' be the two endpoints of the portion of the corresponding Voronoi edge e within A. Note that e separates the cell $V(i)$ containing A and an adjacent cell $V(j)$, so that when $V(j) \cap K$ is decomposed into its subparts, one of them will consist of another sector A' bounded by e and by the two segments x_iy,x_iy'. Hence, instead of computing the areas of A,A' separately, we can compute directly the area of $A \cup A'$, which is simply the quadrangle x_iyx_jy'.

We have thus shown that K can be decomposed into $O(n)$ subparts, each being either a circular sector or a quadrangle. It is also plain that this decomposition can be accomplished in linear time by a simple traversal of the Voronoi diagram, once the diagram has been calculated. Hence we have

THEOREM 6.1:   The area of the union of n arbitrary discs can be calculated in time $O(n \log^2 n)$, as the sum of $O(n)$ circular sectors and quadrangles.

Remarks:

(1) It is also conceivable that this area could be calculated efficiently by some sweeping technique. This is suggested by the fact that the number of points of intersection of pairs of circles in S, which lie on the boundary of K, is $O(n)$ (indeed, each such point lies on one of the $O(n)$ edges of $Vor(S)$, and each of these edges can contain at most two such points). Thus if we could perform a sweep in which only those intersection points are traced, while the other $O(n^2)$ ''inner'' intersection points are ignored, we could obtain the area of K as the sum of the areas of $O(n)$ vertical strips, each bounded by an upper and a lower circular arc. However, we do not know how to achieve this without calculation of the associated Voronoi diagram.

(2) A recent result of Spirakis (1983) gives a Monte-Carlo algorithm for calculating the area of K in expected linear time.

7.   CONCLUSION

In this paper we have introduced the notion of generalized Voronoi diagram for a set of n possibly intersecting circles in the plane, described an $O(n \log^2 n)$ algorithm for constructing this diagram, and presented several applications of this diagram, for detecting intersections and calculating proximities on one hand, and for calculating the area of the union of these circles on the other.

Other properties of these generalized Voronoi diagrams deserve study. Some additional problems concerning applications of this diagram have been noted in Section 5. Another interesting problem is to find efficient techniques for dynamic maintainance of the generalized Voronoi diagram of a set S of circular objects, when one or more objects move continuously along specified trajectories, e.g., along straight lines. An efficient solution of this latter problem would facilitate efficient procedures for checking a prescribed motion of a robot system for collisions; the current alternative is to apply an appropriate static collision-detecting procedure, as described in Section 5, to a sequence of configurations of the system, sufficiently close to each other, along the specified trajectory.

It would also be useful to generalize the techniques described in this paper is to the case of a set S of objects other than circles, e.g., general convex bodies, or special forms of convex bodies, such as "cigar-shaped" displacements of straight segments, etc. A major problem in attempting such generalizations is that if (as we would like to do) we allow objects in S to intersect each other, then the corresponding Voronoi diagram may contain up to $O(n^2)$ cells (consider, e.g., the case of n straight segments intersecting each other at $O(n^2)$ points). Note here that it is a remarkable property of circular bodies that the size of their Voronoi diagram is always linear, even though they can cut each other into $O(n^2)$ regions. A final open problem is whether the generalized Voronoi diagram for n circular discs can be constructed in time $O(n \log n)$ (even in the simpler case in which the discs are disjoint from each other).

CHAPTER 9

Efficient Detection of Intersections among Spheres

JOHN E. HOPCROFT

Department of Computer Science
Cornell University

JACOB T. SCHWARTZ

Courant Institute of Mathematical Sciences
New York University

MICHA SHARIR

Department of Mathematical Sciences
Tel Aviv University

We present an algorithm for detecting intersection between n given spheres in 3-space which requires $O(n \log n)$ space and $O(n \log^2 n)$ time.

Let a collection S of n spheres $s_1, s_2, \ldots, s_n$ be given. In this note we consider the problem of determining whether any two of these spheres intersect each other. This problem is motivated by the following practical problem in robotics: Let B be a robot system in a 3-dimensional space bounded by various walls and other obstacles. Given a position of B, we wish to determine whether some subpart of B intersects a wall, or whether two subparts of B which normally do not meet one another, intersect. As an initial approximation, we enclose each obstacle and each separate subpart of B in a sphere, and ask whether there exists an intersection between any two of these spheres. (Note, however, that a more refined approach would enclose each rigid subpart of B by a collection of spheres, color all spheres in that collection by a unique color corresponding to the subpart enclosed, and ask whether there exists a pair of differently-colored spheres which intersect each other. Since intersection between two equally-colored spheres does not correspond to a collision, this leads to a more difficult problem, not discussed in this paper. However, see Sharir (1983) for a treatment of the two-dimensional version of this more general problem by a method using generalized Voronoi diagrams.

In the problem studied here, we wish to determine whether *any* pair of spheres in the given collection S intersect. Note that, if instead of spheres we consider rec-

Work on this paper by the first author has been partly supported by the National Science Foundation Grant MCF 81-01220; work on this paper by the second and third authors has been partly supported by ONR Grant N00014-82-K-0381 and by a grant from the U.S.-Israeli Binational Science Foundation.

tangular parallelipipeds (which we will call *boxes* for short) whose faces are parallel to the standard coordinate planes, then the problem of detecting intersection between some pair of these boxes can be solved in time $O(n \log^2 n)$ by using known 2-D dynamic range query techniques (cf. Edelsbrunner, 1982).

To handle the more difficult case of spheres, we need the following geometric lemma.

Lemma 1: Let s_1, s_2 be two intersecting spheres having centers c_1, c_2 and radii r_1, r_2 respectively, and suppose that $r_1 \geq r_2$. Let K be a cube inscribed in s_1 and having faces parallel to the coordinate planes. Then if the angle θ between $c_1 c_2$ and the line l parallel to the z-axis and passing through c_2 is $\leq \alpha$, where $\sin \alpha = \frac{1}{2\sqrt{3}}$, then l intersects K.

Proof: Define θ_{min} to be the smallest possible angle that can be formed between $u_1 u_2$ and l such that l lies either exterior to K or on the boundary of K. Then

$$\sin \theta_{min} = \frac{d(u_1, l)}{d(u_1, u_2)} \geq \frac{r_1/\sqrt{3}}{r_1 + r_2} \geq \frac{r_1/\sqrt{3}}{2r_1} = \frac{1}{2\sqrt{3}} = \sin \alpha$$

Hence if $\theta \leq \alpha$ we also must have $\theta \leq \theta_{min}$, from which it easily follows that l intersects K. Q.E.D.

The algorithm we propose is roughly as follows. Let Ψ be a finite set of orientations having the property that for each orientation θ there exists another orientation $\psi \in \Psi$ such that the angle between θ and ψ is $\leq \alpha$. Let M be the size of Ψ. For each $\theta_0 \in \Psi$, we apply the following space-sweeping technique. Assume without loss of generality that θ_0 points in the direction of the positive z-axis. Sweep a plane P parallel to the x,y plane from $z = -\infty$ to $z = +\infty$. As P is swept upwards we maintain a data structure T (to be described below) containing some of those spheres in S whose centers lie below P. Each time P passes through the center u of a sphere $s \in S$, we perform the following steps:

(a) Let r be the radius of s. Perform a range query which enumerates all spheres σ in T whose radius ϱ is not larger than r and whose center projects into a point of P which belongs to the square of side $2r/\sqrt{3}$ centered at u. Check each such sphere σ for intersection with s. If an intersection is found the algorithm reports this intersection and halts. Otherwise delete the sphere σ from T.
(b) Add the sphere s to T.

It is clear that the sweeping procedure just described performs n range queries and n insertions into T. Moreover, each sphere $\sigma \in S$ can appear in at most one range query, since it is deleted from T immediately after any such query. Also the number of sweeps performed by the algorithm is a constant independent of n. We will show that T can be maintained in $O(n \log n)$ space, in such a way that individual insertions and deletions from T require time $O(\log^2 n)$, while each range query needed can be carried out in time $O(\log^2 n + t)$, t being the number of spheres satisfying the range condition of the query. The preceding remarks show that use of such a structure T will lead to an algorithm requiring $O(n \log n)$ space and $O(n \log^2 n)$ time.

The data structure T that we use represents a collection $s_1, \ldots, s_k$ of spheres. For each such sphere s_i let x_i, y_i be the x and y coordinates of its center and r_i its radius. T is simply a balanced binary range tree with a certain amount of auxiliary data stored at its nodes. The leaves of T correspond to the spheres $s_1, \ldots, s_k$, which are sorted by the x coordinates of their centers. Each internal node ξ of T represents an interval $[\xi_{low}, \xi_{high}]$ of x-values, and to each such ξ we attach a *priority search tree* Q_ξ (as defined by McCreight, 1982) which represents the set of all spheres whose centers have x coordinates lying in this range; each such sphere is represented in Q_ξ by the pair $[y, r]$ of the y coordinate of its center and its radius. As shown by McCreight, each such priority search tree Q has size $O(m)$ and supports query operations of the form $E(y_{min}, y_{max}, r_{max})$, which enumerate all pairs $[y, r]$ in Q for which $y_{min} \leq y \leq y_{max}$ and $r \leq r_{max}$. Each such query requires time $O(\log m + t)$, where m is the number of pairs in Q, and where t is the number of pairs satisfying the query conditions to handle each query. Insertions and deletions of elements in such a tree require time $O(\log m)$ each. Note that the space occupied by the range tree T to whose nodes these priority search trees are attached is $O(n \log n)$.

It is easy to apply the operations we require to T. To insert a sphere s (represented by a triple $[x, y, r]$ giving the x and y coordinates of its center and its radius) into T we insert it into each of $O(\log n)$ priority search trees, Q_ξ, for ξ lying on the path from the root of T to an appropriate leaf of T, inserting the pair $[y, r]$ into each of these Q_ξ. Each of these updating operations can be accomplished in $O(\log n)$ time, so that the overall cost of inserting a sphere into T is $O(\log^2 n)$. Deletions are performed in a symmetric and similar fashion. Each of the queries that we apply has the form $EN(x_1, x_2, y_1, y_2, r_0)$, and calls for the enumeration of all spheres (i.e., triples $[x, y, r]$) in T satisfying $x_1 \leq x \leq x_2$, $y_1 \leq y \leq y_2$, and $r \leq r_0$. To apply such a query, we find $O(\log n)$ nodes in T representing disjoint intervals which collectively span the interval between x_1 and x_2. For each such node ξ, we can apply the query $E(y_1, y_2, r_0)$ to Q_ξ in time $O(\log n + t_\xi)$, thus enumerating all the t_ξ spheres satisfying the conditions of the original query whose centers have x-coordinates lying in $[\xi_{low}, \xi_{high}]$. Overall the query $EN(x_1, x_2, y_1, y_2, r_0)$ is handled in time $O(\log^2 n + t)$, where t is the total number of spheres satisfying the query conditions.

This description of the data structure T completes the definition of our algorithm and the analysis of its complexity. We now turn to prove the correctness of the algorithm. It is plain that the algorithm never reports a spurious intersection. Hence we only need to show that if any two spheres in S intersect each other, then at least one such intersection will be detected in at least one sweep performed by the algorithm. In fact we will show that the intersection between the two spheres s_1, s_2 for which $\min(r_1, r_2)$ is maximal must be detected in at least one sweep of the algorithm. Suppose that these two spheres s_1, $s_2 \in S$ intersect each other, and let their centers be c_1, c_2 and their radii be r_1, r_2 respectively, so that $r_1 \leq r_2$, and that no two spheres both having radii greater than r_1 intersect one another (in what

follows we assume for simplicity that the radii of the spheres in S are all distinct).
Let $\theta_0 \in \Psi$ be a direction which forms an angle $\theta \le \alpha$ with the directed segment
$c_1 c_2$, and consider the sweep in the θ_0-direction which the algorithm performs.
Rotating axes if necessary, assume that θ_0 points along the positive z-axis, and that
c_1 is lower than c_2. By Lemma 1, the line l parallel to the z-axis and passing through
c_1 intersects the cube K inscribed in s_2 with faces parallel to the coordinate planes,
i.e., the projection c_1' of c_1 onto the x,y plane falls inside the projection of K onto
that plane. Hence it suffices to show that by the time the swept plane P passes
through c_2, the sphere s_1 is still in T, because then the set produced in response to
the range query performed in step (a) of the algorithm at c_2 will include s_1, so that
the intersection between s_1 and s_2 will be detected by the algorithm. Suppose to the
contrary that s_1 has been deleted from T prior to the processing of c_2. Then there
must exist a third sphere $s \in S$ during whose processing s_1 was deleted from T. It
follows that the radius r of s is $> r_1$, that the center c of s lies above c_1 but below c_2,
and that c_1' also lies in the projection onto the x,y plane of the cube K^1 associated
with the sphere s. Moreover, we can also assume that s and s_1 do not intersect each
other, for otherwise this intersection would have been detected by the algorithm at
the time c was processed. Also by assumption the spheres s and s_2 do not intersect
each other because both their radii are greater than r_1. However, such a configuration of spheres is impossible. Indeed, let u be a point on l having the same z
coordinate as c_2. Then we have

$$c_2 u = c_1 c_2 \sin\theta < (r_1 + r_2) \sin \alpha < 2r_2 \frac{1}{2\sqrt{3}} = \frac{r_2}{\sqrt{3}}$$

Let J, J_1, J_2 be the intervals in which l intersects the spheres s, s_1, s_2 respectively.
Since s intersects neither s_1 nor s_2, and since the center c of s lies above c_1 but below
c_2, it follows that J is disjoint from both J_1 and J_2 and lies between these two
intervals. It follows that

$$c_1 u \ge |J| + \frac{1}{2}|J_1| + \frac{1}{2}|J_2|$$

But $|J_1| = 2r_1$, $|J_2| = 2\sqrt{r_2^2 - c_2 u^2}$, and $|J| \ge \frac{2r}{\sqrt{3}}$ (because l intersects K').
Since $r > r_1$ we have

$$c_1 u > \frac{2r_1}{\sqrt{3}} + r_1 + \sqrt{r_2^2 - c_2 u^2}$$

Squaring the last inequality, we obtain

$$c_1 u^2 > r_1^2 (1 + \frac{2}{\sqrt{3}})^2 + r_2^2 - c_2 u^2 + 2r_1 (1 + \frac{2}{\sqrt{3}})\sqrt{r_2^2 - c_2 u^2}$$

But $c_1 u^2 + c_2 u^2 = c_1 c_2^2 < (r_1 + r_2)^2$ because s_1 and s_2 intersect each other.
Hence

$$(r_1 + r_2)^2 = r_1^2 + 2r_1r_2 + r_2^2 > r_1^2(1 + \frac{2}{\sqrt{3}})^2 + r_2^2 + 2r_1(1 + \frac{2}{\sqrt{3}})\sqrt{r_2^2 - c_2u^2}.$$

By the first inequality above we have

$$\sqrt{r_2^2 - c_2u^2} > \sqrt{r_2^2 - (r_2/\sqrt{3})^2} = \frac{\sqrt{2}}{\sqrt{3}}r_2$$

Substituting this in the preceding inequality, and dividing by r_1, we obtain

$$r_1 + 2r_2 > \left(\frac{7}{3} + \frac{4}{\sqrt{3}}\right)r_1 + \left(2 + \frac{4}{\sqrt{3}}\right)\frac{\sqrt{2}}{\sqrt{3}}r_2,$$

i.e.,

$$\left(\frac{4}{3} + \frac{4}{\sqrt{3}}\right)r_1 + \left(\frac{4\sqrt{2}}{3} + \frac{2\sqrt{2}}{\sqrt{3}} - 2\right)r_2 < 0,$$

which is impossible, since the coefficients of r_1 and of r_2 are both positive. This contradiction implies that the assumed configuration of spheres is impossible, which in turn implies that the intersection between s_1 and s_2 will be detected, proving the correctness of our algorithm, thus justifying the following summary theorem.

THEOREM: The existence of an intersection among n arbitrary spheres in 3-space can be tested by an algorithm requiring $O(n \log n)$ space and $O(n \log^2 n)$ time.

Remark: The algorithm described above can easily be generalized to any number $d \geq 3$ of dimensions. The appropriate generalization of the data structure T is a $(d - 2)$-dimensional range tree with a priority search tree attached to each of its (innermost) nodes. The set Ψ of directions in which sweeps are to be performed must be dense enough so that for each orientation θ in the $(d - 1)$-dimensional unit sphere S^{d-1} there exists an orientation $\psi \in \Psi$ such that the planar angle between θ and ψ is $\leq \alpha_d$, where $\sin \alpha_d = \frac{1}{2\sqrt{d}}$. The correctness proof uses straightforward generalizations of the calculations given above, in which $\sqrt{3}$ is replaced by $\sqrt{d}$. This gives

COROLLARY: The existence of an intersection among n arbitrary d-spheres in Euclidean d-space can be tested by an algorithm requiring $O(n \log^{d-2} n)$ space and $O(n \log^{d-1} n)$ time.

CHAPTER 10

Precise Implementation of CAD Primitives Using Rational Parametrizations of Standard Surfaces

S. OCKEN

Mathematics Department
City College of New York

JACOB T. SCHWARTZ

Courant Institute of Mathematical Sciences
New York University

M. SHARIR

Department of Mathematical Sciences
Tel Aviv University

We discuss the problem of computing the intersection curve of two algebraic surfaces, each of which possesses rational parametrization. The special case where the two surfaces are quadric is analyzed in detail, using a general decomposition theorem which guarantees the existence of a simultaneous canonical reduction of two quadratic forms in Euclidean n-space. In homogeneous 4-space, this yields a classification and simple parametrization of all possible intersections between two quadric surfaces. Using these results, we treat the problem of analyzing the structure of solid bodies defined by Boolean combinations of half-spaces bounded by arbitrary quadric surfaces. The analysis given leads to fast versions of some of the procedures required for geometric modeling systems which admit general quadric surfaces into their vocabulary of basic shapes.

1. INTRODUCTION

In his survey (Wesley, 1980) of geometric modeling systems, M. Wesley makes the following cogent remarks: 'There are some fundamental problems to be solved in the field of geometric representation. Firstly, curved surfaces are required; in the short term, cylinders are probably sufficient for many applications, to be followed by cones and then more general curved surfaces. Secondly, the geometric representation has to be absolutely robust under all conditions. The system can be expected to be used to generate arbitrary objects without human intervention. It is essential that the objects always be numerically and topologically valid; e.g. it must handle

---

Work on this paper has been supported by ONR Grants N00014-75-C-0571 and N00014-82-K-0381, and by a grant from the U.S.-Israeli Binational Science Foundation.

situations where surface details become arbitrarily small without generating invalid topologies. Thirdly, the system must allow computation of necessary results with optimum efficiency in terms of computation time and storage efficiency. As the applications leave the stylized blocks world of the laboratory and approach the full complexity of real engineering situations, the limits of computational capacity can be expected to be reached, and choices between modeling systems may be based on cost rather than the ability to perform the operation at all.'

If for the moment we ignore the computational cost factor on which Wesley lays a very justified stress, it becomes possible for an ideal geometric modeling system to deal with arbitrary bodies and surfaces. There do exist general algorithmic techniques for deciding all the questions about such bodies which a geometric modeling system needs to ask. These techniques derive from Tarski's general method for deciding quantified formulae in the elementary theory of reals, which has been improved substantially by various other authors in the years since Tarski's original paper: see Tarski (1951), Collins (1975), Arnon (1981), Lazard (1977) and Schwartz and Sharir (1983b) for a review of some of this material.

However, these general methods, which are based upon systematic but unoptimized use of elimination theory and of Sturm's location theorem for the roots of real polynomials, are too expensive computationally for use in a practical geometric modeling system. For this reason, one must focus on a class of surfaces whose Boolean combinations can be handled with considerably greater efficiency than can algebraic surfaces in general. The following definition specifies a class of surfaces which is advantageous (but not necessarily sufficiently advantageous) in this regard.

DEFINITION 1:   A surface S is said to be *rational* if
(a) S is defined by a polynomial equation $P(x,y,z) = 0$ having rational (or at least real algebraic) coefficients;
(b) There exists a parametrization $x = X(u,v)$, $y = Y(u,v)$, $z = Z(u,v)$ of the surface S by functions X,Y,Z which are quotients of polynomials in u,v having rational (or at least real algebraic) coefficients. (We suppose this parametrization to be defined for u,v in some domain D which is itself defined in a simple way by polynomial inequalities in u,v, and suppose that, except on a few equally simple curves and points, this parametrization is 1-1.)

The advantage of working with such surfaces is that their intersections can be determined relatively efficiently, since if $P_1 = 0$ is the polynomial equation for one such surface S_1, and X_2,Y_2,Z_2 give the parametric representation for another surface S_2, then the intersection $S_1 * S_2$ is represented by the set of pairs satisfying the simple equation $P_1(X_2(u,v),Y_2(u,v),Z_2(u,v)) = 0$. Although a related algebraic condition can be written even if the surfaces S_1 and S_2 are not rational in the sense defined above, resultants would have to be used to derive this condition, substantially degrading the efficiency of the calculation required to deal with $S_1 * S_2$.

The class of real rational surfaces is extensive. Examples are as follows. Any surface of the form $z = P(x,y)$, where P is a polynomial or a quotient of polynomials, is evidently rational. This includes the paraboloid and the hyperboloid of one sheet. The sphere $x^2+y^2+z^2 = 1$ is rational, since it has the (stereographic) parametrization

$$(x,y,z) = \left(\frac{1 - u^2 - v^2}{1 + u^2 + v^2}, \frac{2u}{1 + u^2 + v^2}, \frac{2v}{1 + u^2 + v^2}\right)$$

Similarly, the cone $z^2 - x^2 - y^2 = 0$ is rational, since it has the parametrization

$$(x,y,z) = v \cdot \left(\frac{1 - u^2}{1 + u^2}, \frac{2u}{1 + u^2}, 1\right)$$

The cylinder $x^2 + y^2 = 1$ is rational and has the parametrization

$$(x,y,z) = \left(\frac{1 - u^2}{1 + u^2}, \frac{2u}{1 + u^2}, v\right)$$

The double-sheeted hyperboloid $z^2 - x^2 - y^2 = 1$ has the parametrization

$$(x,y,z) = \left(\frac{2u}{1 - u^2 - v^2}, \frac{2v}{1 - u^2 - v^2}, \frac{1 + u^2 + v^2}{1 - u^2 - v^2}\right)$$

where u and v vary in the open unit circle. The single-sheeted hyperboloid $x^2 + y^2 - z^2 = 1$ is a ruled surface and has the parametrization

$$(x,y,z) = \left(\frac{1 - u^2 + 2ut}{1 + u^2}, \frac{2u - t(1 - u^2)}{1 + u^2}, t\right).$$

The torus formed by sweeping a circle of radius $r < R$ and center $(R,0,0)$ around the circle $x^2 + y^2 = R^2$ in the $x-y$ plane has the equation

$$\left| (x,y,z) - R\frac{(x,y,0)}{(x^2 + y^2)^{1/2}} \right|^2 = r^2$$

or

$$(x^2 + y^2 + z^2 + R^2 - r^2)^2 = 4R^2(x^2 + y^2),$$

and the parametrization

$$(x,y,z) = \left(r\frac{1 - v^2}{1 + v^2} + R\right) \cdot \left(\frac{1 - u^2}{1 + u^2}, \frac{2u}{1 + u^2}, 0\right) + \left(0, 0, \frac{2vr}{1 + v^2}\right).$$

Many other surfaces are rational also. For example, any surface of revolution of a parametrizable curve $(x,y,z) = (f(t),0,g(t))$ about the z-axis is rational, and has the parametrization

$$(x,y,z) = \left(f(t)\frac{1 - u^2}{1 + u^2}, f(t)\frac{2u}{1 + u^2}, g(t)\right),$$

and a defining equation which can easily be obtained by elimination from the parametric representation of the initial curve. Note also that any affine or projective image of a rational surface is also rational.

2.   INTERSECTION OF QUADRIC SURFACES; ALGEBRAIC ANALYSIS OF THE n-DIMENSIONAL CASE

It is easy to see from the above remarks that all real quadric surfaces are rational. Indeed, if we proceed projectively, writing the polynomial equations for all our surfaces as homogeneous polynomial equations in four variables, then the equation for any quadric can be written as $La \cdot a = 0$, where L is a 4×4 symmetric matrix, and where a is a 4-vector (x,y,z,w). We can diagonalize L, which corresponds to carrying out an appropriate projective transformation of our surface, and normalize the transformed L so that each of its diagonal entries is either ± 1 or 0. Then this normalized L can be brought to a form in which it has at least as many positive as negative entries, and in which the diagonal entries of L are arranged so that the zero entries precede the positive ones which in turn precede the negative entries. Moreover, we can assume that L contains one negative and at least two positive diagonal entries, for otherwise it would represent a point, a line, a plane, or a pair of planes, all of whose intersections with another quadric surface may be easily computed.

Proceeding in this way, the transformed L will have one of the following forms: Either

   (i) L has three positive and one negative entries, in which case it represents the sphere $x^2 + y^2 + z^2 - w^2 = 0$; or
  (ii) L has two positive and two negative entries, in which case it represents the single-sheeted hyperboloid $x^2 + y^2 - z^2 - w^2 = 0$; or
 (iii) L has two positive, one negative and one zero entry, in which case it represents the cylinder $y^2 + z^2 - w^2 = 0$.

Since we have noted in the preceding that all these canonical surfaces are rational, it follows that every quadric surface is rational. These surfaces have relatively simple structure, which makes it easy to deal with their intersection curves. This will be done in the present section, ultimately to yield a relatively simple technique for the analysis of general Boolean combinations of bodies bounded by quadric surfaces.

Let Λ and Σ be two quadric surfaces, represented respectively by the 4×4 matrices L, S in the manner described above. Applying the same projective transformation to both surfaces, we may assume that L has one of the three canonical forms listed above, and that S is still symmetric. Thus L represents either a sphere, a single-sheeted hyperboloid, or a cylinder, and we are left with the problem of finding the intersection curve of such a surface with a second, arbitrary quadric surface. This problem was studied by J. Levin (1976), who gave a fairly systematic account of these intersections. However, since the situation is simpler when viewed projectively than when only affine simplifications are used, we will analyze these intersections from a projective viewpoint. To obtain the simple classification of the intersection curves which we seek, we transform the homogeneous 4-space projectively so as to bring both quadratic forms (Lx,x) and (Sx,x) to simple form, from which the intersection curves can easily be obtained.

First we consider the case in which L is nonsingular (corresponding to the preceding cases (i) and (ii)), but abandon the restriction that the space X in which L and S act is 4-dimensional, since it is not hard to treat the general n-dimensional case.

Our task is to reduce the matrix S to its simplest equivalent form while maintaining some equally simple form of L, i.e., to replace S by a simpler matrix $T'ST$ where T is some real transformation for which $T'LT = L$, or for which $T'LT$ retains some equally convenient form. We will always keep L in a form satisfying $L^2 = I$, and then the eigenvalues of LS are invariant under such transformations T, since

$$LT'ST = (LT)^{-1}ST = T^{-1}LST.$$

In what follows, we will show that these are essentially the only invariants of the pair L, S.

Write $[x,y]$ for the nonsingular indefinite inner product $Lx \cdot y$. Then

$$[LSx,y] = Sx \cdot y = x \cdot Sy = Lx \cdot LSy = [x,LSy],$$

so that LS is self-adjoint relative to the bilinear form $[x,y]$.

The (generalized) eigenspace X_λ belonging to an eigenvalue λ of LS is the set of all vectors x such that $(LS-\lambda I)^n x = 0$ for some integer n. The space X decomposes into the direct sum of such eigenspaces, each of which is invariant under LS. It is plain that for each eigenvalue $\mu \neq \lambda$, the restriction of $LS - \lambda I$ to X_μ is nonsingular. Hence for each $y \in X_\mu$ there exists a $z \in X_\mu$ such that $(LS-\lambda I)^n z = y$. Thus for each $x \in X_\lambda$, $[x,y] = [x,(LS-\lambda I)^n z] = [(LS-\lambda I)^n x,z] = 0$. This shows that eigenspaces X_λ, X_μ belonging to different eigenvalues λ,μ are orthogonal to each other in the bilinear form $[x,y]$.

To reduce L and S simultaneously to the desired canonical form, it is sufficient to analyze the restriction of LS to each of the spaces X_λ, λ real, and to each pair of spaces $X_\lambda, X_{\bar\lambda}$, λ complex. First consider a generalized eigenspace X_λ with λ real. In this space, the operator $R = LS - \lambda I$ is still symmetric relative to the nonsingular bilinear form $[x,y]$ and satisfies $R^n = 0$ for some (smallest) n, so that $R^{n-1} \neq 0$. Thus there exist x and y in X_λ such that $[y,R^{n-1}x] \neq 0$. This implies that there exists a $z \in X_\lambda$ such that $[z,R^{n-1}z] \neq 0$, since otherwise x, y would satisfy

$$0 = [(x+y),R^{n-1}(x+y)] = [y,R^{n-1}x] + [x,R^{n-1}y]$$
$$= 2[y,R^{n-1}x] \neq 0.$$

Suppose without loss of generality that $[x,R^{n-1}x] \neq 0$. Then the space Y_1 spanned by $\{x,Rx, \ldots ,R^{n-1}x\}$ is invariant under R (i.e., under LS), and the restriction of $[x,y]$ to this space is nonsingular. Indeed, if

$$u = \alpha_j R^j x + \alpha_{j+1} R^{j+1}x + \cdots + \alpha_{n-1}R^{n-1}x,$$

where $\alpha_j \neq 0$ is such that $[u,R^m x] = 0$ for all $m \geq 0$, then

$$0 = [u,R^{n-1-j}x] = \alpha_j[R^j x,R^{n-1-j}x] = \alpha_j[x,R^{n-1}x] \neq 0,$$

a contradiction.

Since the space Y_1 is invariant under R and since $[x,y]$ is nonsingular on Y_1, its orthogonal complement Y_1' (relative to the inner product $[x,y]$) has these same properties. Thus the argument given in the preceding paragraph can be applied to Y_1', from which it plainly follows that X_λ decomposes into a direct sum of subspaces $Y_1 + Y_2 + \cdots + Y_k$, mutually orthogonal relative to the inner product $[x,y]$, in each of which the operator R acts as a shift (i.e., each of these spaces has a basis $\{x, Rx, \ldots, R^{n-1}x\}$) for which $[x, R^{n-1}x] \neq 0$.

Consider a single such space $Y = Y_j$, and let its dimension be n, so that $R^n y = 0$ for all $y \, \varepsilon \, Y$. By modifying x appropriately, we can assume without loss of generality that $[x, R^j x] = 0$ except for $j = n-1$. Indeed, suppose that this is false, and consider the largest $j < n-1$ for which $\alpha = [x, R^j x] \neq 0$. Then for each β we have

$$[x - \beta R^{n-j-1}x, \; R^j(x - \beta R^{n-j-1}x)]$$

$$= [x, R^j x] - \beta[x, R^{n-1}x] - \beta[x, R^{n-1}x]$$

$$= \alpha - 2\beta[x, R^{n-1}x].$$

Hence if we put $\beta = \alpha/(2[x, R^{n-1}x])$ and $x' = x - \beta R^{n-j-1}x$, we have $[x', R^j x'] = 0$. Moreover, for each $j < i \leqslant n-1$ we have

$$[x', R^i x'] = [x - \beta R^{n-j-1}x, R^i(x - \beta R^{n-j-1}x)] = [x, R^i x]$$

so that $[x', R^i x'] = 0$ for $j < i < n-1$ and $[x', R^{n-1}x'] \neq 0$.
It is plain that $\{x', Rx', \ldots, R^{n-1}x'\}$ spans the same space Y as $\{x, Rx, \ldots, R^{n-1}x\}$. Thus, if we replace x by x', j is reduced by at least 1, proving that there exists an x as above, but with $[x, R^j x] = 0$ for all $0 \leqslant j < n-1$.

Multiplying x by a suitable real constant, we can suppose without loss of generality that $[x, R^{n-1}x] = c = 1$ or $= -1$. Note then that $[R^j x, R^k x] = c$ if $j+k = n-1$ but 0 otherwise. Thus the bilinear form $[x,y]$ has one of at most two canonical forms on the space Y (on which R has a fixed canonical form). Specifically, in the basis $\{x, Rx, \ldots, R^{n-1}x\}$ the quadratic form $[y,y]$

(where we write $y = \sum\limits_{m=0}^{n-1} y_m R^m x$), is

$$F(y) = \pm(2y_0 y_{n-1} + \cdots + 2y_m y_{n-m-1}), \qquad m = (n-2)/2 \text{ if } n \text{ is even};$$

$$F(y) = \pm(2y_0 y_{n-1} + \cdots + 2y_{m-1}y_{n-m} + y_m y_m), \qquad m = (n-1)/2 \text{ if } n \text{ is odd.}$$

Moreover, in this same basis, the form $(Sy,y) = [LSy,y] = [(R+\lambda I)y,y]$ is

$$G(y) = \pm(2y_0 y_{n-2} + \cdots + 2y_m y_{n-m-2}) + \lambda F(y),$$
$$m = (n-3)/2 \text{ if } n \text{ is odd;}$$

$$G(y) = \pm(2y_0 y_{n-2} + \cdots + 2y_{m-1}y_{n-m-1} + y_m y_m) + \lambda F(y),$$
$$m = (n-2)/2 \text{ if } n \text{ is even.}$$

It is clear that, if n is even, the signature of F is $(n/2,n/2)$, while, if n is odd, the signature of F is either $((n-1)/2, (n+1)/2)$ or $((n+1)/2, (n-1)/2)$. The differing signatures distinguish the two cases in which n is odd, but, even if n is even, the cases $[x,R^{n-1}x] = 1$ and $[x,R^{n-1}x] = -1$ are easily seen to be nonisomorphic. It is convenient to introduce symbolic designations for these various cases; we will use the symbols $[\lambda,m,m+1]$ and $[\lambda,m+1,m]$ to designate the two cases in which n is odd, and $[\lambda,m,m,+]$, $[\lambda,m,m,-]$ to designate the two cases in which n is even.

Next consider a pair λ, λ of mutually conjugate complex eigenvalues of LS, and the generalized eigenspaces $X_\lambda, X_{\bar\lambda}$ associated with them. Since LS is a real matrix, it is clear that $\bar x \in X_{\bar\lambda}$ if $x \in X_\lambda$ (where $\bar x$ denotes the complex conjugate vector of x.) The analysis that we have given applies without change to the space X_λ even if λ is complex (except that for λ complex the vectors which appear in this analysis will not be real.) Hence X_λ can be written as a direct sum of mutually orthogonal subspaces $X_\lambda = Y_1 + \cdots + Y_k$, each of which has a basis $\{x,Rx, \ldots , R^{n-1}x\}$, where $R = LS - \lambda I$, $R^n x = 0$, and $[R^j x, R^m x] = 1$ if $j+m=n-1$ but 0 otherwise. Consider one of these spaces, e.g., Y_1, and as a real basis for the direct sum $Y_1 + \bar Y_1$ take the vectors $a_j = 1/\sqrt2 (x_j + \bar x_j)$ and $b_j = 1/i\sqrt2 (x_j - \bar x_j)$, where $x_j = R^j x$, $j=0, \ldots ,n-1$. Since the vectors $x_i, \bar x_j$ belong to different eigenspaces, they are orthogonal in the form $[u,v]$. It follows easily that

$$[a_j,b_m] = 0, \qquad \text{for all } j,m,$$

$$[a_j,a_m] = -[b_j,b_m] = 1, \qquad \text{if } j+m = n-1, \text{ but 0 otherwise.}$$

Moreover, since $LSx_j = x_{j+1} + \lambda x_j$, and since λ can be written in terms of its real and imaginary parts as $\lambda = \alpha_1 + i\alpha_2$, we have

$$LSa_j = 1/\sqrt2\, LS(x_j + \bar x_j) = a_{j+1} + 1/\sqrt2\, (\lambda x_j + \bar\lambda \bar x_j)$$

$$= a_{j+1} + \sqrt2\, Re(\lambda x_j) = a_{j+1} + \alpha_1 a_j - \alpha_2 b_j,$$

and similarly

$$LSb_j = b_{j+1} + \sqrt2\, Im(\lambda x_j) = b_{j+1} + \alpha_2 a_j + \alpha_1 b_j$$

If the dimension of Y_1 is n, it is plain that the signature of $[x,y]$ on $Z_1 = Y_1 + \bar Y_1$ is (n,n). In the basis $\{a_0, \ldots , a_{n-1}, b_0, \ldots , b_{n-1}\}$ for Z_1, the quadratic form (Lu,u) has the following representation (derived by writing $u = \sum_{j=0}^{n-1} (y_j a_j + z_j b_j)$):

$$F(y,z) = 2y_0 y_{n-1} + \cdots + 2y_m y_{n-m-1} - 2z_0 z_{n-1} - \cdots - 2z_m z_{n-m-1},$$
$$m = (n-2)/2, \text{ if } n \text{ is even;}$$

$$F(y,z) = 2y_0 y_{n-1} + \cdots + 2y_{m-1} y_{n-m} + y_m y_m - 2z_0 z_{n-1} - \cdots - 2z_{m-1} z_{n-m}$$
$$- z_m z_m,$$
$$m = (n-1)/2, \text{ if } n \text{ is odd.}$$

Moreover, in this same basis $(Su,u) = [LSu,u] = [(R + \lambda I)u,u]$ has the representation

$$G(y,z) = 2y_0y_{n-2} + \cdots + 2y_my_{n-m-2} - 2z_0z_{n-2} - \cdots - 2z_mz_{n-m-2}$$
$$+ \alpha_1 F(y,z) + 2\alpha_2(y_0z_{n-1} + \cdots + y_{n-1}z_0),$$
$$m = (n-3)/2, \text{ if } n \text{ is odd;}$$

$$G(y,z) = 2y_0y_{n-2} + \cdots + 2y_{m-1}y_{n-m-2} + y_my_m - 2z_0z_{n-2} - \cdots$$
$$- 2z_{m-1}z_{n-m-2} - z_mz_m$$
$$+ \alpha_1 F(y,z) + 2\alpha_2(y_0z_{n-1} + \cdots + y_{n-1}z_0),$$
$$m = (n-2)/2, \text{ if } n \text{ is even.}$$

These pairs of quadratic forms are defined uniquely by the complex eigenvalue λ and by the dimension n, and we can therefore designate them by $[\lambda,n,n]$, λ complex.

The following theorem, which should be compared to the corresponding result for the complex case, given in Hodge and Pedoe (1952, Vol. II, Chap. XIII, Section 10), summarizes the preceding analysis:

THEOREM 1: Let L and S be a pair of symmetric matrices acting in a real vector space X, and suppose that L is nonsingular. Then X can be decomposed as a direct sum of canonical subspaces Y which are simultaneously orthogonal to each other in the two bilinear forms (Lx,y) and (Sx,y). A descriptor characterizing the representation in a standard basis for Y of each of these two forms is associated with each such Y. The following table shows the dimension of the space Y and the signature of the form (Ls,y) on Y for every possible descriptor:

Descriptor	Dimension of Y	Signature of (Lx,y) on Y
$[\lambda,m,m+1]$, λ real	$2m + 1$	$(m, m+1)$
$[\lambda,m+1,m]$, λ real	$2m + 1$	$(m+1, m)$
$[\lambda,m,m,+]$, λ real	$2m$	(m,m)
$[\lambda,m,m,-]$, λ real	$2m$	(m,m)
$[\lambda,m,m]$, λ complex	$2m$	(m,m)

The representations which the forms (Lx,y) and (Sx,y) take on in standard bases for spaces having these descriptors are detailed in the preceding paragraphs.

To complete our analysis, it remains to handle the situation in which L is singular. For this, we have only to carry over the analysis described in Hodge and Pedoe (1952, pp. 281–286), which we review for the reader's convenience. As previously, let X be the finite dimensional space on which L and S act. We shall show that, if L is singular, X can be decomposed into at least two subspaces, mutually orthogonal with respect to both L and S, in one of which the quadratic forms $F = Lv \cdot v$ and $G = Sv \cdot v$ have one of the four following representations:

Case $[2n, 2n+1]$:

$$F(x) = 2x_2x_3 + \cdots + 2x_{2n}x_{2n+1}, \qquad n \geq 0$$

$$G(x) = 2x_1x_2 + \cdots + 2x_{2n-1}x_{2n} \pm x_{2n+1}{}^2$$

(simplest case: $0, x^2$)

Case $[2n-2, 2n]$:

$$F(x) = 2x_2x_3 + \cdots + 2x_{2n-2}x_{2n-1}, \qquad n \geq 1$$

$$G(x) = 2x_1x_2 + \cdots + 2x_{2n-1}x_{2n}$$

(simplest case: $0, 2xy$)

Case $[2n-1, 2n]$:

$$F(x) = 2x_2x_3 + \cdots + 2x_{2n}{}_2x_{2n-1} \pm x_{2n}{}^2, \quad n \geq 1$$

$$G(x) = 2x_1x_2 + \cdots + 2x_{2n-1}x_{2n}$$

(simplest case: $y^2, 2xy$)

Case $[2n, 2n]$:

$$F(x) = 2x_2x_3 + \cdots + 2x_{2n}x_{2n+1}, \quad n \geq 1$$

$$G(x) = 2x_1x_2 + \cdots + 2x_{2n-1}x_{2n}$$

(simplest case: $2xy, 2yz$)

We can use the symbols $[2n, 2n+1]$, $[2n-2, 2n]$, $[2n-1, 2n]$, $[2n, 2n]$ respectively as designators for these linear spaces. The following table gives the dimensions and ranks of L and S.

Designator	Dimension	Rank of L	Rank of S
$[2n, 2n+1]$	$2n+1, n \geq 0$	$2n$	$2n+1$
$[2n-2, 2n]$	$n, n \geq 1$	$2n-2$	$2n$
$[2n-1, 2n]$	$2n, n \geq 1$	$2n-1$	$2n$
$[2n, 2n]$	$2n+1, n \geq 1$	$2n$	$2n$

We will use the symbol $[0]$ to designate a 1-dimensional space in which both the quadratic forms L and S are 0.

Note that S is nonsingular in all these spaces with the exception of the space $[2n, 2n]$. Thus each of the spaces $[2n, 2n+1]$, $[2n-2, 2n]$, $[2n-1, 2n]$ can be regarded as a kind of 'reverse' of spaces (or sums of spaces) that we have already encountered, obtained by interchanging L and S. By considering the form of SL, it is easily seen that in this sense $[2n+1, 2n]$ is the reverse of either $[0, n, n+1]$ or $[0, n+1, n]$; that $[2n-2, 2n]$ is the sum of the two isomorphic spaces, each of which is the reverse of one of the spaces $[0, k, k+1]$, $[0, k+1, k]$, $[0, k, k, +]$, or $[0, k, k, -]$;

and that $[2n-1,2n]$ is the reverse of either $[0,n,n,+]$ or $[0,n,n,-]$. The exceptional space $[2n,2n]$ is its own reverse.

In view of those facts, it is convenient to introduce the notations $[\lambda,m,m+1]^*$, . . . for the reverse of $[\lambda,m,m+1]$, . . .

To handle the case in which L can be singular, we proceed by induction on the dimension of X. If L is nonsingular, there is nothing to prove, so that we can assume that the nullspace $N=N_L$ of L, i.e., the set $N_L = \{x \in X: Lx=0\}$, is nonempty. If $x \cdot Ly = 0$ for all y then $Lx = 0$ and conversely; hence the range $R = R_L$ of L is the orthocomplement of N_L, giving the direct sum decomposition $X = N_L + R_L$. If there exists a nonzero $x \in N_L$ such that $Sx = 0$, then we can write X as the sum of the 1-dimensional space (x) and its orthocomplement Y, and these spaces are orthogonal in both forms L and S. Hence in this case we achieve a complete canonical decomposition of X by decomposing Y inductively. Thus we need only consider the case in which S is nonsingular on N_L. If there exists any $x \varepsilon N_L$ such that $x \cdot Sx \neq 0$, we can let Y be the S-orthocomplement of the 1-dimensional space (x), and then again X is the direct sum $Y + (x)$ of two spaces orthogonal in both forms L and S, so once more we can proceed inductively. In all remaining cases we will have $x \cdot Sx = 0$ for all $x \in N$, which implies that $y \cdot Sx = 0$ for all $x,y \in N$, so that S is a 1-1 mapping of N into its orthocomplement R.

Take some $x_0 \in N$. Let Z be the space spanned by x_0 and Sx_0. The form $(u \cdot Sv)$ is nonsingular in Z, since if $z \cdot S(ax_0 + bSx_0) = 0$ for all $z \in Z$, then putting $z = x_0$ we have $b|Sx_0|^2 = 0$ since $x_0 \cdot Sx_0 = 0$, and therefore $b = 0$, but now putting $z = Sx_0$ gives $a = 0$. It follows that X can be written as a direct sum $X = Z + Z^*$, where Z^* is the S-orthocomplement of Z. Note that Z and Z^* are S-orthogonal, but not necessarily L-orthogonal.

Since $x_0 \cdot Sx_0 = 0$, the nonsingular form S is neither positive definite nor negative definite on the space Z, and hence it must have signature $(1,1)$. Thus in appropriate coordinates it has the representation x^2-y^2, and in these coordinates x_0 will appear as a 2-dimensional vector (a,b) satisfying $a^2-b^2 = 0$. Put $e_1 = x_0$, and let e_2 be the vector with coordinates $c(a,-b)$ in these same coordinates, where $c = (2a^2)^{-1}$; then $e_1 \cdot Se_1 = e_2 \cdot Se_2 = 0$ and $e_1 \cdot Se_2 = 1$. Let $e_3, \ldots e_n$ be a basis for Z^*. In the basis for Z consisting of the vectors e_1 and e_2, the form S has the matrix A_2 appearing just below. Moreover, the symmetric linear transformation representing the form L satisfies $Le_1 = 0$. In the following discussion we will make use of this basis, and write $u \cdot v$ for a positive-definite inner product in which the vectors e_j are orthonormal.

At this point we begin a subsidiary induction. Let n be the dimension of X, let A_2 designate the 2×2 matrix

$$\begin{bmatrix} 0 & 1 \\ 1 & 0 \end{bmatrix}$$

and for each even m let A_m designate the $m \times m$ block diagonal matrix built from $m/2$ copies of A_2. For m odd, let A_m designate the $m \times m$ matrix

$$\begin{bmatrix} 0 & \\ & A_{m-1} \end{bmatrix}$$

Let B,C designate either L,S or S,L. Suppose that for some k we have found a basis in which B,C have the respective forms

$$\begin{bmatrix} A_k & \\ & B' \end{bmatrix} \quad \text{and} \quad \begin{bmatrix} A_{k+1} & \\ & C' \end{bmatrix} \tag{1}$$

where B' is $(n-k)\times(n-k)$ and C' is $(n-k-1)\times(n-k-1)$. (Note that in the preceding paragraph, we have in fact defined such a basis for $k = 1$.) Then we will show that there also exists a basis in which C has the same form, but in which B has one of the three following forms:

$$\begin{bmatrix} A_k & & \\ & 0 & \\ & & B'' \end{bmatrix} \quad \text{or} \quad \begin{bmatrix} A_k & & \\ & \pm 1 & \\ & & B'' \end{bmatrix} \quad \text{or} \quad \begin{bmatrix} A_{k+2} & \\ & B''' \end{bmatrix} \tag{2}$$

where B'' is $(n-k-1)\times(n-k-1)$ and B''' is $(n-k-2)\times(n-k-2)$. In the first two of these three cases, there evidently exists a pair of spaces of respective dimensions $k+1$ and $n-k-1$ which are orthogonal to each other in both forms L,S, and it is also plain that, for the first of these spaces, the forms L,S have one of the four previously listed canonical representations. In the third case we simply interchange B,C, increase k by 1, and continue our subsidiary induction on k, which must eventually terminate.

For the inductive step we argue as follows. Let the basis in which B,C have the forms (1) be $e_1, \ldots ,e_k,e_{k+1}, \ldots e_n$. If $B'e_{k+1} = 0$, we have the first case (2). If not, modify the basis so that $e_j, \ldots e_n$, where $j \geq k+1$, is a basis for the nullspace of B'. This gives B the form

$$\begin{bmatrix} A_k & & \\ & B_0 & \\ & & 0_m \end{bmatrix} \tag{3}$$

where 0_m designates an $m\times m$ square matrix of zeroes, $m = n-j+1$, and B_0 is nonsingular, say of dimension (and thus rank) r. Use an auxiliary positive definite inner product in which the basis $e_1 \ldots e_n$ is orthonormal, and let E and F denote the orthogonal projections of X onto $e_{k+2}, \ldots ,e_{j-1}$ and onto $e_k, \ldots ,e_{j-1}$ respectively. Then EBE omits one row and one column of B_0, and hence its rank is either $r-1$ or $r-2$.

Suppose first that the rank of EBE is $r-1$, i.e., identical to the rank of EB, so that these two matrices have the same range. Then there exists a $u = Eu$ such that $EB(e_{k+1}-u) = 0$. Plainly $e_{k+1}\cdot u = 0 = (Ce_k)\cdot u$ since $Ce_k = e_{k+1}$. Define a linear transformation T on X by putting

$$Tx = x - (e_{k+1}\cdot x)u + (Cu\cdot x)e_k - \tfrac{1}{2}(e_{k+1}\cdot x)(Cu\cdot u)e_k.$$

Then if $Tx = 0$, we have $e_{k+1} \cdot Tx = e_{k+1} \cdot x$, so $Tx = x + (Cu \cdot x)e_k$. Hence $0 = Cu \cdot Tx = Cu \cdot x + (Cu \cdot x)(Cu \cdot e_k) = Cu \cdot x$, and therefore $x = 0$, showing that T is nonsingular. Plainly $Te_i = e_i$ for $i = 1 \ldots k$. Moreover $FBFe_k = 0$, so that

$$(Ty) \cdot FBF(Tx) = (y - (e_{k+1} \cdot y)u + ((Cu \cdot y) - \tfrac{1}{2}(e_{k+1} \cdot y)(Cu \cdot u))e_k)$$

$$\cdot FBF(x - (e_{k+1} \cdot x)u + ((Cu \cdot x) - \tfrac{1}{2}(e_{k+1} \cdot x)(Cu \cdot u))e_k)$$

$$= ((e_{k+1} \cdot y)(e_{k+1} - u) + Ey) \cdot FBF((e_{k+1} \cdot x)(e_{k+1} - u) + Ex)$$

$$= y \cdot EBEx + (e_{k+1} \cdot y)(e_{k+1} \cdot x)(e_{k+1} - u) \cdot B(e_{k+1} - u).$$

Thus in the basis $Te_1, \ldots Te_n$, the form B has the representation

$$\begin{bmatrix} A_k & & & \\ & c & & \\ & & B_0' & \\ & & & 0_m \end{bmatrix}$$

where B_0' has one less row and column than B_0. Plainly, $c \neq 0$, since B_0 is nonsingular; hence we can normalize and take $c = \pm 1$.

Since $Ce_k = e_{k+1}$ and $e_k \cdot Ce_k = u \cdot Ce_k = e_k \cdot Cu = 0$, we have also

$$Ty \cdot C(Tx) = (y - (e_{k+1} \cdot y)u + ((Cu \cdot y) - \tfrac{1}{2}(e_{k+1} \cdot y)(Cu \cdot u))e_k)$$

$$\cdot C(x - (e_{k+1} \cdot x)u + ((Cu \cdot x) - \tfrac{1}{2}(e_{k+1} \cdot x)(Cu \cdot u)) e_k)$$

$$= y \cdot Cx - (e_{k+1} \cdot y)u \cdot Cx + ((Cu \cdot y) - \tfrac{1}{2} (e_{k+1} \cdot y)(Cu \cdot u))e_k \cdot Cx$$

$$- (e_{k+1} \cdot x)(y \cdot Cu - (e_{k+1} \cdot y)u \cdot Cu) + ((Cu \cdot x) - \tfrac{1}{2}(e_{k+1} \cdot x)(Cu \cdot u))y \cdot Ce_k$$

$$= y \cdot Cx - (e_{k+1} \cdot y)u \cdot Cx + (e_{k+1} \cdot x)(u \cdot Cy) - (e_{k+1} \cdot x)(y \cdot Cu)$$

$$+ (e_{k+1} \cdot y)(u \cdot Cx) - \tfrac{1}{2}(e_{k+1} \cdot y)(e_{k+1} \cdot x)Cu \cdot u$$

$$+ (e_{k+1} \cdot x)(e_{k+1} \cdot y)u \cdot Cu - \tfrac{1}{2}(e_{k+1} \cdot y)e_{k+1} \cdot x)Cu \cdot u$$

$$= y \cdot Cx$$

In this new basis, B plainly has the second of the forms (2), completing our treatment of the case in which the rank of EBE is $r-1$.

Next suppose that the rank of EBE is $r-2$, so that there exists a nonzero vector $w \in EX$ such that $EBEw = 0$, i.e., $EBw = 0$. Change the basis to make $w = e_{k+2}$. Then in (3) the submatrix B_0 will have the form

$$\begin{bmatrix} * & c & . & . & . & * \\ c & 0 & . & . & . & 0 \\ . & . & & & & \\ . & . & & B_0' & \\ * & 0 & & & & \end{bmatrix} \qquad (4)$$

where the asterisks denote elements that may be nonzero, and B_0' is an $(r-2)\times(r-2)$ matrix which is plainly nonsingular. Define F and E as before. Since B_0' is nonsingular, there clearly exists a $u = Eu$ satisfying $EB(e_{k+1}-u) = ce_{k+2}$. (The vector u is determined by this condition only up to addition of a multiple of e_{k+2}.) Computing just as before, we find that

$$(Ty)\cdot FBF(TX) = y\cdot EBEx +$$
$$c((e_{k+1}\cdot y)\,(e_{k+2}\cdot x) + (e_{k+1}\cdot x)\,(e_{k+2}\cdot y))$$
$$+ a(e_{k+1}\cdot y)(e_{k+1}\cdot x),$$

where $a = (e_{k+1}-u)\cdot B(e_{k+1}-u)$, and we also have
$$Ty\cdot C(Tx) = y\cdot Cx.$$

We must have $c \neq 0$, since B_0 is nonsingular; changing the scale of e_{k+2}, we can therefore assume that $c = 1$. Write $u = u_0 + be_{k+2}$, where b is at our disposition, since, as already noted, u is determined only up to addition of a multiple of e_{k+2}. Then, since $Be_{k+2} = ce_{k+1} = e_{k+1}$ we have

$$a = (e_{k+1} - u_0 - be_{k+2})\cdot B(e_{k+1}-u_0 - be_{k+2})$$

$$= (e_{k+1} - u_0)\cdot B(e_{k+1}-u_0) + 2b(e_{k+1}-u_0)e_{k+1}$$

$$= (e_{k+1} - u_0)\cdot B(e_{k+1}-u_0) + 2b.$$

Choosing $b = -\tfrac{1}{2}(e_{k+1}-u_0)\cdot B(e_{k+1}-u_0)$ gives $a = 0$. Hence in the basis Te_1, . . . ,Te_n, B has the form

$$\begin{bmatrix} A_k & & \\ & \begin{smallmatrix}0 & 1\\ 1 & 0\end{smallmatrix} & \\ & & B'' \end{bmatrix}$$

where B'' is an $(n-k-2)\times(n-k-2)$ matrix. This is the third case of (3), and simply continues our induction, which must eventually terminate with one of the first two cases of (3).

This completes the proof of the following theorem, which extends Theorem 1 to the singular case:

THEOREM 2:   Let L and S be a pair of symmetric matrices acting in a real vector space X. Then X can be decomposed as a direct sum of canonical subspaces Y which are simultaneously orthogonal to each other in the two bilinear forms (Lx,y) and (Sx,y). A descriptor characterizing the representation in a standard basis for Y of each of these two forms is associated with each such Y. Each such descriptor, and its associated space and form, is either one of those appearing in Theorem 1, or one of the spaces described by $[0,m,m+1]^*$, $[0,m+1,m]^*$, $[0,m,m,+]^*$, or $[0,m,m,-]^*$, where D^* designates the result of taking the space and pair of canonical forms described by D and interchanging the two forms, or is the 1-dimensional space with both forms zero, whose descriptor is [0], or is an odd-dimensional space with descriptor $[n,n]$ and forms

$$x_1x_2+\cdots+x_{2n-1}x_{2n} \text{ and } x_2x_3+\cdots+x_{2n}x_{2n+1}.$$

3. APPLICATION TO COMPUTATION OF INTERSECTION CURVES

Now we can apply the preceding analysis of the general n-dimensional case to the 4-dimensional case required for the geometric application which interests us. We use the notations introduced at the beginning of the preceding section. For the moment we assume that L is nonsingular. Suppose first that LS has a nonsimple real eigenvalue λ. Make $\lambda=0$ by subtracting λL from S, and let n be the dimension of one of the spaces Y appearing in the canonical decomposition of the generalized eigenspace corresponding to λ, so that n equals either 2, 3, or 4. If $n=2$, then Y has coordinates in which (Lu,u) and (Su,u) have the respective forms $\pm 2zw$ and $\pm z^2$. Thus, if we pass to inhomogeneous coordinates for the whole of our original Euclidean 3-space by dividing by z, the equations $(Lu,u) = 0 = (Su,u)$ take on the form

$$Q_1(x,y) + 2w = 0 = Q_2(x,y) + 1$$

where Q_1 and Q_2 are quadratic forms in two variables. The second of these equations defines a quadratic plane curve to which we can easily give a rational parametrization, and then the first equation gives the w coordinate for the corresponding points on the intersection of Λ and Σ. Thus, in this case, the intersection curve of Λ and Σ has rational parametrization.

Next suppose that $n=3$. Then on Y the forms (Lu,u) and (Su,u) can be represented as $\pm(2yw + z^2)$ and $\pm 2yz$. Pass to inhomogeneous 3-space coordinates by dividing by z, so that $(Lu,u) = 0 = (Su,u)$ takes on the form

$$\pm x^2 + 2yw + 1 = 0 = ax^2 + 2y$$

In this case, the second of these equations can be used to eliminate y from the first, giving the equation of an elementary curve in the x,w plane, and then the second equation gives the y coordinate for the corresponding points on the intersection of Λ and Σ. Hence the intersection curve of Λ and Σ is simply $(x, -ax^2/2, (1\pm x^2)/(ax^2))$.

Finally, suppose that $n=4$. Then (Lu,u) and (Su,u) can be represented as $\pm(2xw + 2yz)$ and $\pm(2xz + y^2)$. Here we can pass to inhomogeneous coordinates by dividing by y, which gives the equation

$$xw + z = 0 = 2xz + 1$$

whose elementary parametrization is $(x,w,z) = (1/t, t^2/2, -t/2)$.

It is also possible that LS should have complex eigenvalues, either simple or nonsimple. If LS has a nonsimple complex eigenvalue, then by subtracting a real multiple of L from S, and multiplying S by an appropriate constant, we can suppose without loss of generality that this eigenvalue is i. In this case, the preceding analysis shows that the equations $(Lu,u) = 0 = (Su,u)$ reduce to

$$2(xy - zw) = 0 = x^2 - z^2 + 2xw + 2yz$$

We can then pass to inhomogeneous coordinates by dividing by z, which gives the equations

$$xy - w = 0 = x^2 - 1 + 2xw + 2y$$

Using the first equation to eliminate w from the second and solving for y yields $y = (1-x^2)/(2x^2+2)$. Thus, the intersection of $\varLambda$ and $\varSigma$ consists of a nonsingular rational curve.

Next suppose that LS has a simple complex eigenvalue. If its other eigenvalues are real, then either the preceding analysis or the analysis given below will apply, so the only case that need concern us specially is that in which LS has two pairs of conjugate complex eigenvalues. Again, we can suppose without loss of generality that one of these eigenvalues is i, and then the representation theorem derived above implies that there exists a pair of 2-dimensional spaces orthogonal in both forms (Lu,u) and (Su,u), on one of which these two forms have the representations $z^2 - w^2$ and $2zw$ respectively. Thus $(Lu,u) = 0 = (Su,u)$ can be written as

$$Q_1(x,y) + z^2 - w^2 = 0 = Q_2(x,y) + 2zw,$$

where Q_1 and Q_2 are homogeneous nonsingular quadratic forms in the variables x and y. Passing to inhomogeneous coordinates by dividing by w, these equations become

$$Q_1(x,y) + z^2 - 1 = 0 = Q_2(x,y) + 2z.$$

Eliminating z from the first equation gives $4Q_1 + Q_2^2 - 4 = 0$. Since Q_1 and Q_2 describe two simple conjugate eigenspaces on which LS has a pair of conjugate complex eigenvalues we can find coordinates in which they appear as $a(x^2 - y^2) + 2bxy$ and $2xy$ respectively, so that this equation has the form

$$0 = a(x^2 - y^2) + 2bxy + x^2y^2 - 1 = (a+y^2)x^2 + 2bxy - (1+ay^2)$$

If $a = 0$, this equation is solvable in rational terms, giving an intersection curve which is a hyperbola parallel to the x,y plane. Thus, only the case $a \neq 0$ requires special consideration. In this case, we can regard the preceding equation as a quadratic equation in x, and then its discriminant is $\varDelta = b^2y^2 + (a+y^2)(1+ay^2)$. Suppose first that $a > 0$. Then it is easy to see that $\varDelta > 0$ for all values of y, so that the equation describes a curve consisting of two infinite branches, which can be parametrized in terms of y using rational functions of y together with a single square root.

If $a < 0$, then $\varDelta$ is negative for $y = 0$ and for large y. Moreover, unless $b = 0$ and $a = -1$, there always exists a nontrivial interval I such that $\varDelta > 0$ whenever $y^2 \in I$. Thus the curve defined by the original equation is a pair of symmetric ovals which can again be parametrized using rational functions of y and a single square root.

In all remaining cases, LS has eigenvalues which are all simple and at most one pair of these eigenvalues is complex. Suppose first that all four eigenvalues λ_1,

. . . ,λ_4 of LS are real and simple. Then the homogeneous 4-space decomposes as a direct sum $Y_1 + \cdots + Y_4$ of four corresponding 1-dimensional orthogonal eigenspaces. In this decomposition, the forms (Lu,u) and (Su,u) are $\pm x^2 \pm y^2 \pm z^2 \pm w^2$ and $\pm\lambda_1 x^2 \pm\lambda_2 y^2 \pm\lambda_3 z^2 \pm\lambda_4 w^2$ respectively. Subtracting an appropriate multiple of L from S and passing to inhomogeneous coordinates by dividing by w, we can bring S to the form

$$ay^2 + bz^2 = c,$$

so that the intersection of Λ and Σ is the intersection of Λ (which is either a sphere or a hyperboloid) with an elliptic or hyperbolic cylinder parallel to the x-axis.

Finally, suppose that LS has two real simple eigenvalues and one complex conjugate pair of eigenvalues. We can assume without loss of generality that the complex eigenvalues are $\pm i$. Hence we can decompose the 4-space into the direct sum $Z + Y_1 + Y_2$, where Y_1, Y_2 are the 1-dimensional eigenspaces corresponding to the two real eigenvalues λ_1, λ_2, and where Z is a two-dimensional space corresponding to the eigenvalues $\pm i$. In this decomposition, the forms (Lu,u) and (Su,u) are $x^2 - y^2 \pm z^2 \pm w^2$ and $2xy \pm\lambda_1 z^2 \pm\lambda_2 w^2$, respectively. Subtracting an appropriate multiple of L from S and passing to inhomogeneous coordinates by dividing by w, we can bring S to the form

$$a(x^2 - y^2) + 2bxy = c,$$

so that the intersection curve of Λ and Σ is the intersection of Λ with a hyperbolic cylinder parallel to the z-axis.

In each of the last two cases, the coordinate system that we have introduced represents the intersection curve as the intersection of the unit sphere or the single-sheeted hyperboloid $x^2 + y^2 - z^2 = 1$ with a vertical elliptic or hyperbolic cylinder (or with one or two planes) whose base quadratic curve Q lies in the x,y plane and has center at the origin. First consider the case in which Λ is a sphere. Depending on the relationship of Q to the unit circle, this intersection curve can either consist of one or two disjoint closed loops, four branches meeting at two symmetrically situated critical points, or two symmetric isolated points. All these curves can be conveniently (though not quite rationally) parametrized by introducing a rational parametrization $x = x(t)$, $y = y(t)$ for the curve Q, and noting that $z = (1 - x^2(t) - y^2(t))^{1/2}$.

Next consider the case in which Λ is the single-sheeted hyperboloid $x^2 + y^2 - z^2 = 1$. If the cylinder base Q is an ellipse, the intersection can be either a circle, a symmetrical pair of closed loops, four branches meeting at two symmetrically situated critical points, or two symmetric isolated points. If Q is a hyperbola, the intersection will be either a set of four symmetrically situated infinite curves symmetric in the x,y plane, or two sets of four such curves meeting at two diametrically opposite critical points. These curves can be parametrized by introducing a rational parametrization $x = x(t)$, $y = y(t)$ for Q, and noting that $z = (x^2(t) + y^2(t) - 1)^{1/2}$.

With the exception of the spaces with designators [0] and [2,2], the quadratic forms corresponding to $\alpha S + \beta L$ are nonsingular for all but finitely many pairs (α, β) in every canonical space appearing in the preceding discussion. But when there exists (α, β) such that $\alpha S + \beta L$ is nonsingular, we can analyze the intersection of Σ and Λ in the manner explained in the preceding paragraphs by forming the intersection of Λ with the nonsingular surface defined by $(\alpha S + \beta L)u \cdot u = 0$. Thus only cases in which the designators [0] or [2,2] appear require special consideration. In the first of these cases, Σ and Λ can be represented simultaneously as vertical cylinders with quadratic curves as bases, and their intersections are simply vertical lines. On the other hand, if the space with designator [2,2] appears, then the intersection of Λ and Σ can be represented as the set of points satisfying

$$\pm x^2 + 2yw = 0 = ax^2 + 2zw.$$

Dividing by w to pass to inhomogeneous coordinates, we see that the intersection is just the rational plane curve $(x, \pm x^2/2, -ax^2/2)$.

4. BOOLEAN COMBINATIONS OF QUADRIC SURFACES

For a solid geometry modeling system, one will need to use the detailed information concerning quadric surface intersections developed in the preceding section to represent the geometry of an arbitrary boolean combination B of volumes bounded by such surfaces. The solid B is formed by applying intersection, symmetric difference and union operators to halfspaces H_i, each defined by an equation of the form

$$P_i(x,y,z) \geq 0 \text{ or } P_i(x,y,z) \leq 0$$

where each P_i is a quadratic polynomial. Let Σ_i be the quadric boundary surface defined by $P_i(x,y,z) = 0$. If necessary, we can apply infinitesimal translations to the surfaces Σ_i to ensure that they are in general position, i.e., that any two such surfaces which meet intersect transversally. Thus the boundary of the solid B is the union of closed subregions R_i of Σ_i, where R_i is the intersection of Σ_i with the boundary of B. In a geometric modeling system which admits general quadric surfaces to its vocabulary of basic shapes, we will have to determine the precise geometry of the region R_i for each quadric surface Σ_i.

To do this, we consider a fixed surface Σ and study the intersection curves C_i of Σ and Σ_i ($i \geq 1$). This network of curves divides Σ into a collection of 2-dimensional cells k_j, each of which is bounded by finitely many piecewise differentiable arcs α lying on various curves C_i, the arcs lying along C_i being separated by the intersection points of C_i with other curves C_j. The region $R = R_i$ we seek is a union of certain of the cells k_j. We shall show how to determine the geometry of the region R by dividing it into patches, each a union of one or more cells. Intuitively speaking, these patches are the connected subregions of R. More precisely, patches which meet only at isolated points will be considered to be distinct, i.e., 'patches' are defined as closures of the connected components of the interior of R. Thus distinct patches have disjoint interiors and either are disjoint or intersect at finitely many boundary points.

We now show how algorithms for manipulating algebraic numbers, like those reviewed in Schwartz and Sharir (1983b), can be used to determine the ordering along a fixed parametrized intersection curve C_i of this curve's intersections with all other curves C_j. (To avoid discussion of 'exceptional' coincidences, it is assumed in the following discussion that where necessary we have applied an infinitesimal perturbation to these curves to put them into general position, so that any curves which intersect meet transversally.) Recall from the previous section that each curve C_i can be parametrized by coordinate functions $x_i(u)$, $y_i(u)$, and $z_i(u)$, each of which is a rational combination of polynomials, or square roots of polynomials, in the real parameter u. This remains true even after we apply projective transformations to the parametrizations given in the last section in order to transform all of them into a common coordinate system. Let S and S_i be the quadratic forms which represent the surfaces Σ and Σ_i respectively. Examination of the intersection parametrizations listed above shows that except for the case in which the product SS_i has two pairs of complex eigenvalues, each such coordinate function either has the form

$$\frac{p(u) + d(u)^{1/2}}{1 + u^2}$$

where p and d are of degrees at most two and four respectively, or a similar form with denominator u^2, or is a rational fraction with denominator u and numerator at most quadratic, or is a rational fraction with denominator $1+u^2$ and numerator at most biquadratic, or is a simple quadratic polynomial. In the above formula, the polynomial d which appears under the radical is the same, up to a constant factor, for all three coordinate functions x, y, and z. The intersection points of C_i and some other curve C_j are the same as the intersections of C_i and the surface Σ_j, and may be determined by substituting the coordinate functions above into the defining polynomial $P_j(x,y,z) = 0$ of the surface Σ_j. This yields a polynomial equation $q_j(u) = 0$, of degree at most eight.

In the exceptional case that the product SS_i has two pairs of complex eigenvalues, one of the coordinate functions is of the form shown above, but with numerator multiplied by u. In this case, which can only occur when both quadratic forms have signature (2,2), substitution of the intersection parametrizations in the polynomial P_j may yield a polynomial of degree twelve. If many such cases occur in a particular application, it may be desirable to use the following somewhat different approach so as not to have to work with polynomials of degree more than eight. (This approach may in fact be superior whenever a hyperboloid is dealt with.) Choose a basis in which the surface Σ is given by the formula $x = yz$ in inhomogeneous coordinates. Then the intersection of Σ and any other surface Σ_i can be obtained by substituting yz for x in the equation of Σ_i, yielding an equation of the form

$$f_i(y,z) = P_i(z)y^2 + Q_i(z)y + R_i(z) = 0$$

where the coefficient polynomials are at most quadratic in z. This makes it clear that the intersection curve can be parametrized by z, at most one square root being involved. Two such curves intersect if and only if, for some real z, the polynomials f_i and f_j, viewed as polynomials in y, have a common real root. They have a common (possibly complex) root if and only if their resultant, which is the determinant of the 4×4 matrix

$$\begin{bmatrix} P_i & Q_i & R_i & 0 \\ 0 & P_i & Q_i & R_i \\ P_j & Q_j & R_j & 0 \\ 0 & P_j & Q_j & R_j \end{bmatrix}$$

of quadratic polynomials in z, vanishes. Note that this resultant is of degree at most eight. One can eliminate nonreal roots y by determining the sign of the discriminant of f_i, which is only of fourth degree, at each root of the resultant.

It follows that the intersection points of the fixed parametrized curve C_i and the various other curves C_j have coordinates $(x_i(t), y_i(t), z_i(t))$, where t is a real root of some q_j. In order to determine the sequence of arcs into which the collection of all such intersection points divides the curve C_i, we can make use of algorithms like those described in the Appendix of Schwartz and Sharir (1983b), which perform exact computations with algebraic numbers, i.e., roots of arbitrary rational polynomials. Given a polynomial q, these algorithms show how to find disjoint intervals on the real line, each having rational endpoints and containing precisely one root of q. By combining such root-isolating intervals corresponding to different polynomials q_j, one may determine the relative ordering of the collection of roots of all the polynomials q_j of our discussion. Thus we may assume that each intersection curve C_i has been divided into arcs α, each bounded by intersection points of C_i with other curves, and that the ordering of these arcs and points is known.

Our next step is to determine how the various cells into which the network of curves C_i divides the surface Σ combine to form surface patches on Σ. To do this, we map Σ to the u,v plane by using the inverse of the parametrization of the surface Σ, which corresponds to one of the three canonical diagonal quadratic forms enumerated at the beginning of Section 2 above. For instance, if Σ is the unit sphere, the inverse of the parametrization

$$x = \frac{2u}{1 + u^2 + v^2}; \quad y = \frac{2v}{1 + u^2 + v^2}; \quad z = \frac{1 - u^2 - v^2}{1 + u^2 + v^2}$$

is given by

$$u = x/(1+z); \quad v = y/(1+z).$$

This transforms the network of curves on Σ to a topologically equivalent network in the plane, whose natural orientation facilitates determination of the structure of the patches P_i. From now on, we shall therefore treat the collection of curves, arcs, intersection points, and patches as lying in the u,v plane rather than on the surface Σ.

For each arc α, it is first necessary to determine whether the cell immediately to the left (or right) of this arc lies in the region R, i.e., on the boundary of the solid B. To do this, find a rational number s such that the point $p = (u(s),v(s))$ lies on α. Working in the u,v plane and using polynomial calculations only, we can determine the equation of the normal line to the arc α through the point p. Now choose a point p' infinitesimally to the left (or right) of p on this normal line. The cell containing p' is a subset of the solid B if and only if an appropriate Boolean combination of conditions $P_i(x(p'),y(p'),z(p')) \geq 0$ is satisfied, where the P_i are the defining polynomials of the bounding surfaces Σ_i of the solid B. In order to exclude cells which are interior to B, we perform a similar test on two points q which lie close to Σ on the normal line to Σ passing through p'. If both points tested lie in B, the cell containing p is interior to the solid B, hence does not lie on R. Otherwise, precisely one of the points q lies in B, hence the cell lies in R and on the boundary of B. Note that at least one point q must lie in B, for otherwise B would contain a degenerate two-dimensional component; this possibility is ruled out by our prior assumption that the various surfaces Σ_i intersect only transversally.

Thus we may assume that the curves C_i in the u,v plane are divided into sequences of arcs, and that the region to the left of each of these arcs or to its right (but not both) is specified as lying on the boundary of the solid B. From this information, we wish to reconstruct the global geometry of the region R in the u,v plane which represents the intersection of S with the surface of B.

In order to do this, divide the collection of arcs α into a set of closed loops l_i which bound the patches of R. The patches so obtained will not be disjoint, but nondisjoint patches will meet only at a finite collection of vertices which lie at the intersection of different curves. Each simply-connected patch will be surrounded by a single boundary loop, but patches with n holes will have $n+1$ boundary loops. Thus, to understand the topological situation, we need to know whether a given patch lies to the left or to the right of each of its (oriented) boundary loops, and must also determine the manner in which different loops are nested.

For this, we can proceed as follows. Let A denote the collection of arcs α_i. Begin by discarding from A all arcs which are interior to R. These are detected by the fact that points near the arc on both halves of its normal line lie in R. Next, recall that each arc is oriented by the parametrization of the curve on which it lies, so that we may legitimately refer to the tail and head of the arc. (Of course, we will have to reorient some of the arcs in order to link them into the loops which we seek to construct.)

Begin the algorithm by picking an arbitrary arc α_1. Assume that the region to the right of this arc lies on R; if not, reverse the orientation of the curve on which the arc α_1 lies. The first loop will begin with α_1; we will find successive arcs which border the patch to the right of α_1 and add these arcs to the loop which we are constructing.

For this, choose the successor α_2 of the initial arc α_1 from among the collection of arcs in A whose tail or head is the head h_1 of α_1. Reorient α_2 so that its tail is h_1. Since R lies to the right of α_1, α_2 will be that arc which is first encountered when we rotate α_1 counterclockwise while holding its head fixed. Note that the clockwise ordering of the arcs with endpoint h_1 can be determined precisely from the parametrizations $(u_i(t), v_i(t))$ of the intersection curves C_i: the slope of each curve at h_1 is the quotient of derivatives of u and v with respect to t, evaluated at a suitable algebraic number t, and hence these slopes can be ordered by the polynomial algorithms described in Schwartz and Sharir (1983b). Once the ordering of the slopes is known, it is easy to determine the clockwise order of the arcs around h_1. No two slopes can be equal, since all intersections have been made transversal.

This procedure for finding the successor to an arc on a boundary loop of R should be iterated until the initial arc a_1 is again encountered. At this point, we can delete the collection of arcs on this loop from the set of arcs A. Then we pick a new initial arc from A, and continue in this manner until A is exhausted and the complete collection of loops which bound patches of R has been constructed. Note that, during its construction, each loop has been oriented so that the patch of R which it bounds lies its right. It remains to determine whether the patch lies inside or outside this loop; for this we need to establish the sequence of nesting of the various loops.

In order to do this, consider a collection of loops l_i in the u,v plane and define the distance from infinity d_i of each loop l_i as follows. If l_i is an outermost loop, define d_i to be 1. Inductively, if l_i is contained inside a loop l_j whose distance from infinity is n, but if no loop lies between l_i and l_j, define d_j to be $n + 1$. Note that l_i runs clockwise if and only if d_i is odd and that each patch in R is bounded by one clockwise (outer) boundary loop and by a number of clockwise (inner) boundary loops equal to the number of holes in the patch.

In order to compute the distances d_i for the collection of loops l_i, proceed as follows. Choose a point p on a loop l and let r be an infinite ray with endpoint p. Assume that r is in general position with respect to the loops l_i, i.e., that at any intersection, r crosses non-tangentially between the outside and inside of l_i. Of course, the number of intersections of this ray with loops l_i is finite, although the ray may intersect a given loop several times. Order the list of intersections of r with the loops so that the first intersection point lies on an outermost loop. Define integers m_i by the condition that l_{m_i} is the loop which contains the i-th intersection point on the ray. Thus d_{m_1} is 1. Let L be the ordered list of integers m_i. Construct a sequence of sublists P_i of L, beginning with $P_i = [m_1]$, which will record the passage of a point q traversing the loop diagram along r, as follows. Given P_i, let P_{i+1} be the list formed from P_i by deleting the last element m_j from P_i if m_{i+1} equals m_j; otherwise by appending m_{i+1} to P_i. In the second case, record the fact that $l(m_{i+1})$ is nested immediately inside $l(m_j)$, and set $d(m_{i+1})$ to $d(m_j) + 1$, if it has not yet been defined. The first case occurs when q exits $l(m_j)$, the second when q moves deeper into the nest of loops. This procedure concludes by setting the last list P_{last} to the empty list when q exits the outermost loop.

If loops remain whose distance from infinity has not yet been calculated, draw a ray from one of these loops and repeat the algorithm, continuing until $d(l_i)$ has been calculated for all loops. These distances, together with the nesting information which was recorded during the algorithm, completely determine the geometry of the region R, and hence determine the intersection of the boundary of the solid B with the quadric surface Σ. This geometric information can be used for various computations involving the boundary of R, e.g., in connection with a drawing or surface area algorithm. A geometric modeling system supporting these applications is currently being programmed.

CHAPTER 11

Complexity of the Generalized Mover's Problem

JOHN H. REIF

Aiken Computation Laboratory
Harvard University

This paper concerns the problem of planning a sequence of movements of linked polyhedra through three-dimensional Euclidean space, avoiding contact with a fixed set of polyhedra obstacles. We prove this generalized mover's problem is polynomial space hard. Our proof provides strong evidence that robot movement planning is computationally intractable, i.e., any algorithm requires time growing exponentially with the number of degrees of freedom.

1. INTRODUCTION

1.1. The Movers Problem

The classical *mover's problem* in d-dimensional Euclidean space is:

Input: (O,P,p_I,p_F) where O is a set of polyhedral *obstacles* fixed in Euclidean space and P is a rigid polyhedron with distinguished initial position p_I and final position p_F. The inputs are assumed to be specified by systems of rational linear inequalities.

Problem: Can P be moved by a sequence of translations and rotations from p_I to p_F without contacting any obstacle in O?

For example, P might be a sofa[1] which we wish to move through a room crowded with obstacles. Figure 1 gives a simple example of a two-dimensional movers problem.

The mover's problem may be *generalized* to allow P (the object to be moved) to consist of multiple polyhedra freely linked together at various distinguished vertices. (A typical example is a robot arm with multiple joints.) Again, the input is specified by systems of rational linear inequalities. (A precise definition of the generalized problem is given in Section 2.)

This work was supported by the Office of Naval Research Contract N00014-80-C-0647.

Thanks to Martin Cherez, Christoph Freytag, Debby Joseph, and Paul Spirakis for helpful comments on this paper.

[1] The author first realized the nontrivial mathematical nature of this problem when he had to plan the physical movement of an antique sofa from Rochester, New York to Cambridge, Mass.

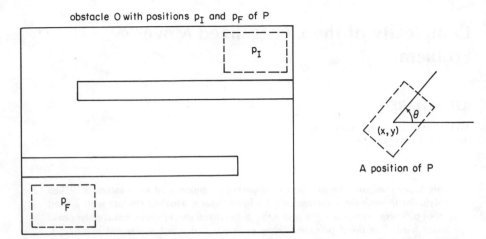

obstacle O with positions p_I and p_F of P

A position of P

Figure 1.   A 2-D Mover's Problem: Can rectangle P be moved from p_I to p_F without contacting an obstacle in $O?$

1.2.   Lower Bounds for Generalized Mover's Problems

Our main result, first presented in Reif (1979) (and given in full detail in Section 2) is that the generalized mover's problem in three dimensions is polynomial space hard. That is, we prove that the generalized mover's problem is at least as hard as any computational problem requiring polynomial space. (Polynomial space problems are at least as hard as the well-known NP problems; see Garey & Johnson, 1979.)

This was the first paper investigating the inherent computational complexity of a robotics problem, and in fact was the first polynomial space hardness result for any problem in Computational Geometry. Our proof technique is to use the degrees of freedom of P to encode the configuration of a polynomial space bounded Turing maching M, and to design obstacles which forced the movement of P to simulate the computation of M.

This work was originally motivated by applications to robotics: The author felt it was important to examine *computational complexity issues in robots* given the recent development of mechanical devices autonomously controlled by micro and minicomputers, and the swiftly increasing computational power of these controllers. However, it took a number of years before computational complexity issues in robotics became of more general interest. Recently there has been a flurry of papers in the now emerging area which we might term *Computational Robotics.*

Recent investigations in lower bounds have provided some quite ingenious lower bound constructions for restricted cases of the generalized mover's problem. For example, Hopcroft, Joseph, and Whitesides (1982) showed that the generalized mover's problem in three dimensions is also polynomial space hard, and Hopcroft and Sharir (1984) show that the problem of moving a collection of disconnected

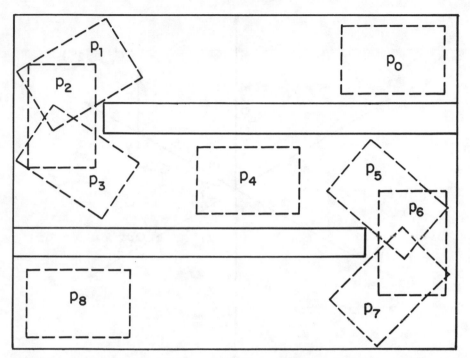

Figure 2. **A solution to the 2-D Mover's Problem of Figure 1.** P **may be moved through** positions $p_I = p_0, p_1, \cdots, p_8 = p_F$.

polyhedra in a two-dimensional maze is polynomial space hard. The problem of moving a collection of disks in two dimensions is known to be NP-hard (Sparakis & Yap, 1985), but it remains open to show this problem polynomial space hard.

1.3. Upper Bounds for Mover's Problems

Our lower bounds for the generalized mover's problem provide evidence that time bounds for algorithms for movement planning must grow exponentially with the number of degrees of freedom. We next give a brief description of known algorithms for mover's problems. In our original paper (Reif, 1979) we also sketched a method for efficient solution of the classic mover's problem where P, the object to be moved, is rigid. In spite of considerable work on this problem by workers in the robotics fields and in artificial intelligence (for example, Nilson, 1969; Paul, 1972; Udupa, 1977; Widdoes, 1974; Lozano-Pérez & Wesley, 1979), no algorithm guaranteed to run in polynomial time had previously appeared. Our approach was to transform a classic mover's problem (O, P, p_I, p_F) of size n in d dimensions to an apparently simpler mover's problem (O', P', p_I', p_F') of dimension d', where P' is a single part and d' is the number of degrees of freedom of movement in the original problem. The transformed problem is thus to find a path in d'-dimensional space

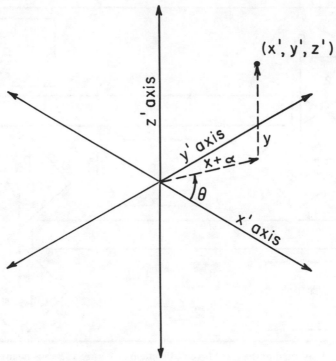

Figure 3.   The mapping $f(x,y,\theta) = (x',y',z')$.

avoiding the transformed obstacles O. The fundamental difficulty is that the induced obstacles may be non-linear constraints. (Lozano-Pérez & Wesley, 1979, did not construct O', but instead approximated the induced obstacles O' by linear constraints. Unfortunately, an exponential number of linear constraints were required to approximate even a quadratic constraint within accuracy 2^{-n}. Thus their method required exponential time—i.e., 2^{cn} time for some $c>0$—even if the original movers problem was two-dimensional.)

   Example.   Consider a classical mover's problem (O,P,p_I,p_F) restricted to *dimension $d=2$*, with the obstacles O consisting of a set of line segments and P a single polygon. A *position* of P can be specified by a triple (x,y,θ) where (x,y) are the cartesian coordinates of some fixed vertex of P and θ is the angle of rotation around this vertex. We define a mapping f from the position of P to 3-space. Let $f(x,y,\theta) = (x',y',z')$ where $y=z'$, $tan(\theta)=x'/y'$, and $x = (x')^2 + (y')^2-\alpha$, for some sufficiently large constant $\alpha\geqslant 0$. (α may be taken as the diameter of a circle enclosing P.) See Figure 3.

   In this case, we define a *1-contact set* to be a maximal set of positions of P where a vertex of P contacts a line segment of O, or a vertex of O contacts a line segment of P. (See Figure 4.) The transformed obstacles O' are the union of these 1-contact

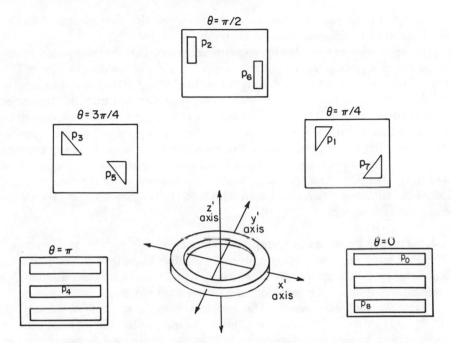

Figure 4. Transformed mover's problem from Figure 1. The obstacles of the transformed problem define a torus with cross-sections illustrated for $\theta j = 0$, $\pi/4$, $\pi/2$, $3\pi/4$, π. P may be moved through positions $p_I = p_0, p_1, \cdots, p_8 = p_F$ as in Figure 2.

sets. Thus each obstacle in O' is a quadratic surface patch which may be easily constructed from the input, there are at most $O(|O||P|)$ such obstacles and their $O(|O|^2|P|^2)$ intersections can easily be computed within accuracy 2^{n-c} for any $c>0$, by known polynomial time procedures (Comba, 1968) for intersection of quadratic surface patches. Hence in this simple example the connected regions bounded by O' can be explicitly constructed in polynomial time within accuracy 2^{-nc} which is sufficient for solution of this mover's problem.

In the case of a classical mover's problem (O,P,p_I,p_F) of dimension $d=3$, the transformed problem (O',P',p_I',p_F') has dimension $d' =6$. In this case we define a 1-contact set to be a maximal set of positions of P where an edge of P contacts a face of O or an edge of O contacts a face of P. Again, the 1-contact sets are constant degree polynomials. The transformed obstacles O' are the union of the 1-contact sets. The connected regions defined by O' can again be explicitly constructed by intersecting these constraints. In Reif (1979), we briefly suggested a method for this construction, but Schwartz and Sharir (1983a) later gave a complete detailed description of a method for explicit construction of such a transformed mover's problem in three dimensions in polynomial time. (O'Dunlaing, Sharir, & Yap,

1983, further improved this construction by observing that movement of P can be restricted to be equidistant from the obstacles.)

This approach was extended by Schwartz and Sharir (1983b) to solve any generalized mover's problem of input size n with d' degrees of freedom in time $n^{2^{O(d')}}$. They make use of the algebraic decomposition of Collins (1975) (previously used to decide formulas of the theory of real closed fields) to construct the connected regions bounded by O'. Note that their upper bounds grow doubly exponentially with d', where as our polynomial space lower bounds suggest only single exponential time growth with d'. It remains a challenging problem to close the gap between those lower and upper bounds for generalized mover's problems. Further progress will likely depend on improvements to decision algorithms for the theory of real closed fields; recently Ben-Or, Kozen, and Reif (1984) gave a single exponential space decision algorithm.

1.4. Further Problems in Computational Robotics

There are some very challenging problems remaining in the field of Computational Robotics beyond the complexity of the mover's problem and its generalization. We mention below three such problems and some recent progress.

(1) *Frictional Movement.* The problem here is to plan movement for (O, P, p_I, p_F) in the case contact is allowed in the presence of friction between surfaces. (An unpublished manuscript of Burridge, Rajan, and Schwartz analyzes some aspects of this surprisingly complex problem.) Miller and Reif (1985) prove undecidability of planning frictional movement. What natural subclass of frictional movement problems is decidable?

(2) *Minimal Movement.* The problem is, given a set O of k polyhedral obstacles in d space defined by a total of n linear constraints, and points p_I, p_F, find a minimal length path from p_I to p_F avoiding the obstacles in O. Chazelle (1982) gives a $O(n \log n)$ algorithm in the case $d=2$ and $k=1$. Sharir and Schorr (1984) give a $2^{2^{O(n)}}$ algorithm for $d=3$. Reif and Storer (1985) gave a $O(nk \log n)$ algorithm for $d=2$ and $n^{k^{O(1)}}$ time and $n^{O(\log k)}$ space algorithms for $d=3$. Recently Canny and Reif (1986) have shown that the minimal movement problem for $d=3$ dimensions is NP-hard. Is this problem NP-complete for $d=3$?

(3) *Dynamic Movement.* The problem is to plan the movement of a polyhedron in d dimensions with bounded velocity modulus from point p_I at time $t=0$ to p_F at some given time t' so as to avoid contact with a set O of k polyhedral obstacles (defined by a total of n linear constraints) moving with fixed, known velocity. Reif and Sharir (1985) give the first known investigation of the computational complexity of planning dynamic movement. They show that the problem of planning dynamic movement of a single ($k=1$) disk P in $d=3$ dimensions is polynomial space hard. (This result is somewhat surprising, since P in this case has only three degrees of freedom. Our key new idea is to use time to encode a configuration of a polynomial space bounded Turing machine.) Is this problem polynomial space hard for dimension $d=2$?

Asteroid avoidance problems are a natural subclass of dynamic mover's problems where each obstacle is convex and does not rotate. Reif and Sharir (1985) give a polynomial time algorithm for dimension $d=2$ with a bounded number $k=O(1)$ of obstacles and give $n^{n^{O(1)}}$ time and $(k+n)^{O(\log n)}$ space algorithms for dimension $d=3$ with an unbounded number k of obstacles. Is the asteroid avoidance problem polynomial in the case $d=3$?

1.5. Organization of the Paper

In Section 2.1 we give a precise definition of the generalized mover's problem. In Section 2.2 we define symmetric Turing machines. In Section 2.3 we give the relevant complexity theoretic definitions and results. In Section 2.4, we give our proof that the generalized mover's problem is polynomial space hard.

2. THE GENERALIZED MOVER'S PROBLEM IS PSPACE-HARD

2.1. Definition of the Generalized Mover's Problem

We let a convex polyhedron in 3 space be specified by a finite set of linear inequalities with rational coefficients. We let a *(rational) polyhedron* be specified by a finite union of such convex polyhedra. Such a polyhedron P can be encoded by some fixed convention as a finite binary string $\langle p \rangle$.

We will formally specify the three-dimensional generalized mover's problem (O,P,p_I,p_F) as follows:

1. the *obstacle set* O consists of a finite set of (rational) polyhedra $O_1, \ldots, O_{n_1}$
2. the *object to be moved*, P, consists of a finite set of (rational) polyhedra $P_1, \ldots, P_{n_2}$ which are freely linked at distinguished *linkage vertices* $v_1, \ldots v_{n_3}$
3. p_I, p_F are distinguished *initial* and *final* rational positions of P.

Hence we may encode (O,P,p_I,p_F) as the string $(\langle O_1 \rangle, \ldots, \langle O_{n_1} \rangle) (\langle P_1 \rangle, \ldots, \langle P_{n_2} \rangle, v_1, \ldots, v_{n_3}) (\langle p_I \rangle, \langle p_F \rangle)$. The *size* of (O,P,p_I,p_F) is the length of this encoding.

A *legal position* of P is any position where each polyhedron p_i of P intersects no obstacle of O and furthermore intersects no other polyhedron of p except at its specified linkage vertices. We assume, of course, that p_I and p_F are both legal positions. A *legal movement* of P is a continuous sequence of simultaneous translations and rotations of the polyhedra of P through only legal positions. The *generalized mover's problem* is to determine the existence of a legal movement from p_I to p_F.

It is important to observe that any generalized mover's problem is *reversible* in the sense that if there is a legal movement of P from p_I to p_F, then the movement can always be reversed so as to begin at p_F and end at p_I. This reversibility property

imposes a constraint on the class of computation problems which can be simulated
by generalized movement problems; in particular the simulated machine must be
symmetric in a sense precisely defined below.

2.2. Symmetric Computations

A *symmetric Turing machine* is defined (see also Lewis & Papadimitriou, 1982,
for an equivalent definition) as $M = (\Gamma, \Sigma, Q, q_I, q_F, \Delta)$ where

i. Γ is the *tape alphabet* with distinguished *pad symbol* $\$ \in \Gamma$ and *blank symbol*
     $\# \in \Gamma$
ii. $\Sigma \subseteq \Gamma - \{\$, \#\}$ is the *input alphabet*
iii. Q is the state set with distinguished *initial state* $q_I \in Q$ and *accepting state* q_F
     $\in Q$
iv. $\Delta \subseteq (Q \times \Gamma^2 \times \{-1, 1\})^2$ is the *transition relation,* where we require that for each
     transition $((q, L, R, D), (q', L', R', D')) \in \Delta$
     $D' = -D$
     if $L = \$$, then $D \neq 1$. Alternatively, if $R = \$$, then $D \neq 1$
     also $((q', L', R', D'), (q, L, R, D)) \in \Delta$.

We will also be given a *space bound* $s = s(n)$ which is a function of the input
length n such that $s(n) \geq n$. M has a single read/write tape with $s + 2$ tape cells. This
tape has contents $t = t_0 t_1 \cdots t_s t_{s+1}$ where $t_0 = t_{s+1} = \$$ and $t_1, \ldots, t_s \in \Gamma - \{\$\}$.

M has a single read/write tape head which simultaneously scans the tape cell
under the current head position, as well as the tape cell immediately to the left or
right of the current head position depending on the direction of the next move of the
tape head (this convention is used to allow for reversibility). Restriction (b) insures
M never moves its head off the end of the tape. Restriction (c) implies that the
transition relation is a symmetric relation.

More precisely, a *configuration* of M is a tuple $ID = (q, h, t)$ where $q \in Q$ is the
current state, $h \in \{1, \ldots, s\}$ is the current position of the tape head, and $t = t_0 t_1$
$\cdots t_s t_{s+1} \in \$(\Gamma - \{\$\})^s \$$ is the current tape contents. The *next move relation* $\vdash$ is a
relation on configurations such that $(q, h, t) \vdash (q', h', t')$ iff there exists a transition
$((q, L, R, D), (q', L', R', D')) \in \Delta$ the new head position is $h' = h + D$, the new tape
contents t' are identical to the previous tape contents t except at positions h and
$h + D$.

1.   if $D = 1$ then $t_h = L$, $t_{h+1} = R$, $t'_h = L'$, and $t'_{h+1} = R'$
2.   if $D = -1$ then $t_{h+1} = L$, $t_h = R$, $t'_{h+1} = L'$, and $t'_h = R'$.

Given the input string $w = w_1 \cdots w_n \in \Sigma^n$, the *initial configuration* is $ID_I(w) =$
$(q_I, 1, \$w_1 \cdots w_n \#^{s-n}\$)$. We define $ID_F = (q_F, 1, \$\#^s\$)$ to be the accepting configuration of M. Let $\vdash^*$ be the transitive closure of $\vdash$. M accepts input w iff $ID_0(w)$
$\vdash^* ID_F$. Let $L(M)$ be the language accepted by M.

2.3. Complexity Definitions

For some space bound $s=s(n)\geq n$ let $DSPACE(s)$, $SSPACE(s)$, $NSPACE(s)$ denote the class of language accepted by deterministic, symmetric, and nondeterministic Turing machines, respectively. Savitch (1970) shows:

PROPOSITION 1. $NSPACE(s) \subseteq DSPACE(s^2)$.

Lewis and Papadimitriou (1982) show

PROPOSITION 2. $DSPACE(s) \subseteq SSPACE(s) \subseteq NSPACE(s)$.

Let

$$PSPACE = \bigcup_{c\geq 1} DSPACE(n^c).$$

The above imply;

PROPOSITION 3. $PSPACE = \bigcup_{c\geq 1} SSPACE(n^c)$.

A *log-space reduction* from a language $L' \subseteq \Sigma^*$ to a language L is a mapping f computable by a $O(\log n)$ space bounded deterministic Turing machine such that for each input $w \in \Sigma^*$, $w \in L'$ iff $f(w) \in L$. In this case, we say L' is *log-space reducible* to L. Note that any log-space reduction requires only time bound $2^{O(\log n)} = n^{O(1)}$.

Given a language class L, a language L is L-*hard* if each language $L' \in$ L is log-space reducible to L.

2.4. The Simulation of a Symmetric Turing Machine

We now prove:

THEOREM. *The generalized mover's problem is* PSPACE-*hard*.

Proof. Let $M=(\Gamma,\Sigma,Q,q_I,q_F,\Delta)$ be a symmetric Turing machine with polynomial space bound $s(n)=n^c$ for some constant $c \geq 1$. We will construct a log-space reduction from $L(M)$ to the generalized mover's problem. In particular, given an input $w=w_1 \ldots w_n \in \Sigma^n$, we construct in $O(\log n)$ space a mover's problem $f(w) = (O,P,p_I,p_F)$ such that P has a legal movement from p_I to p_F iff M accepts input w.

We can assume, without loss of generality, that $s=s(n)$ is constructible in deterministic $O(\log n)$ space.

It will be useful to consider the tape alphabet Γ to be the integers $\{1, \ldots, \gamma\}$, where $\gamma = |\Gamma|$.

We begin by defining P, the object which is to be moved. P will contain a sequence $A_0, \ldots, A_{s+1}$ of triangular pyramids of identical size which will be called *arms*. For each $i=0, \ldots, s+1$ arm A_i has a distinguished *apex vertex* v_i. A_i has an equilateral triangular base with base sides of length $a = 1/(4(\gamma+1))$. Each of

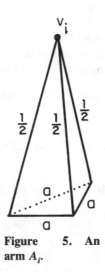

Figure   5.   An arm A_i.

the vertices of the base is of length ½ from the apex vertex v_i (see Figure 5). For each $i=0,\ldots,s$ there is also a straight (one-dimensional) link of length l from v_i to v_{i+1} which freely links A_i to A_{i+1} (see Figure 6).

It will be useful to define a *cutout polygon* Q consisting of the union of a rectangle and a set of triangles $\{Q_{ij}\}$ of identical size for $i=0,\ldots,2s+1$ and $j=1,\ldots,\gamma$. The rectangle is of horizontal length $2s+1$ and vertical height $\epsilon=a/10$. Each triangle Q_{ij} has a distinguished vertex u_i, connected to two sides of length $½+\epsilon$, and a base side of length $a+\epsilon$ opposite u_i (see Figure 7). On the upper side of the rectangle is the sequence of vertex $u_0,\ldots,u_{2s+1}$ spaced at distance l between each other. For each $i=0,\ldots,2s+1$ the triangles $Q_{il},\ldots,Q_{i\gamma}$ each share vertex u_i but are otherwise disjoint, and arranged in cyclic order (as in Figure 8).

Let *TUNNEL(Q)* be a cylinder with perpendicular cross-section Q. Therefore, the interior of *TUNNEL(Q)* is formed by sweeping Q in a direction perpendicular to the plane in which Q is contained. We will call the region swept out by triangle $Q_{i,j}$ the $Q_{i,j}$-*slot*.

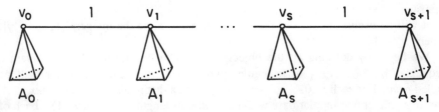

Figure 6.   P, the object to be moved.

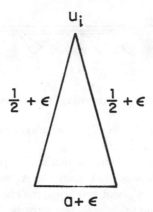

Figure 7. A triangle Q_{ij}.

The basic idea in our construction will be to use the $s+2$ degrees of freedom of P to encode a given configuration of M.

Let $h \in \{1, \ldots, s\}$ be a head position and let $t = t_0 t_1 \ldots t_s t_{s+1} \in \$(\Gamma - \{\$\})^s \$$ be the contents of the tape. We say P *encodes* (h, t) if P is positioned in the interior of $TUNNEL(Q)$ so that for $i = 0, 1, \ldots, s+1$ arm A_i is in the Q_{s-h+i, t_i}-slot (see Figure 9). We say P is *properly positioned* if P encodes some (h, t). We shall define obstacles and the initial position in such a way that P is always properly positioned.

Observe that we have defined $TUNNEL(Q)$ so that if P is properly positioned in its interior and P encodes (h, t), then P always encodes (h, t) on any legal movement of P within the interior of $TUNNEL(Q)$ since the arms of P remain in the same slots.

A *segment* of $TUNNEL(Q)$ is a copy of the cylinder $TUNNEL(Q)$ bounded by two planes perpendicular to the cylinder (see Figure 10). We will allow separate segments of $TUNNEL(Q)$ to be merged into a single copy of a $TUNNEL(Q)$ segment. This can be done as in Figure 11, so that if P encodes (h, t) on an entrance, P encodes (h, t) on the exit. Note that of course, P can also move from the exit back to either entrance, without modifying the encoding (h, t). Thus this construction can also be viewed as the branch of a segment of $TUNNEL(Q)$ into two segments of $TUNNEL(Q)$.

Next we require a construction of obstacles which force P to modify its position so as to simulate next moves of the symmetric machine M.

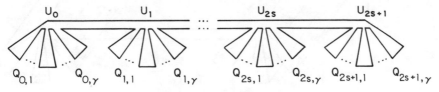

Figure 8. The polygon Q.

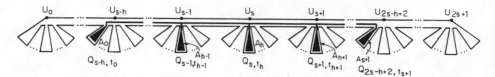

Figure 9.   A position of P encoding (h,t) where $t=t_0 t_1 \ldots t_{s+1}$.

For any $L,R \in \{1, \ldots, \gamma\}$, let $Q[L,R]$ be the polygon derived from Q by deleting all triangles Q_{s,j_0} and Q_{s+1,j_1} for all $j_0 \in \{1, \ldots, \gamma\} - \{L\}$ and $j_1 \in \{1, \ldots \gamma\} - \{R\}$ (see Figure 12). Observe that if P is positioned in the interior of *TUNNEL($Q[L,R]$) and P encodes* (h,t), then arm A_h must be in the $Q_{s,L}$-slot and arm A_{h+1} must be in the $Q_{s+1,R}$-slot and hence the encoded tape symbols in the h and $h+1$ position are $t_h=L$ and $t_{h+1}=R$, respectively.

Let $\hat{Q}$ be the figure derived from Q by adding two semidisks with radius $\frac{1}{2}+\epsilon$, and with centers at u_s and u_{s+1} (see Figure 13). Note that if P is positioned in the interior of *TUNNEL($\hat{Q}$)* so that P encodes (h,t) except at t_h and t_{h+1}, then the arms A_h and A_{h+1} are each free to move within the interior region swept out by a semidisk.

Let $\delta \in \Delta$ be a transition, where

$$\delta = ((q,L,R,D),(q',L',R',-D)).$$

We will define an obstacle B_δ with a connected interior region with distinguished *entrance* and *exit,* and with the property that if P enters the interior of B_δ encoding (h,t), then when P exits B_δ P encodes (h',t'), where

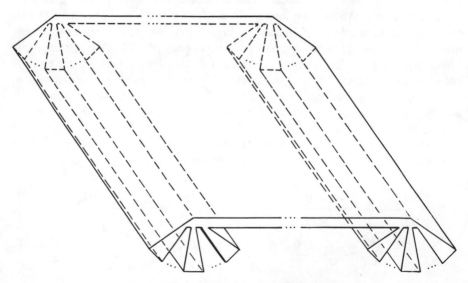

Figure 10.   A segment of *TUNNEL(Q)*.

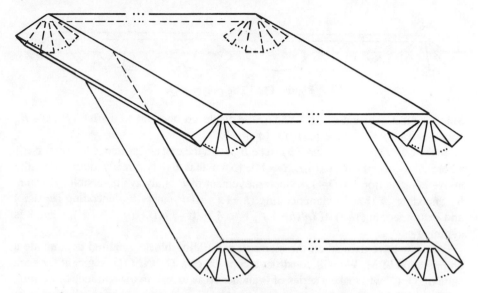

Figure 11.   The merge of two segments of *TUNNEL(Q)*.

$$(q,h,t) \vdash (q',h',t').$$

We first consider the case $D=1$. Then we let B_δ consist of a concatenation of unit length symbols of the following:

1. *TUNNEL(Q)*
2. *TUNNEL(Q[L,R])*
3. *TUNNEL($\hat{Q}$)*
4. *TUNNEL(Q[L',R'])*
5. *TUNNEL(Q)*, which is displaced one unit to the left with respect to segments (1)–(4).

(See Figure 14.)

Suppose P enters B_δ encoding (h,t). Then P can move through *TUNNEL(Q[L,R])* only if $t_h=L$ and $t_{h+1}=R$. After moving through *TUNNEL($\hat{Q}$)*, P encodes (h,t'), where t' is identical to t except t'_h and t'_{h+1} are arbitrary elements of $\{1, \ldots, \gamma\}$. However, P can move through *TUNNEL(Q[L',R'])* only if $t'_h=L'$ and $t'_{h+1}=R'$.

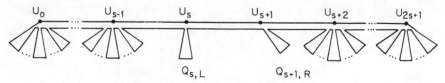

Figure 12.   The polygon Q[L,R].

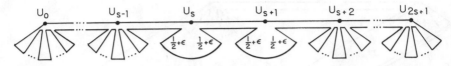

Figure 13.   The polygon $\hat{Q}$.

Since the last segment of *TUNNEL(Q)* is displaced one unit to the left, P exits B_δ encoding $(h+1,t')$, where $(q,h,t) \vdash (q',h+1,t')$.

In the case $D = -1$, we take B_δ to be $B_{\delta'}$ with the exit and entrance face reversed, where $\delta' = ((q',L',R',1),(q,L,R,-1))$. (Note that $B_{\delta'}$ is already defined by the above construction for $D=1$.) Since movement of P is always reversible, P enters B_δ encoding (h,t) and exits encoding $(h-1,t')$ iff P enters $B_{\delta'}$ encoding $(h-1,t')$ and exits encoding (h,t) iff $(q',h-1,t') \vdash (q,h,t)$ iff $(q,h,t) \vdash (q',h-1,t')$, since $\vdash$ is symmetric.

We now have defined all the elementary building blocks required to simulate a computation of M. We will construct a copy C_q of a *TUNNEL(Q)* segment for each state $q \in Q$. C_q will make a series of branches so as to lead to the entrance of each B_δ such that $\delta \in \Delta$ is a transition from state q. Also C_q will make a series of branches in the opposite direction, so as to lead to the exit of each $B_{\delta'}$ such that $\delta' \in \Delta$ is a transition to state q. Note that the construction is of polynomial size and can easily be done by a $O(\log n)$ space deterministic Turing machine.

For the proof of our construction, it will be useful to extend our definition of encoding so that if P is located in the interior of C_q encoding (h,t), we also then say that P encodes configuration $ID=(q,h,t)$.

Given input $w=w_1 \ldots w_n \in W^n$, we define the *initial position* p_I to be a rational position of P encoding the initial configuration $ID_0(w)=(q_I,1,\$w_1 \ldots w_n\#^{s-n}\$)$.

The *final position* p_F is defined to be a rational position of P encoding the accepting configuration $ID_F=(q_F,1,\$\#^s\$)$.

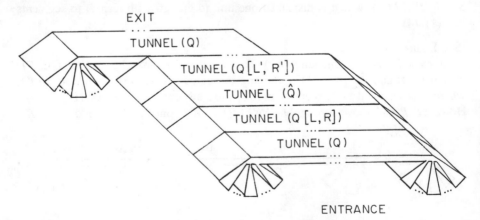

Figure 14.   The obstacle B_δ.

LEMMA. *P has a legal movement from $p_I(w)$ to a position encoding configuration ID iff $ID_0(w) \vdash^* ID$.*

Proof. $ID_0(w) \vdash^* ID$ iff $\exists$ a sequence of configurations $ID_0(w)=ID_0, ID_1,$ $\ldots, ID_k=ID$ where $ID_0 \vdash ID_1, \ldots, ID_{k+1} \vdash ID_k$ iff $\exists$ a sequence of transitions $\delta_1,$ $\ldots, \delta_k \in \Delta$ where $ID_i = \delta_i(ID_{i-1})$ for $i=1, \ldots, k$. We now claim that this holds iff P has a legal movement from $p_I(w)$ through $B_{\delta_1}, \ldots, B_{\delta_k}$ (in this order) to a position p_k encoding $ID_k=ID$. In the case $k=0$, the claim obviously holds since $p_I(w)$ encodes $ID_0(e)$. Suppose the claim holds for all $k'<k$. Then P has a legal movement from $p_I(w)$ through $B_{\delta_1}, \ldots B_{\delta_{k-1}}$ to position p_{k-1} encoding ID_{k-1} iff $ID_0(w) \vdash^* ID_{k-1}$. But our above construction of B_{δ_k} insures that there exists a legal movement of P from p_{k-1} through B_{δ_k} to a position encoding ID_k iff $ID_k=\delta_k(ID_{k-1})$. Hence the claim holds.

The Lemma then implies: P has a legal movement from initial position $p_I(w)$ to final position p_F iff $ID_0(w) \vdash^* ID_F$, where ID_F is the accepting configuration. This completes the proof of our theorem.

CHAPTER 12

Movement Problems for 2-Dimensional Linkages

JOHN HOPCROFT

Department of Computer Science
Cornell University

DEBORAH JOSEPH

Computer Science Department
University of Wisconsin, Madison

SUE WHITESIDES

School of Computer Science
McGill University, Montreal

This paper is motivated by questions concerning the planning of motion in robotics. In particular, it is concerned with the motion of planar linkages from the complexity point of view. There are two main results. First, a planar linkage can be constrained to stay inside a bounded region whose boundary consists of straight lines by the addition of a polynomial number of new links. Second, the question of whether a planar linkage in some initial configuration can be moved so that a designated joint reaches a given point in the plane is PSPACE-hard.

1. INTRODUCTION

This paper is concerned with the motion of linkages from the computational complexity point of view. The research was motivated by earlier work in robotics, particularly that of Lozano-Perez and Wesley (1979), Lozano-Perez (1980), Reif (1979), and Schwartz and Sharir (1983a,b). There are two natural ways in which linkage movement problems arise in robotics. First, a linkage can model a robot arm. A frequently encountered model consists of a sequence of links connected together consecutively at movable joints. Second, linkages can also model hinged objects being moved by an arm or other type of manipulator. In both cases, it is essential to plan collision-avoiding paths of motion, as the manipulator and the object it is moving are generally required to lie within regions whose boundaries are determined by walls and the presence of other objects in the work space.

A *linkage* is a collection of rigid rods called *links* (see Figure 1.1). The endpoints of various links are connected by joints, each joint connecting two or more links.

This research was supported in part by ONR contract N00014-76-C-0018, NSF grant MCS81-01220, an NSF Postdoctoral Fellowship, and a Dartmouth College Junior Faculty Fellowship.

(a)

(b)

Figure 1.1.   (a) A planar linkage (b) An arm in the plane

The links are free to rotate about the joints. In a *planar* linkage, links are allowed to cross over one another, and the linkage may be fastened to the plane so that the locations of certain joints are fixed (the fixed joints are indicated by ''*'''s in the figures).

In a physical realization of a planar linkage, each link could move in a separate plane parallel to the ground. If links were joined together or to the ground by pins, then a link in one plane might collide with a pin joining links in two other planes. However, it is not difficult to design simple devices that function like pins but that do not interfer with the motions of the linkage. Thus the mathematical model in which links cross over one another and in which the locations of some joints are fixed can be physically realized.

An *arm* is a simple type of linkage consisting of a sequence of links joined together consecutively with the location of one end fixed.

Suppose that an arm is required to stay inside some given region R of the plane. It is a natural question to ask whether new links can be adjoined to the arm in such a way that the original links in the arm automatically stay inside R. The new links may move outside R, and some of the links may have an endpoint fixed in the plane. The key requirement is that no motion of the arm inside R be prevented by the addition of the new links. Figure 1.2 shows how this can be done if R is a circular region.

The reason that we are interested in reductions of this sort is that the motions of

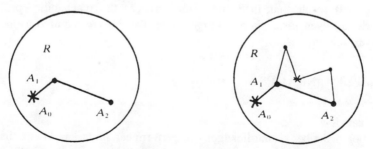

Figure 1.2.   An arm confined to a circular region. Connecting joints A_1 and A_2 to the center of the circle by two-link ''elbows'' keeps the arm in R.

the new linkage can be studied without reference to the region R. The first main result of our paper is that for any compact connected, but not necessarily simply connected, region R whose boundary consists of a finite set of straight line segments and any linkage L positioned within R, there is a reduction of the type we have just described. What's more the number of new links that must be added is bounded by a polynomial in the number of original links and the number of sides of R. Also the lengths and placements of the new links are easy to compute.

Our second result is that the reachability question for planar linkages is PSPACE-hard. In other words, given an initial configuration of an arbitrary planar linkage L, a joint J in that linkage, and a point p in the plane, the question of whether L can be moved so that J reaches p is PSPACE-hard.

The main technique used throughout the paper is to build complex linkages by connecting together simpler special purpose linkages. Some of these simpler linkages date from the 19th century and are described in Section 2. These include Peaucellier's straight line motion device, which is a linkage containing a joint whose locus is *exactly* a straight line segment, and linkages that translate and rotate vectors and multiply distances.

Section 3 contains an easy demonstration that a linkage required to move inside a closed bounded *convex* polygonal region R can be embedded in a more complex linkage that enforces the boundary constraint for the original linkage. The extension of this result to a linkage L constrained to move inside a nonconvex bounded region R with straight line boundaries appears in Section 4. We obtain this result by triangulating the complement of R in its convex hull H, designing a linkage that contains a joint whose locus is a triangle, and then using this device to build a linkage that can keep a link entirely outside a triangle. By keeping each link of L outside each triangle in $H-R$ while requiring each joint of L to remain inside H, we keep L inside R. We do this in such a way that the motion of L is not restricted in any other way.

Section 5 contains our other main result, that the reachability question for planar linkages is PSPACE-hard. We obtain this result by designing a linkage that can simulate a linear bounded automaton (LBA). The result should be compared to Reif's (1979) result that in 3-dimensional space, the reachability problem is PSPACE-hard even for a simple, hinged, tree-like linkage required to move in a nonconvex region.

2. SIMPLE LINKAGES

2.1. Overview

This section describes planar linkages that perform certain tasks. After a discussion in Section 2.2 of Peaucellier's straight line motion linkage, we show in Section 2.3 how to use this device to build linkages that can translate and rotate vectors. Then in Section 2.4 we use these devices to give a modified version of Kempe's construction of a linkage that "solves" a multivariable polynomial equation (Kempe, 1876). This linkage has certain joints whose positions represent values of variables $x_1, \ldots, x_n$, and the only constraint that the linkage puts on the motion of these joints is that the implied values of the x_i stay within given bounded domains and satisfy a given polynomial equation.

The linkage for solving a polynomial equation plays an important role in both the main results. We use it to keep links outside of triangular regions when we show how to build boundary constraints into a linkage in Section 4. We also use it to synchronize the motions of the LBA simulator given in Section 5.

Two important subtleties arise in designing special purpose linkages. First, we often want to construct a linkage L having a joint J whose locus is some specified set of points. It is important to understand that in such a case, L must be able to move to all points in the set but to *no* other points. Historically, some linkages that have been proposed for performing certain tasks have been faulty because, while they are able to move in some desirable way, they can also move to "configurational singularities" at which they can begin undesired motions. Hence for the sake of completeness we include redesigned versions of these linkages that avoid this problem.

The second important subtlety is this. Suppose that the locus of some joint J in linkage L is a set of points S and that the locus of some joint J' in linkage L' is a set S'. Now suppose that J and J' are identified. It is not necessarily true that the joint $J=J'$ can then reach all points in $S \cap S'$. Indeed $S \cap S'$ need not be connected! We have been careful to avoid this pathology in designing our linkages. The crucial observation is the following. Suppose that $x(t)$ and $y(t)$ are given continuous functions of time. When a new linkage is formed from L and L' by identifying joints J and J', this new linkage can move so that the position of $(J=J')$ is given by $(x(t), y(t))$ if and only if L can move so that the position of J is given by $(x(t), y(t))$ and L' can move so that the position of J' is also given by the same $(x(t), y(t))$. This observation should be kept in mind when checking that the linkages we build up from smaller pieces function as claimed.

2.2. Peaucellier's Straight Line Motion Linkage

In 1864, Peaucellier designed a linkage, shown in Figure 2.2.1, that converts circular motion to linear motion. Links AD, AB, DC, and BC have equal length as do links EA and EC. The length of FD equals the distance from E to F. The locations of joints E and F are fixed points in the plane, but the linkage is allowed to rotate about these points. As it does, the joint B traces out the line segment XY. This can be seen by observing two facts. First, joints D and B always lie on a ray through E. Second, the distances h, r and t shown in Figure 2.2.2. satisfy $h^2 = a^2 - t^2 = b^2 - (r+t)^2$, where r is the distance between D and E. Consequently the distance s between E and B is such that rs is equal to the constant $b^2 - a^2$. (Here h, r, s and t are functions of the position of D.) Hence, this device can be thought of as performing the well-known mapping called "inversion with respect to a circle" (Eves, 1963). In this mapping, the image

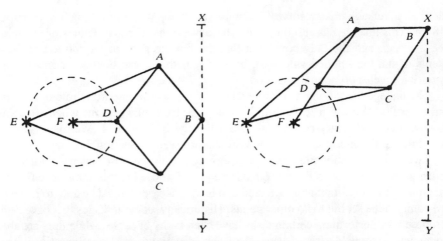

Figure 2.2.1.   The Peaucellier straight line motion linkage

of a point $p = (r, \theta)$ is the point $p' = (r', \theta)$ where rr' is some given constant. It is known that this mapping takes circles to circles, where a straight line is regarded as a circle of infinite radius. Suppose that the joint E of the Peaucellier device is at the origin of the polar coordinate system and that the given constant is $b^2 - a^2$. Then the device computes the images of the points that D can reach. Since the circle of radius $|FD|$ about F goes through the origin, this circle is mapped to a straight line, in particular the line through X and Y. The points X and Y represent the extremes that B can reach.

The relative lengths of the links are not important provided that the linkage can

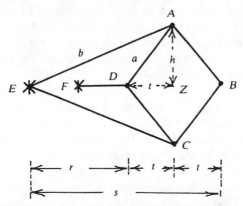

Figure 2.2.2. Consideration of triangles
EAZ **and** *DAZ* **shows that** $h^2 = a^2 - t^2 = b^2 - (r+t)^2$, **where** h **is half the distance between** A **and** C, t **is half the distance between** D **and** B, **and** r **is the distance between** E **and** D.

be assembled as shown in Figure 2.2.1 with E,F,D and B on a straight line. In order to argue that the Peaucellier device works correctly we must demonstrate that joint B cannot reach joint D, for if this could occur, the joint B could leave the line segment XY and trace out part of the circle that D traces. Similarly we must demonstrate that joint A cannot reach joint C. Joint B cannot reach joint D since the line XY does not intersect the circle of radius $|FD|$ centered at F. To see that joint A cannot reach joint C, suppose that B moves along XY toward X, say. The parallelogram $ABCD$ begins to collapse: diagonal DB lengthens, and diagonal AC shortens. Also, link EA moves counterclockwise about E, thereby increasing the distance from A to F. Consequently, the collapsing of the parallelogram is stopped when angle FDA straightens, preventing further counterclockwise rotation of EA. This occurs when B reaches X, as shown in Figure 2.2.1. The reader is referred to Eves (1963) for a more detailed discussion.

Now we point out some consequences of a simple modification of the Peaucellier device that we will need in Section 3, where we will describe how to confine a linkage to the inside of a convex polygon.

Suppose that the Peaucellier linkage is modified by adding a new link BG. As joint B travels up and down line segment XY, link BG can rotate freely about B. Clearly the set S of points that joint G can reach is the union of a rectangle and two discs (see Figure 2.2.3). Note that G can follow any curve that stays inside S but avoids the discs centered at X and Y. (G cannot move inside the discs when B is at X or Y.) Consequently, when we are faced in Section 3 with the problem of designing

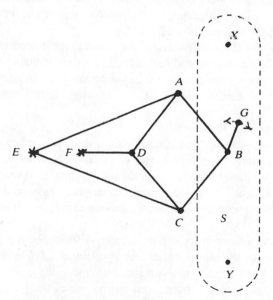

Figure 2.2.3. The region of points reachable by the modified Peaucellier linkage.

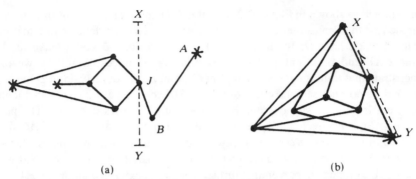

Figure 2.2.4.   (a) An arm *ABJ* whose endpoint *J* is moving in a slot. (b) A slot on
a platform.

a modified Peaucellier linkage whose joint *G* must be able to move freely inside
some given polygonal region *R,* we simply scale the linkage shown in Figure 2.2.3
by an appropriate constant so that region *R* fits inside set *S* with the discs removed.

We will take advantage of the fact that *S* has a straight line segment in its
boundary by placing the modified Peaucellier linkage so an edge of the polygon *R*
lies along a boundary line of *S*. This will keep joint *G* from crossing that edge of *R*.

We will frequently use the Peaucellier linkage to constrain a joint *J* of some other
linkage to a line. This can be done by identifying the joint *J* with the joint of the
Peaucellier linkage that moves on a line. When we do this identification, we say that
J is moving in a *slot*. The geometry and positioning of the Peaucellier linkage
determine the length and position of the slot.

Also observe that the two points of the Peaucellier device that are normally
attached to the plane could instead be attached to a rigid structure made up of links
that is free to move in the plane. In this situation the slot itself has allowable
motions, and we say that the slot is on a ''platform''. (See Figure 2.2.4.)

2.3. Translators and Rotators

We will need linkages to perform certain basic tasks. Since many of the previously
published constructions have deficiencies of the sort described earlier, we include
correct versions of these linkages. We do not attempt to construct the simplest
linkage for a task, but rather one that is conceptually easy to understand and to
prove correct. Throughout this section, we assume that *R* is a given closed bounded
planar region.

The first device we construct is a *translator*. A translator is a linkage such that
the only restriction on the movement of four of its joints *S,T,U* and *V* in the region *R*
is that the position of *T* relative to *S* remains the same as the position of *V* relative to
U. Alternatively, any three of these joints can be moved freely, and the position of
the fourth joint is uniquely determined by the above relation and the position of the
other three.

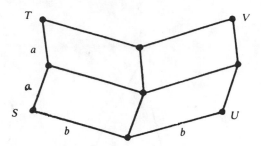

Figure 2.3.1. A faulty translator

The linkage consisting of four parallelograms shown in Figure 2.3.1 is a natural candidate for a translator. Joints T, U, and V can be moved to any three points in the plane, provided the distance between S and T does not exceed $2a$ and the distance between S and U does not exceed $2b$. At first it appears that the position of V relative to U is always the same as the position of T relative to S, i.e., that the vector ST is equal to the vector UV. The difficulty is that one or more of the parallelograms may convert to a contraparallelogram (see Figure 2.3.2), and thus other motions are possible.

One might attempt to overcome this difficulty by attaching to each diagonal of the parallelograms a sufficiently short two-link segment. This would keep a parallelogram from straightening. Unfortunately, this also prevents the movement of T to S when S is held fixed, and this motion is essential in a construction of Kempe's that we use. We solve the problem by using a more complex device involving nine parallelograms. The linkage shown in Figure 2.3.3 will be part of this device. The lengths of the links A_1B_1, B_1C_1 and C_1D_1 can be chosen long enough so that, no matter where A_1 is positioned inside the bounded region R, D_1 can move freely in R while A_1 is kept fixed and C_1 is constrained to move on a line l through A_1 by means of a *slot* (see Section 2.2). In fact, if the links are sufficiently long, then the slot can

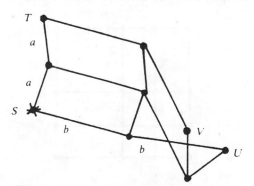

Figure 2.3.2. Conversion of a parallelogram to a contraparallelogram

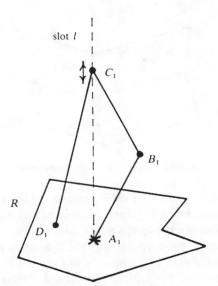

Figure 2.3.3. Keeping links nearly "vertical". (The two-link connection between A_1 and D_1 is not shown.)

be constructed so that the angles between l and links A_1B_1 and B_1C_1 do not exceed 30° no matter how D_1 moves in R. (Of course, the joints B_1 and C_1 are outside R, but this does not concern us, as we will never need to attach them to the joints of a linkage required to stay inside R.) Also note that the angle between C_1D_1 and l can be kept to at most 30° by the addition of a two-link segment connecting A_1 to D_1 and having length equal to the diameter of R. A similar linkage with joints A_1, A_2, A_3, and A_4 can be constructed so that A_4 can move freely in R while the angles between A_1A_2, A_2A_3, and A_3A_4 and another line l' through A_1 are kept within 30°. It is convenient to choose l' perpendicular to l since the links in the segments between A_1

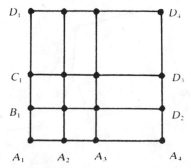

Figure 2.3.4.   The predecessor of a translator.

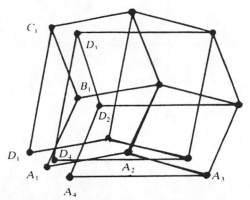

Figure 2.3.5.   A translator. (The constraining devices attached to A_1, . . . ,D_1 and A_1, . . . ,A_4 are not shown. The picture is planar, not 3-dimensional.)

and D_1 and between A_1 and A_4 will appear as sides of parallelograms in our nine-parallelogram translator. The fact that these links can be kept nearly parallel to l and l' will prevent any of the nine parallelograms from straightening. In this way, we avoid the flaw in the faulty four-parallelogram translator.

To construct the main body of the translator, begin with the nine-parallelogram linkage shown in Figure 2.3.4. The three-link segments connecting A_1,B_1,C_1, and D_1 and A_1,A_2,A_3, and A_4 are not yet constrained as described above. Notice (by applying the parallelogram law of vector addition) that it is possible to move these segments independently of each other without breaking links or creating contraparallelograms, although parallelograms may straighten. As long as no contraparallelograms are created, the position of D_1 relative to A_1 is the same as the position of D_4 relative to A_4. Now, move D_1 and A_4 (and hence D_4) to A_1, as shown in Figure 2.3.5, and then attach the constraining devices described in the discussion of Figure 2.3.3 to A_1,B_1,C_1, and D_1 and to A_1,A_2,A_3, and A_4. Joints D_1 and A_4 can still move freely in R, but the links in the segments between A_1 and D_1 and between A_1 and D_4 must remain nearly parallel to l and l', preventing the formation of contraparallelograms.

The next device we construct is called a *rotator*. A rotator is a linkage such that the only restriction on the movement inside of R of three of its joints A, I, and H is that the distance from A to I be equal to the distance from A to H. In this construction, we begin with the quadrilateral linkage $ABCD$ shown in Figure 2.3.6a. The lengths of the sides of $ABCD$ satisfy $|AD|=|AB|<|CD|=|CB|$. Then we constrain C to a slot through A. We want to insure that the slot through A always bisects the angle DAB. We also want to insure that links AD and AB can rotate freely about A, so it is necessary that $ABCD$ be able to straighten to allow links AB and AD to cross over each other. However, when B coincides with D, B and D must not be allowed

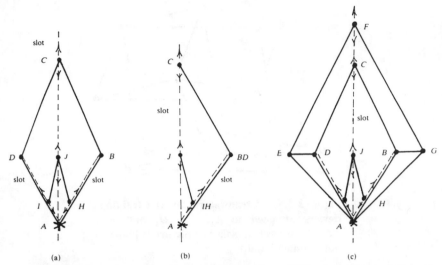

(a)   (b)   (c)

Figure 2.3.6.   A distance rotator. Without the addition of the quadrilateral
AGFC, *D* **could move to** *B* **and the two superimposed joints could then move to**
the same side of the slot through *A* **and** *C*. **Hence the slot through** *A* **and** *C* **would**
no longer bisect angle *DAB*.

to simultaneously move off the line *AC* in the same direction (see Figure 2.3.6b). If
this happens, the slot through *A* would no longer bisect the angle *DAB*. To solve this
problem, we construct another quadrilateral *AGFE* with link lengths satisfying
$|AG|=|AE|<|FG|=|FE|$ and, also, $|AE|+|EF|>|AD|+|DC|$. Then we constrain *F* to
move in the slot through *A* in which *C* moves. Finally, we join the quadrilaterals by
adding links *ED* and *BG* (see Figure 2.3.6c). Now *D* and *B* can rotate freely about *A*
(for an appropriately designed slot), but *ABCD* must be straight whenever *B* and *D*
coincide. Hence, *B* and *D* cannot move to the same side of the slot through *A*, and
the slot remains the bisector of angle *DAB*.

Now we attach "platforms" to *AD* and *AB* (as shown in Figure 2.2.4), and slots
that coincide with *AD* and *AB*. We then add links *IJ* and *HJ*, where *I* is constrained
to move in the slot along *AD*, *H* is constrained to move in the slot along *AB*, and *J* is
constrained to the slot in which *C* and *F* move. Since triangles *AIJ* and *AHJ* are
always congruent, the distance between *A* and *I* must equal the distance between *A*
and *H*. Note that this is the only constraint on the motion of *I* and *H*. This completes
the construction of a rotator.

Now we combine a translator that keeps the relative position of *T* to *S* equal to the
relative position of *V* to *U* (but does not otherwise restrict their motions inside *R*)
with a rotator that keeps the distance between *A* and *H* equal to the distance between
A and *I* (but does not otherwise constrain their motions inside *R*). We do this simply
by identifying *A* with *U* and *H* with *V*. The result is a linkage containing four joints
S,*T*,*A*=*U* and *I* whose motions inside R must satisfy only one requirement, that the

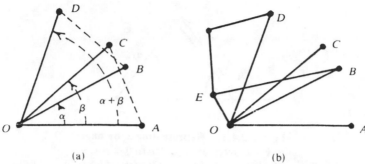

(a)   (b)

Figure 2.3.7.   An angle adder. In b), the right triangle *EOB* has been added, together with a two-link segment connecting *E* to *D* of length $|EB|$.

distance between S and T be equal to the distance between $A=U$ and I. We call this device a *distance copier*.

A distance copier can be used to construct an *angle adder*. An angle adder is a linkage containing four equal-length links OA, OB, OC and OD whose motions are constrained only by the requirement that angle AOD be equal to angle AOB plus angle AOC. We will only need a device that correctly adds angles AOB and AOC when angle AOC is less than π. To construct such a device, we take four equal-length links OA, OB, OC and OD and attach a distance copier to A, B, C and D that keeps the distance between A and C equal to the distance between B and D. Then to insure that angle AOC is added to angle AOB rather than subtracted from it, we add two links OE and EB to form a triangle with a right angle at O. Now we connect E to D with a two-link segment of length $|EB|$ (see Figure 2.3.7). These additional links constrain D to be on the correct side of the line OC.

2.4.   Linkages for Multiplication

In the late 1800's, Kempe (1876) showed how to construct linkages to "solve" multivariable polynomial equations. We will make important use of a modified version of his construction. Given a set of variables $x_1, x_2, \ldots, x_n$ with bounded domains and a polynomial equation $p(x_1, x_2, \ldots, x_n) = 0$, we can design a linkage that will force the x_i to satisfy the equation.

Consider the links AB and BC of equal length shown in Figure 2.4.1. Joint A is fastened to the plane, and joint C moves in a slot on the x-axis. The position of C represents the value of a variable x whose domain is determined by the slot. The length of the slot is such that B cannot straighten. Additional links, whose description we omit, can be added to insure that AB remains vertical when $x=0$. Then $x=a$ cos α. Thus for a fixed value of a greater than max $|x|$ the variable x can be represented by the angle α, where $0 < \alpha < \pi$.

We can now rewrite the polynomial equation, expressing each x_i as $a\cos\alpha_i$. Replace products of cosines by cosines of sums of angles using the formula

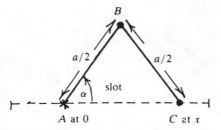

**Figure 2.4.1.   Representing x by angle
α: $x = a \cos \alpha$. Max $|x| < a$ so $0 < \alpha < \pi$.
The linkage that keeps AB vertical when
$x = 0$ is not shown.**

$$\cos\alpha \, \cos\beta = \frac{1}{2}(\cos(\alpha + \beta) + \cos(\alpha - \beta))$$

thereby reducing the equation to the form

$$a_0 + \Sigma a_i \cos\theta_i = 0,$$

where each θ_i is a sum of α_i's.

Using the technique for adding angles described in the previous section, we can design a linkage that constructs each θ_i from the α_i's. Recall that the construction for adding angles works correctly as long as the second summand is in the range $[0, \pi]$, and since joint B cannot straighten, this condition is satisfied by the α_i's. The terms $a_i \cos\theta_i$ can be summed by constructing a sequence of links $L_1, L_2 \ldots$ of lengths $a_1, a_2, \ldots$ connected together at their end points and making each link L_i form the angle θ_i with the horizontal by using a translator. The translator is attached to the end points of L_i and to the end points of another link of length a_i that is rigidly attached to the moving side of the angle θ_i. Finally, the free joint of the last link is constrained to a slot on the vertical line $x = -a_0$.

Note that, for all motions of the linkages, $p(x_1, \ldots ,x_n)=0$. Furthermore, for each choice of x_i' s solving the equation, the linkage can move to a configuration that represents this choice. The number of links in the straightforward implementation of Kempe's idea can be exponential in n because the summation may have exponentially many terms. However, we need the Kempe construction to enforce only two particular equations, so this problem does not concern us.

One of the equations that we are interested in is $x_1x_2x_3=0$. This equation states that at least one of x_1, x_2 or x_3 must be zero. Substituting $\cos\alpha_i$ for x_i reduces the equation to

$$\cos(\alpha_1+\alpha_2+\alpha_3)+ \cos(\alpha_1+\alpha_2-\alpha_3)+ \cos(\alpha_1-\alpha_2+\alpha_3)+ \cos(\alpha_1-\alpha_2-\alpha_3)=0.$$

Note that in terms of the previous notation, $a_0=0$ and $a_1=a_2=a_3=a_4=1$. Figure 2.4.2. is a simplified picture of how this equation can be mechanically solved. If

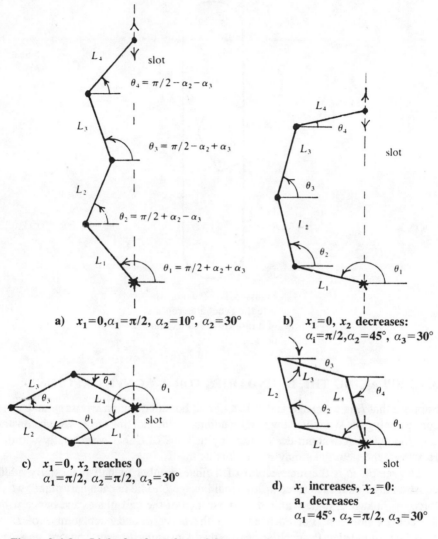

a) $x_1=0, \alpha_1=\pi/2,\ \alpha_2=10°,\ \alpha_2=30°$

b) $x_1=0,\ x_2$ decreases:
$\alpha_i=\pi/2, \alpha_2=45°,\ \alpha_3=30°$

c) $x_1=0,\ x_2$ reaches 0
$\alpha_1=\pi/2,\ \alpha_2=\pi/2,\ \alpha_3=30°$

d) x_1 increases, $x_2=0$:
a_1 decreases
$\alpha_1=45°,\ \alpha_2=\pi/2,\ \alpha_3=30°$

Figure 2.4.2. Links for the polynomial $x_1x_2x_3=0$
$\theta_1 = (\alpha_1 + \alpha_2 + \alpha_3),\ \theta_2 = (\alpha_1 + \alpha_2 - \alpha_3),\ \theta_3 = (\alpha_1 - \alpha_2 + \alpha_3),\ \theta_4 = (\alpha_1 - \alpha_2 - \alpha_3)$

$x_1=0$, meaning that $\alpha_1=\pi/2$, then the links L_1 and L_4 must be oriented so that $\theta_4 = \pi-\theta_1$. Similarly links L_2 and L_3 must be oriented so that $\theta_3=\pi-\theta_2$. If $x_2=0$ then L_1 and L_3 must be parallel as must L_2 and L_4. All these conditions are met when $x_1=0$ and $x_2=0$. At this point, the shape of the figure changes from that in (a) and (b) to a parallelogram, shown in (c).

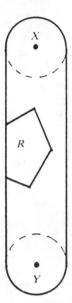

**Figure 3.1.   Polygon in-
side the reachable region
of a modified Peaucellier
device.**

3.   REPLACING THE BOUNDARIES FOR A CONVEX REGION

Suppose that L is a linkage and R is a closed bounded region whose boundary is a
convex polygon. We will show that by adding additional links to L we can constrain
L to the region R without destroying any motions of L that were totally within R.
However, the new links may move outside R.

The region R is the intersection of a finite number of half-planes. By adding
constraining linkages to force the original linkage L to lie in each half-plane, we can
force the linkage L to lie within the intersection of the half-planes and hence within
the convex polygon. For each edge of the polygon and each joint J of L, we
construct a modified Peaucellier device that constrains J to the appropriate side of
the edge (see Figures 2.2.3 and 3.1). The device does not interfere with the motion
of J inside the polygon provided that the polygon avoids the discs centered at X and
Y. Clearly, this will constrain the linkage L to remain within R. However, we must
show that we have not restricted the allowable motions of L. As pointed out earlier
at the end of Section 2.1, identifying joint J_1 of one linkage with joint J_2 of another
may restrict the movement of $J_1 = J_2$ to a region smaller than the intersection of the
original reachable regions of J_1 and J_2. In fact the intersection may not even be a
connected region. Recall that the subtle point that one must consider is that even
when the intersection is connected, the joints still may not be able to reach all points

in the intersection since the possible paths the joints can follow may not be compatible.

However, in this particular case, each Peaucellier device constraining J to a region bounded in part by a side of R allows J to move along any curve in R. Hence, the allowable motions of a joint J of L are not restricted by the addition of the devices.

The number of Peaucellier devices needed is equal to the product of the number of joints of L and the number of sides of the polygon. The lengths of the links in each Peaucellier device can be determined simply by choosing an appropriate scaling factor for Figure 2.2.3. Consequently, the description of the new linkage is polynomial in the size of the description of the original linkage L and region R.

4. REPLACING THE BOUNDARIES FOR A NONCONVEX REGION

4.1. Overview

In this section, we show how to incorporate into a linkage L the boundaries of an arbitrary bounded region R whose boundary consists of a finite number of straight-line segments. Here two problems arise. First, the region is not simply the intersection of half-planes. Second, constraining the end-points of a link to be in a region does not necessarily constrain the entire link to be in the region. In Figure 4.1.1, even if A and B are constrained to lie within the region R, the link AB may be partially outside the region.

To handle these problems, let H be the convex hull of the region R. The region H-R can be quickly partitioned into a small set of triangles, the number of triangles being polynomial in the number of line segments in the boundary. (See Eves, 1963.) If we can exclude a link from a triangle without otherwise restricting its

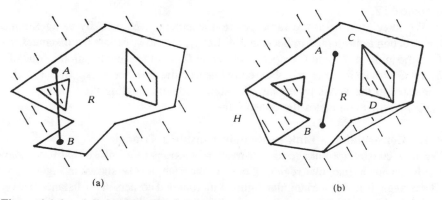

(a)   (b)

Figure 4.1.1.   A link with endpoints in a nonconvex region. The link must be kept out of each triangle in the triangulation of $H-R$. In b), CD is an edge of the triangulation.

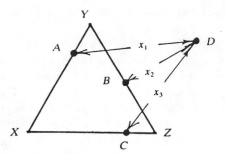

Figure 4.2.1. Forcing D to trace the boundary of a triangle XYZ

motion, then we can restrict a link to R without restricting its motion. (To keep a link from sliding along an edge of the triangulation, we can cover the edge with two additional triangles that lie in $H\text{-}R$ and then require that the link remain outside those triangles also.) Applying the construction to each link of L will solve the problem.

In Section 4.2, we describe a linkage for tracing a triangle and then use this construction in Section 4.3 for constraining a link to remain outside a triangle.

4.2. A Linkage for Tracing a Triangle

In order to construct a linkage that can reach all points in the closed exterior of a triangle, we begin by constructing a linkage that traces the boundary of a triangle. Suppose that we are given a triangle XYZ. Then we can construct three straight-line motion linkages with designated joints A, B, and C such that the joints A, B, and C move along the segments XY, YZ, and XZ respectively (see Figure 4.2.1). We would like to construct a fourth linkage with a designated joint D such that D must be at the same position as either A, B, or C. Then provided D can move freely subject to the above constraint, we will have constructed a linkage that traces the triangle XYZ.

We force D to be at the same position as either A, B, or C by using Kempe's construction as presented in Section 2.4. Let x_1, x_2, and x_3 denote the distances from D to the joints A, B, and C respectively. Joint D is at the same location as one of A, B, or C provided $x_1 x_2 x_3 = 0$. For convenience, the distance x_1, x_2, and x_3 can be translated to the x-axis by means of distance copiers. Adding the linkage to force $x_1 x_2 x_3 = 0$ then completes the construction.

4.3. Constraining a Link to Remain Outside a Triangle

We now construct a linkage to constrain the motions of a link so that it can move freely outside a triangular region. Consider the triangle XYZ shown in Figure 4.3.1. The triangle is inside a triangular figure with rounded corners. The distance between parallel edges of the inner triangle and the outer is d. The corners of the outer triangle have been replaced by circular arcs of radius d centered at the vertices of the inner triangle.

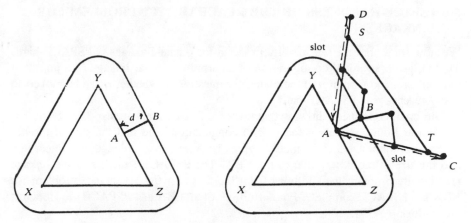

Figure 4.3.1.   Constraining a link to remain outside a triangle

Using the construction given in Section 4.2, we can constrain a joint A to the boundary of XYZ, and, using a similar construction, we can constrain a joint B to the boundary of the outer triangular figure. We can connect A and B with a link AB of length d. The possible motions of the link AB consist of rotating about the inner triangle but always remaining perpendicular to an edge, except at the vertices. At a vertex, the link AB can rotate from a position perpendicular to one edge to a position perpendicular to the other.

We now add two additional links AC and AD at joint A and two-link segments connecting B to AC and AD. The lengths of the segments when fully extended are designed to force the angles BAD and BAC to be in the range $\left[-\dfrac{\pi}{2}, \dfrac{\pi}{2} \right]$. Thus AD and AC are forced to lie outside triangle XYZ.

Attached to the links AD and AC are platforms that contain slots coinciding with AD and AC. The end points of the link ST that we wish to exclude from the triangle XYZ move in these slots. Clearly, ST can never enter the triangle since its end points are always in a half-plane whose boundary is a line through A perpendicular to AB. The triangle XYZ is outside this half-plane and thus, by convexity, ST does not intersect the triangle XYZ.

We must show that the motions of link ST are not further restricted as long as ST does not move far from the triangle. Since ST is completely contained within a half-plane associated with the perpendicular to AB passing through A, we can fix A at any point on the triangle and move T by rotating D about A and sliding T along AD. The movement of S is obtained by analogous use of AC. Since S and T are confined to slots along AD and AC, the lengths of AD and AC must be long enough to allow S and T to move as desired.

5. PSPACE-HARDNESS OF THE REACHABILITY PROBLEM FOR LINKAGES

We now show that the reachability problem for planar linkages is PSPACE-hard. That is, given an initial configuration of an arbitrary planar linkage L, a joint J in that linkage, and a point p in the plane, the question of whether L can be moved so that J reaches p is PSPACE-hard.

Our proof consists of showing that there are linkages that are capable of simulating Linear Bounded Automaton (LBA) computations and that the size of the description of a linkage that simulates a given LBA on inputs of length n is linear in n and the size of the description of the LBA. The PSPACE-hardness of the linkage reachability problem then follows from the fact that the acceptance problem for LBA's is PSPACE-complete. For definitions of an LBA and PSPACE see Hopcroft and Ullman (1979).

5.2. Some Useful Linkages

We begin by building up a collection of simple devices that perform various functions. First, we define a *cell* to be a horizontal slot of some fixed size containing a joint. The joint represents the value of a Boolean variable. The left end of the slot indicates value 0, the right end value 1. Certain cells will be grouped together to form *registers*.

It is convenient to have a device called a *lock* that can be used to force the value of each cell in a register to be 0 or 1 and to prevent the value of any cell from changing during certain time periods. Figure 5.2.1 shows a lock attached to a register. The horizontal rectangular bar is part of the lock. The bar is attached to slots so that it can only move vertically; no rotation is possible. Attached to the bar are a number of platforms carrying vertical slots. Each joint representing a Boolean variable is attached to a link, the other end of which moves in one of the vertical slots. The links are designed so that when the bar is in the lower, unlocked position the Boolean variable joint can move freely in its cell because the other end of the link can move up and down the vertical slot. When the bar is in the upper, locked position each Boolean variable joint is in a 0-1 position. Note that these variable joints cannot move when the bar is up.

In order to coordinate the linkage motions that take place during the simulation of two moves of the LBA, we design a *sequence controller* with five variables s_1, s_2,

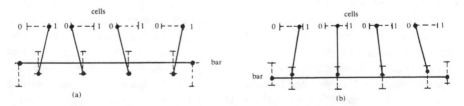

Figure 5.2.1.   A lock on cells of a register (a) locked (b) unlocked

Table 5.2.1. Possible values for variables.
Dashes denote arbitrary values between
0 and 1.

s_1	s_2	s^3	l_1	l_2
0	0	0	1	—
—	0	0	1	1
1	0	0	—	1
—	0	0	0	1
0	0	—	0	1
0	0	1	—	1
0	—	1	1	1
0	1	1	1	—
0	—	1	1	0
0	0	—	1	0

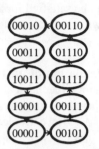

Figure 5.2.2. The allowable sequence of values

s_3, l_1, and l_2. Each variable is represented as a joint in a slot. We restrict the values that the variables can assume by adding a Kempe linkage to force

$$[s_1^2+s_2^2+s_3^2+(1-l_1)^2][s_2^2+s_3^2+(1-l_1)^2+(1-l_2)^2] \cdots [s_1^2+s_2^2+(1-l_1)^2+l_2^2]=0.$$

The consequence of this equation is that the only possible values the variables can assume are those shown in Table 5.2.1. The restriction on the value of the variables allows only one variable to change value at a time, and the variable that can change value is determined by the values of the remaining variables. As a result, the only allowable sequence of changes from one set of 0-1 values to another is that shown in Figure 5.2.2. Of course, the changes can reverse at any time. We will use the values of these variables to control certain events, thereby sequencing the order in which the events can take place. In particular, the variables l_1 and l_2 will control locks, and the s's will sequence the order in which these locks are opened and closed.

Since we represent values of variables by positions of joints, we will often use the words "joint" and "variable" interchangeably. Also, we will denote a variable and the joint that represents it with the same symbol.

The next device we need is a gate for NOT and a gate for AND. To obtain negation, we use the distance copier of Section 2.3 to force the distance of a joint from one end of a slot to be the same as the distance of another joint from the opposite end of its slot. Thus when one cell has value 0, the other has value 1 and vice versa.

To construct an AND gate, we force the product of the distances of two joints from the 0-ends of their slots to equal the distance of a third joint from the 0-end of its slot. Let x_1, x_2, and x_3 be these distances. Then when x_1 and x_2 both have 0-1 values, $x_3 = x_1 AND x_2$.

Using these linkages it should be clear that we can construct a linkage to compute any Boolean function. However, to make it easy to check the correct behavior of the linkage we must be careful not to form a loop by using the output of a gate as an input to one of its predecessors. This might cause the linkage to be rigid since the loop might imply a relationship between the rates of motion of certain joints that would not be satisfied for any nonzero rate. Our design will contain only two loops, and we will use a decoupling mechanism with them to insure that the entire linkage does not jam.

5.3. Simulation of an LBA

One idea for a mechanical simulation of a given LBA, M, is the following. Suppose that we have two registers that can be used for storing instantaneous descriptions (ID's) of M. Since Boolean variables are modeled by joints moving in slots, the contents of a register at a given time will not necessarily be a sequence of 0's and 1's. However, we will design a linkage connecting these two registers so that whenever the contents of both registers are sequences of 0's and 1's (i.e., whenever both registers contain ID's), the ID in one represents the result of a legal move of M from the ID in the other. We would also like the linkage to have the property that as M makes its moves, its ID's appear alternately in one register and then in the other. In this way, we can simulate the operation of an LBA. Since we are only interested in the reachability problem, however, we do not need to build a linkage that actually simulates M; rather, we only need a linkage which is *able* to simulate M. The linkage could make other moves as well, provided that it never moved a certain joint J to a point p representing an accepting state of M by accident. The linkage we are about to construct can simulate M, but in addition, it can undo and then redo sequences of moves. Because of this we will assume that M is deterministic and has no move from any accepting state.

We begin the construction with the two registers R_1 and R_2 used to store the ID's of M. Attached to the cells of the registers are two Boolean circuits constructed from NOT and AND gates. The output f_1 is true whenever the ID in register R_2 follows from the ID in register R_1 by one move of the LBA. The output f_2 is true whenever the ID in register R_1 follows from the ID in the register R_2 by one move of the LBA.

The variables l_1 and l_2 in the sequence controller described in Section 5.2 are connected to locks on registers R_1 and R_2, respectively, with $l_i=1$ when its lock is in the closed position. The variables s_1 and s_2 are connected to the outputs f_1 and f_2 by the linkage in Figure 5.3.1. Joints f_1 and f_2 are free to move when s_1 and s_2 are 0. However, s_1 or s_2 can move to value 1 only if f_1 or f_2, respectively, has value 1.

Initially, R_1 holds the configuration of the LBA at time zero, and $s_1=0$, $s_2=0$, $s_3=0$, $l_1=1$ and $l_2=0$. This corresponds to the first entry in Table 5.2.2.

We now describe a sequence of events that simulates the behavior of M on a given input. Since l_2 is initially 0, the variables in R_2 can move freely. In particular, they can move to the ID of M after its first move. Then l_2 can move to a locked position, i.e., l_2 can take on the value 1. Note that as variables in R_2 were changing values, f_1 and f_2 were also changing, but this is allowed since $s_1=0$ and $s_2=0$. At

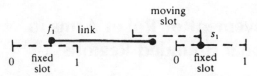

Figure 5.3.1. Decoupling mechanism. The moving slot has one end point attached to s_1, which moves in a fixed slot. s_1 cannot move to 1 unless f_1 is at 1.

this point, the sequence controller has advanced to the second state shown in Table 5.2.2 and can now advance to the third state, with $s_1 = 1$. This is because f_1 must be 1 since the ID in R_2 follows from the ID in R_1 by one move of M. Hence s_1 can move to 1.

Next, R_1 unlocks allowing s_1 to return to zero. (Note that s_i can change to zero independently of f_i's value.) Now the variables in R_1 can change to the next ID of M. Again, f_1 and f_2 must be changing while R_1 is changing, but this is permitted since s_1 and s_2 have value 0. At this point, the variable s_3 can change to 1 and then l_1 can lock. The next step is for s_2 to change value to 1. This is allowed because f_2 has value 1: the configuration in R_1 follows from the configuration in R_2 by one move of M. As soon as s_2 changes value to 1, then l_2 can unlock, and s_2 can change back to 0. Finally, s_3 can change back to 0, completing a cycle of the sequence controller. During the cycle the linkage has simulated two moves of the LBA.

Observe that the simulation may proceed forward or backward. If the simulation proceeds from ID_1 to ID_2 and then reverses, it may back up into an ID other than ID_1 since two IDs may both have the same successor ID. The only concern here is that the simulation might accidentally back into an accepting ID. This can be prevented by modifying the LBA so that no move is possible from any ID with an accepting state and then basing the design of the linkage on the modified LBA. Note that the linkage may back into a configuration corresponding to an ID of M that could not be reached from its initial state. However, since we are only considering deterministic LBAs, the linkage must move forward along the same computational path on which it has just backed up. Of course its forward progress may be interrupted from time to time by additional backing up and retracing of sequences.

Finally, another Boolean circuit is attached to the two registers R_1 and R_2. The Boolean circuit computes a 1 output whenever one of the registers is locked and contains an accepting ID. The output of this circuit is a joint J. There is a motion of the linkage that moves J to 1 if and only if the LBA reaches an accepting ID.

This completes the proof that the reachability problem for planar linkages is PSPACE-hard.

CHAPTER 13

On the Movement of Robot Arms in 2-Dimensional Bounded Regions

JOHN HOPCROFT
Computer Science Department
Cornell University

DEBORAH JOSEPH
Computer Science Department
University of Wisconsin, Madison

SUE WHITESIDES
School of Computer Science
McGill University, Montreal

The classical *mover's problem* is the following: can a rigid object in 3-dimensional space be moved from one given position to another while avoiding obstacles? It is known that a more general version of this problem involving objects with movable joints is PSPACE complete, even for a simple tree-like structure moving in a 3-dimensional region. In this paper, we investigate a 2-dimensional mover's problem in which the object is a robot arm with an arbitrary number of joints. In particular, we give a polynomial time algorithm for moving an arm confined within a circle from one given configuration to another. We also give a polynomial time algorithm for moving the arm from its initial position to a position in which the end of the arm reaches a given point within the circle. Finally, we show that 148 circles suffice to cover the boundary of the reachable region of a joint in an arm enclosed in a circle.

1. INTRODUCTION

With current interests in industrial automation and robotics, the problem of designing efficient algorithms for moving 2- and 3-dimensional objects subject to certain geometric constraints is becoming increasingly important. The *mover's problem* (see Schwartz & Sharir, 1983a,b; Reif, 1979), is to determine, given an object X, an initial position P_i, a final position P_f, and a constraining region R, whether X can be moved from position P_i to position P_f while keeping X within the region R.

In the classical problem, X is a rigid 2- or 3-dimensional polyhedral object, and R is a region described by linear constraints. Recently, several authors (Schwartz & Sharir, 1983a,b; Reif, 1979; Lozano-Perez, 1980) have presented polynomial time algorithms for solving this type of problem.

This work was supported in part by ONR contract N00014-76-C-0018, NSF grant MCS81-01220, an NSF Postdoctoral Fellowship, and a Dartmouth College Junior Faculty Fellowship.

A more difficult problem, which is related to problems in robotics, assumes that the object X has joints and is hence nonrigid. Again, one desires a fast (polynomial time) algorithm for moving X from position P_i to P_f within a region R. Unfortunately, such an algorithm is unlikely, as Reif (1979) has shown that the problem of deciding whether an arbitrary hinged object can be moved from one position to another in a 3-dimensional region is PSPACE complete.

Our paper investigates variants of the mover's problem which we believe are of practical interest. We begin in Section 2 and 3 by considering the problem of folding a *carpenter's ruler*—that is, a sequence of line segments hinged together consecutively. This problem arises because a natural strategy for moving an arm in a confining region is to fold it up as compactly as possible at the beginning of the motion. Unfortunately, deciding whether an arbitrary carpenter's ruler (whose link lengths are not necessarily equal) can be folded into a given length is NP-complete. Because of this, it turns out to be at least NP-hard to decide whether or not the end of an arbitrary *arm* (i.e., a carpenter's ruler with one end fixed) can be moved from one position to another while staying within a given 2-dimensional region.

In Sections 4–6, we consider the problem of moving an arm inside a circular region, and we are able to give polynomial time algorithms for changing configurations and reaching points. Also, we show that circles covering the boundary of the set of points reachable by a joint can be computed in polynomial time.

2. FOLDING A RULER

In this section, we ask how hard it is to fold a carpenter's ruler consisting of a sequence of n links $L_1, \ldots L_n$ that are hinged together at their endpoints. These links, which are line segments of integral lengths, may rotate freely about their joints and are allowed to cross over one another. We assume that the endpoints of the links are consecutively labeled $A_0, \ldots, A_n$ and for $1 \leq i \leq n$, we let $\mathbf{1}_i$ denote the length of link L_i. We define the RULER FOLDING problem to be the following:

GIVEN: Positive integers n, $\mathbf{1}_1, \ldots, \mathbf{1}_n$, and k.
Question: Can a carpenter's ruler with lengths $\mathbf{1}_1, \ldots, \mathbf{1}_n$ be folded (each pair of

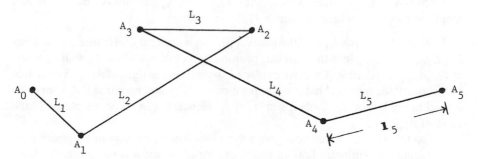

Figure 2.1.   A typical ruler with five links.

consecutive links forming either a $0°$ or $180°$ angle at the joint between them) so that its folded length is at most k?

By a reduction from the NP-complete PARTITION problem (see Garey & Johnson, 1979) we can easily show that the RULER FOLDING problem is also NP-complete. The PARTITION problem asks whether, given a set S of n positive integers $1_1, \ldots, 1_n$, there is a subset $S' \subseteq S$ such that

$$\sum_{1_i \in S'} 1_i = \sum_{1_j \in S - S'} 1_j.$$

Theorem 2.1:   The RULER FOLDING problem is NP-complete.
Proof: Given an instance of the PARTITION problem with $S = \{1_1, \ldots, 1_n\}$, let $d = \sum_{i=1}^{n} 1_i$. Then the desired subset S' of S exists if and only if a ruler with links of length $2d, d, 1_1, \ldots, 1_n, d, 2d$ (in consecutive order) can be folded into an interval of length at most $2d$. To see that this is the case, imagine that the ruler is being folded into the real line interval $[0, 2d]$, and notice that both the initial endpoint A_0 of link L_1 (the third link in our ruler) and the terminal endpoint A_n of link L_n (the third from last link) must be placed at integer d. The set S' in the PARTITION problem then corresponds to the set of links L_i whose initial endpoints A_{i-1} appear to the left of their terminal endpoints A_i in a successful folding of the ruler. $\square$

The RULER FOLDING problem and the PARTITION problem share not only the property of being NP-complete, but also the property of being solvable in pseudo-polynomial time. The time complexity of the RULER FOLDING problem is bounded by a polynomial in the number of links, n, and the maximum link length, m. In fact, it is possible to find the *minimum* folding length in time proportional to nm by a dynamic programming scheme. However, in order to carry out this scheme we need to know that a ruler with maximum link length m can always be folded to have length at most $2m$.

LEMMA 2.1:   A ruler with lengths $1_1, \ldots, 1_n$ can always be folded into length at most $2m$, where $m = \max \{1_i | 1 \leq i \leq n\}$.

PROOF: Place link L_1 into the interval $[0, 2m]$ with A_0 at O. Having placed links $L_1, L_2, \ldots, L_{i-1}$ into the interval, position L_i as follows: Place L_i with A_i to the left of A_{i-1}, if possible. Otherwise, place L_i with A_i to the right of A_{i-1}. To see that this is possible, suppose that p is the position of A_{i-1} and note that if A_i cannot be placed to the left of A_{i-1}, then $p \leq 1_i \leq m$. Hence A_i can surely be placed to the right of A_{i-1}. $\square$

Using this result, we can now give a dynamic $O(mn)$ programming algorithm for determining the minimum folding length of a ruler, where n is the number of links in the ruler and m is the maximum length of any given link.

ALGORITHM 2.1:   *Ruler Folding in Minimum Length*

Given a ruler with links $L_1, \ldots, L_n$, compute the maximum link length m. Then, for each k, $1 \le k \le 2m$, construct a table with rows numbered 0 to n and columns numbered 0 to k. Row i corresponds to endpoint A_i, and column j corresponds to the position j in the interval $[0,k]$. Fill in row 0 by writing a T in each column j for which L_0 fits in $[0,k]$ with A_1 at integer j, and F's in the other columns. Once row $i-1$ has been filled in, fill in row i by writing a T in each column j for which the linkage $L_1, \ldots, L_i$ fits in $[0,k]$ with endpoint A_i at integer j. To do this, examine row $i-1$ to obtain the possible locations for A_{i-1}. The last row of the completed table contains a T if and only if the ruler can be folded into $[0,k]$. Find the smallest k for which the table contains a T in the last row, and read the table from bottom to top to reconstruct the desired folds. $\square$

The next example shows that $2m$ is, in fact, the best upper bound for the minimum folding length.

Example 2.1: A ruler with minimum folding length $2m-\varepsilon$. Consider a ruler (see Figure 2.2) which has $n = 2k-1$ links $L_1, \ldots, L_n$. Suppose that links with odd subscripts have length m and that links with even subscripts have length $m-\varepsilon$, where $\varepsilon = m/k$. It is easy to check that this ruler cannot be folded into length less than $2m-\varepsilon$. $\square$

Having established some basic results about folding rulers, we now return to the original problem of moving such objects.

3.   MOVING AN ARM IN TWO DIMENSIONS

The remainder of this paper is concerned with moving a ruler that has one endpoint, A_0, pinned down. We will refer to such a ruler as an *arm*.

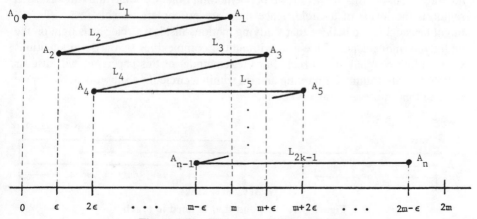

Figure 2.2.   The ruler of Example 2.1.

Unrestricted Movement

It is easy to find out what points can be reached by the free end of an arm placed in the plane. The answer is given in the next lemma, whose simple proof we omit. (The lemma extends readily to three dimensions.)

Lemma 3.1: Let $L_1, \ldots, L_n$ be an arm positioned in 2-dimensional space, and let $r = \sum_{i-1}^{n} 1_i$, the sum of the lengths of the links. Then the set of points that A_n can reach is a disc of radius r centered at A_0—unless some 1_i is greater than the sum of the other lengths. In that case, the set of points A_n can reach is an annulus with center A_0, outer radius r, and inner radius $1_i - \sum_{j \neq i} 1_j$.

Restricted Movement

If an arm is constrained to avoid certain specified objects during its motions, then determining whether A_n can reach some given point p is difficult. In the following example, we use a reduction of RULER FOLDING to show that even for "walls" consisting of a few straight line segments, this problem can be NP-hard.

Example 3.1: A hard decision problem. We want to know whether the arm shown in Figure 3.1 can be moved so that A_n reaches the given point p. The arm consists of a ruler with links of integral lengths attached to a chain of very short links. The chain links are short enough to turn freely inside the tunnel, which is sufficiently narrow that links of the ruler can rotate very little once they are inside. Since the ruler cannot change its shape very much while moving through the tunnel, it must be foldable into length at most k in order to move through the gap of width k. Thus, point p can be reached if and only if the ruler can be folded into length at most k. □

We would like to find natural classes of regions for which questions concerning the movement of arms are decidable in polynomial time. Certainly the simplest such region is the inside of a circle, since there are no corners in which an "elbow" might be caught. We believe that studying motions inside a circle sheds light on the underlying movements of the arm without the complexities that arise in situations where a link can jam in a corner. For the remainder of this paper, we will discuss polynomial algorithms for moving an arm within a circle. In a subsequent paper, we hope to treat more general situations.

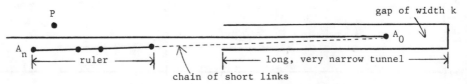

Figure 3.1. A point that is hard to reach.

4.   CHANGING CONFIGURATIONS INSIDE A CIRCLE

In this section, we solve the problem of moving an arm from one given configuration to another inside a circular region. Simply determining whether this can be done turns out to be a matter of checking that links whose "orientations" differ in the two configurations can be reoriented. This checking can be done in time proportional to the number of links. Assuming that it is feasible to change configurations, we show how to move the arm to its desired final position by first moving it to a certain "normal form" and then putting each link into place, correcting its orientation if necessary. Correcting orientation involves destroying and then restoring the positions of previous links. Our algorithm consists of a sequence of "simple motions" (which we are about to define), and the length of this sequence is on the order of the cube of the number of links.

Simple Motions
A definition of a "simple motion" is needed in order to make clear the sense in which our algorithms for moving an arm are polynomial. This definition should not limit the positions the arm can reach nor should it complicate the algorithms and proofs. With these considerations in mind, we define a "simple motion" of an arm as follows. (There are many other definitions which would give similar results.)

DEFINITION 4.1:   A *simple motion* of an arm is a continuous motion during which at most four joint angles change. (The angle between the first link and some reference line through the fixed point A_0 may be one of these.) Moreover, a changing angle is not allowed both to increase and to decrease during one simple motion.

Figure 4.1 illustrates some simple motions of the type we use. Note that in the motions shown, the joints where angles are changing are connected together by straight sections of the arm. This is true of all the simple motions we will use.

Normal Form
It is convenient to begin by showing that any arm positioned within a circle can be moved by a short sequence of simple motions into a normal form that has as many joints as possible positioned *on* the circle. We immediately dispense with the case in which the distance from A_0 to the circle is greater than the length of the entire arm, since in this case the circle is irrelevant.

Definition 4.2:   Suppose A_0 is fixed at some point distance d_0 from the circle, and suppose that j is the smallest integer such that $\sum_{i=1}^{j} \mathbf{1}_i \geq d_0$. Then the arm is in

normal form if and only if $L_1, \ldots, L_j$ contains at most one bent joint, and for each k, $j \leq k \leq n$, A_k is on the circle. Moreover, if $L_1, \ldots, L_j$ is bent, the bend is at joint A_{j-1} (see Figure 4.2). In any event, $L_1, \ldots, L_{j-1}$ lie on a radius.

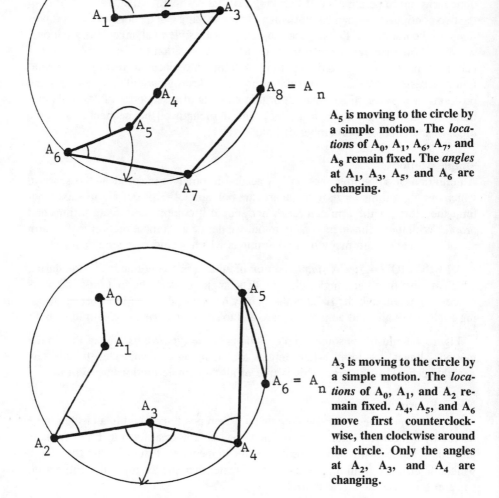

A_5 is moving to the circle by a simple motion. The *locations* of A_0, A_1, A_6, A_7, and A_8 remain fixed. The *angles* at A_1, A_3, A_5, and A_6 are changing.

A_3 is moving to the circle by a simple motion. The *locations* of A_0, A_1, and A_2 remain fixed. A_4, A_5, and A_6 move first counterclockwise, then clockwise around the circle. Only the angles at A_2, A_3, and A_4 are changing.

Figure 4.1.   Examples of simple motions.

LEMMA 4.1 (Normal Form):   For any given configuration of an arm within a circle there is a sequence of $O(n)$ simple motions that moves the arm to normal form. Moreover, this sequence can be computed in $O(n)$ time.

Proof: The process consists of two stages. First, the tail will be straightened until A_n reaches the circle. Then, starting with A_{n-1}, the other joints will be moved one by one onto the circle.

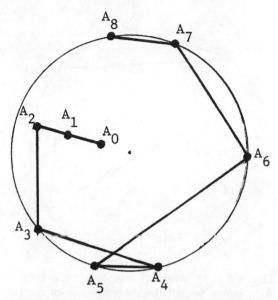

Figure 4.2. An arm in normal form. A_0, A_1, and A_2 lie on a radius. A_3 is the first joint that can reach the circle. The successors of A_3 lie on the circle.

Suppose L_j, L_{j+1}, . . . , L_n form a straight line segment. Move A_n toward the circle by rotating this segment about A_{j-1} until A_n reaches the circle or L_{j-1} is added to the straight segment. In this latter case, rotate the extended straight segment about A_{j-2}. Eventually, A_n reaches the circle or the entire arm becomes a straight segment that can be rotated about A_0 to place A_n on the circle. (Recall that we are assuming that the arm is long enough to reach the circle.) This process requires at most $O(n)$ simple motions and can be computed in $O(n)$ time.

Now assume that A_n, A_{n-1}, . . . , A_j are on the circle, and let L_i, L_{i+1}, . . . , L_j be the maximal straight segment leading back from A_j. Keeping $L_i, L_{i+1}, \ldots,$ L_{j-1} straight and the positions of A_j and A_{i-2} fixed, rotate L_j about A_j, moving A_{j-1} *away* from A_{i-2} (see Fig. 4.3). L_j is rotated until A_{j-1} hits the circle (in which case we have a new joint on the circle), or L_{i-1} is added to the straight segment $L_i, \ldots, L_{j-1}$, or A_{i-1} hits the circle. If L_{i-1} is added to the straight segment, then the process of rotating L_j is continued with the straight segment replaced by a new one containing at least $L_i, \ldots, L_{j-1}$ and L_{i-1}. If A_{i-1} hits the circle, then A_{i-1} is held fixed while the angles at joints A_{i-1}, A_{j-1}, and A_j are adjusted so as to push A_{j-1} to the circle while keeping A_j and its successors on the circle. In this way, one can force onto the circle as many joints as possible (i.e., A_j can be placed on the circle, where j is minimum such that the sum of the lengths of the first j links exceeds the distance from A_0 to the circle). Once these joints are on the circle, it is easy to position the links at the beginning of the arm as desired. This process

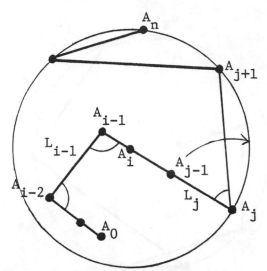

Figure 4.3. Moving an arm to normal form.
A_{j-1} **moves toward the circle** *away* **from** A_{i-2}.
The locations of A_{i-2} **and its predecessors and the**
locations of A_j **and its successors remain fixed.**
Only the angles at A_{i-2}, A_{i-1}, A_{j-1}, **and** A_j **are**
changing.

requires $O(n)$ simple motions and once again, these motions can be computed in $O(n)$ time. Thus, a total of $O(n)$ simple motions is needed to put an arm into normal form, and $O(n)$ time is needed to compute the motions. $\square$

Reorientation of Links

For any given position of an arm inside a circle, we define each link to have either "left" or "right" orientation. This is done by first observing that the straight line extension of a link L_i cuts the circle into two arcs. L_i is said to have *left orientation* if the arc on the left of the extension, viewed from A_{i-1} to A_i, is no longer than the arc on the right. *Right orientation* is defined in a similar manner (see Figure 4.4). Note that a link that is on a diagonal of the circle can be regarded as having either orientation and that a link must move to a diagonal in order to change orientation.

An obvious necessary condition for being able to move the arm from one configuration to another is that it be possible to reorient each link whose orientation differs in the two configurations. (It turns out that this condition is also sufficient.) We are about to show that determining whether a link can be reoriented is simply a matter of determining how far its endpoints can be moved from the circle.

For an arm with A_0 fixed within a circle C, let c_i and d_i denote the minimum and maximum distance that A_i can be moved from C by arbitrary motions of the arm within C. Of course, distance is measured along a radius of C, so $0 \leq c_i \leq d_i \leq d/2$, where d is the diameter of C.

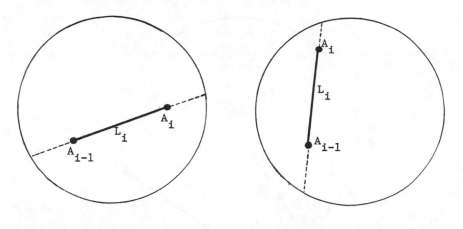

Figure 4.4. Link orientations.

Since A_0 is fixed, c_0 and d_0 are determined by the position of A_0. The Normal Form Lemma (4.1) shows that each successive A_i can get closer to the circle by the amount $\mathbf{1}_i$ until the circle is reached. Thus,

$$c_i = max\ \{c_{i-1} - \mathbf{1}_i,\ 0\}.$$

Computing the d_i's is slightly more complicated. We begin by computing for each i, $0 \le i \le n$, the maximum distance t_i that A_i could move from the circle if it were constrained only by the tail of the arm (i.e., if $L_{i+1}, \ldots, L_n$ were freed from $L_1, \ldots, L_i$ and $L_1, \ldots, L_i$ were discarded). Then we compute d_i from t_i and d_{i-1}.

LEMMA 4.2: For any arm $L_1, \ldots, L_i, \ldots, L_n$ inside a circle of diameter d.

$$t_i = \begin{cases} d/2 \text{ if no link beyond } A_i \text{ is longer than } d/2; \\ \min\{d/2,\ d - \mathbf{1}_k + \sum_{i<j<k} \mathbf{1}_j\}, \text{ where } \mathbf{1}_k \text{ is the length of the first link beyond } A_i \text{ longer} \\ \text{than } d/2\} \text{ otherwise.} \end{cases}$$

Proof: Think of the links beyond A_i as an arm with A_i fixed. Move this arm to normal form. Let A_j be the first joint on the circle. If $j \ge i+2$, the straight section of arm between A_i and A_{j-1} lies on a radius of the circle. (If $j = i$ or $i+1$, this section is just the point A_i.) While changing only the angles at joints A_{j-1} and A_j, one can push this straight section along the radius toward the circle's center while A_j and its successors move around the circle (see Figure 4.5). New links are added to the moving straight section until A_i reaches the center or the first long link L_k prevents further travel because it has folded against the straight section (or reached the diagonal in the case $L_k = L_{i+1}$). $\square$

Now that we have calculated the t_i's, it is easy to calculate the d_i's. For $i > 0$:

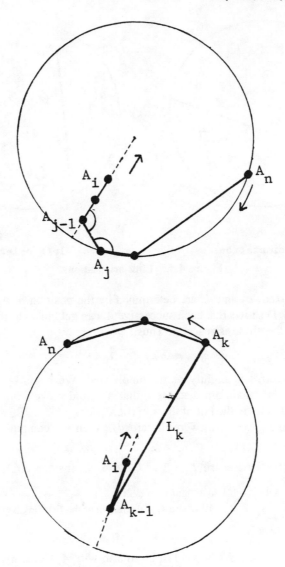

Figure 4.5.   Moving A_i distance t_i from the circle.

Above, $A_0, \ldots, A_{i-1}$ have been removed. $A_i, \ldots, A_{j-1}$ move along the radius while $A_j, \ldots, A_n$ move around the circle. Only the angles at A_{j-1} and A_j are changing.

Below, joint A_{k-1} is about to fold completely, preventing further travel of A_1 along the radius.

$$d_i = \begin{cases} \min\{t_i, d_{i-1} + 1_i\} & \text{if } 1_i < d/2 - d_{i-1}; \\ \min\{t_i, d/2\} & \text{if } d/2 - d_{i-1} \leq 1_i \leq d/2 - c_{i-1}; \\ \min\{t_i, d - 1_i - c_{i-1}\} & \text{if } 1_i > d/2 - c_{i-1}. \end{cases}$$

For any given distance x between c_i and d_i, there is obviously some way to move A_i to a position that is distance x from the circle. The point of the next remarks and lemma, which we need before we can give an algorithm for reorienting the links of an arm, is that this can be done using a short sequence of simple motions.

Remark 4.1: Suppose that the tail $L_{j+1}, \ldots, L_n$ has been detached from the arm $L_1, \ldots, L_n$. Then note that this tail can be moved from its initial position so that the distance between A_j and the circle *monotonicly* increases or decreases. To see this, put the tail (regarded as an arm with initial point A_j fixed) into normal form. Then move the straight segment of links containing A_j along the radius on which it lies, adding or deleting links from the segment as A_j gets closer to or farther from the center of the circle. ☐

Remark 4.2: Consider the arm as a whole, and suppose the tail beginning at A_j is in normal form. Then L_j can be rotated about A_{j-1} to push A_j closer to or farther from the circle while the angles at A_j and two other joints in the tail are adjusted to keep the tail constantly in normal form. In fact, Remark 4.1 shows that any rotation of L_j for which the distance between A_j and the circle is either an increasing or a decreasing function can be carried out in at most $n-j$ simple motions. ☐

LEMMA 4.3: Let A_j be a joint of an n-link arm positioned within a circle. For any x between c_j and d_j, there is a sequence of $O(n^2)$ simple motions that moves the arm from its original position to a position in which A_j is distance x from the circle.

Proof: Compute the c_i and d_i for each predecessor A_i of A_j. Then, given x, compute the sequence of numbers defined by the following recursive formula:

$$x_i = x \text{ for } i = j;$$
$$x_{i-1} = max\{c_{i-1}, x_i - 1_i\} \text{ for } 2 \leq i \leq j.$$

(Note that $c_i \leq x_i \leq d_i$.) To position A_j distance x_j from the circle, first put the entire arm into normal form ($O(n)$ steps). Then, beginning with A_1, move each A_i in turn to a position distance x_i from the circle. This is done by rotating L_i about A_{i-1} while keeping the tail in normal form. All together, at most $(n-1) + (n-2) + \cdots + (n-j)$ additional simple motions are needed, so the entire repositioning sequence contains $O(n^2)$ motions. Note that this sequence can be computed in $O(n^2)$ time. ☐

We are now ready to give the conditions under which links can be reoriented.

LEMMA 4.4: A link L_i can be reoriented if and only if at least one of the following inequalities holds:

(i) $d - 1_i \leq d_{i-1} + d_i;$
(ii) $d_i \geq 1_i + c_{i-1};$
(iii) $d_{i-1} \geq 1_i.$

Furthermore, if L_i can be reoriented, then this can be done with $O(n^2)$ simple motions that can be quickly computed.

Proof: As we noted at the beginning of this subsection, L_i must lie on a diagonal in order to be reoriented. Hence, the above conditions are obviously necessary because (i) holds when L_i is on a diagonal and the center of the circle is between A_{i-1} and A_i, (ii) holds when L_i lies on a radius with A_i closer to the center than A_{i-1}, and (iii) holds when L_i lies on a radius with A_{i-1} closer to the center than A_i.

To prove that the conditions are also sufficient, first suppose that inequality (i) holds. Using the method in the proof of Lemma 4.3, move A_{i-1} to a position distance d_{i-1} from the circle in $O(n^2)$ simple notions. If inequality (iii) holds, move A_{i-1} to a position distance d_{i-1} from the circle, again using $O(n^2)$ simple motions. After this has been done, hold A_{i-1} fixed, and rotate L_i about A_{i-1} to bring L_i to the radius through A_{i-1}. By Remark 4.2, this takes at most $n-i$ simple motions, and these can be quickly computed.

If inequality (ii) holds, then $c_{i-1} \leq d/2 - \mathbf{1}_i \leq d_{i-1}$. Move A_{i-1} distance $d/2 - \mathbf{1}_i$ from the circle, and then rotate L_i to the diagonal. $\square$

We need to make one more observation before we can show how to change configurations.

Remark 4.3: Suppose L_i is a link that can be reoriented. Then starting from any initial configuration of the arm, we can reorient L_i and with $O(n^2)$ additional motions, return $A_1, \ldots, A_{i-1}$ to their starting positions without changing the new orientation of L_i. To see this, bring L_i to a diagonal with $O(n^2)$ simple motions, and then "undo" these motions but with the orientation of L_i reversed. That is, keep the angle at A_{i-1} adjusted so that at corresponding moments before and after L_i reaches the diagonal through A_i, L_i forms the same angle with this diagonal but lies on the opposite side of it. This keeps A_i the same distance from the circle at corresponding times (see Figure 4.6). To check that the tail can be moved in a compatible fashion, note that reversing the changes in the size of the angles in the tail indeed keeps A_i the same distance from the circle at corresponding times. Although the tail does not return to its original *position,* it does return to its original *shape.* $\square$

An Algorithm for Changing Configurations

Suppose we are given an initial configuration and a desired final configuration of an arm within a circle. Using the formulas of the preceding subsection, we can quickly compute the c_i's, d_i's, and t_i's. Using Lemma 4.4, we can then quickly check whether each link with differing initial and final configuration can be brought to the diagonal. If this necessary and sufficient condition holds, then the following motion algorithm shows that the arm can be moved to the desired final configuration with $O(n^3)$ simple motions.

ALGORITHM 4.1: Algorithm for Changing Configuration
Step (i) Move the arm to normal form ($O(n)$ simple motions);
Step (ii) Once the predecessors of A_i are in their final positions, reorient L_i if

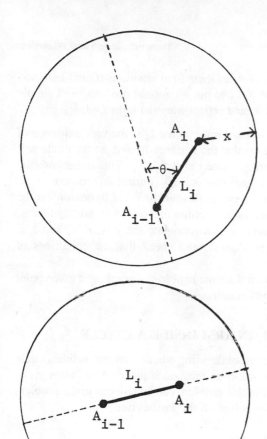

At time $t_0 - t$, L_i forms an angle θ with the diagonal through A_{i-1}, and A_i is distance x from the circle.

At time t_0, L_i reaches a diagonal.

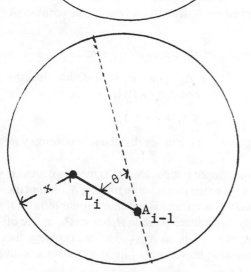

At time $t_0 + t$, A_{i-1} has returned to the position it occupied at time $t_0 - t$. L_i again forms angle θ with the diagonal through A_{i-1}, but has changed orientation. The distance between A_i and the circle is again x.

Figure 4.6. Reorientation of a link L_i with restoration of $A_1, \ldots, A_{i-1}$.

necessary, restoring the predecessors of A_i to their final positions ($O(n^2)$ motions, by Remark 4.3). Then rotate L_i about A_{i-1} to put A_i in final position ($n-i$ simple motions, by Remark 4.2). Increment i, and repeat Step (ii) until $i > n$. $\square$

Notice that, since the c_i's and d_i's depend only on the 1_i's, the very existence of the desired final configuration assures us that the distance from A_i to the circle will stay between c_i and d_i while L_i is being rotated about A_{i-1}. This is because the distance between A_i and the circle changes monotonicly during this rotation.

Notice also that the *question* of whether the desired final configuration can be attained can be answered in linear time on a machine that does real arithmetic ($+$, $-$, $*$, $/2$, $\min(,)$) since it is necessary only to compute the c_i's, d_i's, and t_i's, determine the links which must be reoriented, and check that the conditions of Lemma 4.4 hold for these links.

In the next section, we show how to reduce the *problem* of reaching a given point with A_n to a problem of changing configurations.

5.   REACHING A POINT WITH AN ARM INSIDE A CIRCLE

In this section, we will solve the problem of deciding whether an arm inside a circle can be moved from a given initial position to one which places A_n at some given point p. We will do this by showing that this problem can be reduced to the problem of changing configurations, which we solved in the last section.

Points on the Circle Reached by the A's

We want to compute a *feasible configuration* (i.e., one to which the arm can be moved from its initial configuration) that places A_n at a given point p (inside or on the circle). In order to find such a configuration, we first construct the set R_j of points *on* the circle that can be reached by A_j from the given initial position of the arm.

LEMMA 5.1:   Each R_j consists of at most two arcs of the circle.

Proof: (Induction on j) Clearly, $R_0 = \{A_0\}$ if A_0 is on the circle. Otherwise, the Normal Form Lemma 4.1 shows that the first non-empty R_j is the one for which

$$1_1 + \cdots + 1_{j-1} < c_0 = d_0 \leq 1_1 + \cdots + 1_j,$$

and that all subsequent R_j's are nonempty. It is easy to see that the *first* nonempty R_j consists of at most two arcs.

Now consider a j for which R_{j-1} is nonempty but consists of at most two arcs. If A_j is at some point in R_j, we can move A_{j-1} *to* the circle while moving A_j *around* the circle. (This can be done in the same way that an arm is put into normal form.) Of course, A_j stays in R_j during this process. Thus, each point in R_j belongs to an arc of R_j that contains a point reached by A_j with A_{j-1} in R_{j-1}. Hence, counting the number of arcs in R_j is equivalent to counting how many of its arcs contain a point that A_j can reach with A_{j-1} in R_{j-1}.

Suppose that A_{j-1} and A_j are on the circle and that $d_{j-1} \geq \mathbf{1}_j$. Then we can reorient L_j while moving A_j around the circle, keeping A_j in R_j. Our observation about counting arcs shows that each arc of R_{j-1} gives rise to only one arc in R_j. Thus in this case, R_j consists of at most two arcs.

Now suppose that A_{j-1} and A_j are on the circle and that $d_{j-1} \leq \mathbf{1}_j$. Then we can move A_{j-1} from any point in R_{j-1} to any other point in R_{j-1} without ever taking A_j off the circle or changing the orientation of L_j. Hence, all the points of R_j that are reached from R_{j-1} by L_j with left orientation are in the same arc of R_j. The same is true for L_j with right orientation, so again R_j consists of at most two arcs. $\square$

In our algorithm for reaching a point p, we will need to find for any given point in R_j a feasible configuration of the arm that positions A_j at that point. In the next section, we show how to compute this information quickly.

Determining the R's

First we will show that each set R_j is a union of certain contributions from its predecessors, and then we will describe an algorithm for calculating the R_j's and determining how to reach them.

The following lemma, whose proof we omit, can easily be established using the ideas in the proof of the Normal Form Lemma 4.1.

LEMMA 5.2:   Suppose an arm is positioned inside a circle so that A_j is located at a point p_j on the circle. Then A_j can be kept fixed at p_j while the arm is moved to a position where one of the following conditions holds:

(i)   links $L_1, \ldots, L_j$ form either a straight line (with no folds) or an "elbow" whose only bend is as A_{j-1};
(ii)   for some $i < j$, A_i is on the circle, and links $L_{i+1}, \ldots, L_j$ form either a straight line or an elbow whose only bend is at A_{j-1}.

Given a value for j, we need to find out for each R_i, $i < j$, which points of R_j can be reached from R_i by the straight lines and elbows of Lemma 5.2.

Suppose that p_i is a point in R_i and that $\mathbf{1}_{i+1} + \cdots + \mathbf{1}_j \leq d$. If all the links between A_i and A_j can be given the same orientation, then p_i contributes a point to R_j by means of a straight line. (If both orientations are possible, then p_i contributes two points to R_j.) Contributions of this type from points in R_i form at most four arcs, two for each arc of R_i. These arcs amount to shifts of R_i around the circle.

Now consider the possibilities for joining a point p_i in R_i to a point p_j in R_j by an elbow whose last joint is the one which is bent. Certainly $\mathbf{1}_{i+1} + \cdots + \mathbf{1}_{j-1}$ must be at most d. Since L_j and the straight line from A_i to A_{j-1} might have either orientation, there are four types of elbows to consider. Consider a particular feasible elbow, and note that it must place A_{j-1} somewhere on an arc of a circle of radius $\mathbf{1}_{i+1} + \cdots + \mathbf{1}_{j-1}$ centered at p_i. Since the orientations of the links in the elbow are specified, this arc is bounded by the circle at one end and by the diagonal through A_i at the other. The set of points that can then be reached by L_j in its specified orientation, with A_{j-1} on the arc, forms an arc on the circle. Hence, each feasible elbow type allows R_i to contribute a widened shift of itself to R_j.

The contributions of A_0 to R_j can be determined in a similar fashion.

It is now easy to give an $O(n^2)$ algorithm to do the following: compute the endpoints of the R_j's, and build a table that allows one, given a p_j in R_j, to find in $O(n)$ time (where n is the number of links in the arm) a feasible configuration having A_j at p_j.

ALGORITHM 5.1: Finding R's

First, determine how the links can be oriented ($O(n)$ time). Next, compute the contributions from A_0 of straight lines and elbows whose last joint is the one that is bent. Record these contributions by listing the endpoints of the arcs together with the description of the lines or elbows that generated them ($O(n)$ time). At this stage, the first nonempty R_i has been completely determined, and so its endpoints (of which there are at most four) can be computed ($O(n)$ time). Finally, for each R_i in turn, compute the contribution of R_i to its successors, and then compute the endpoints of R_{i+1} ($O(n)$ time per iteration). $\square$

In the next subsection, we use the information about the R_j's to solve the problem of moving A_n to an arbitrary point inside the circle.

How to Reach a Point

If we want to place A_n at a point p on the circle, we merely compute R_n and test p for membership. If p is in R_n, we use the table generated by Algorithm 5.1 to determine a feasible arm configuration that has A_n at p. Then we can use Algorithm 4.1 to move the arm to this configuration.

Now suppose p is inside the circle. If the arm can be moved to a configuration in which A_n is at p and some other joint is on the circle, then p can be reached by a feasible configuration in which some A_i is on the circle and links $L_{i+1}, \ldots, L_n$ form either a straight line or an elbow with the bend at A_{i+1}. To see whether this happens, we compute the R_j's and then look for an appropriate straight line or elbow reaching from p back to a nonempty R_j. If no such line or elbow can be found, we check to see whether p can be reached by a configuration that does not touch the circle.

LEMMA 5.3: Suppose that an arm $L_1, \ldots, L_n$ can be moved to a configuration in which A_n is at a given point p inside the circle, but that no such feasible configuration can have any joint on the circle. Then the arm can be moved to a configuration in which A_n is at p and at most two joints are bent.

Proof: Consider a feasible configuration with A_n at p. If it has more than two bends, proceed as follows. Let A_i, A_j, and A_k, where $0 < i < j < k < n$, denote the first three bent joints. Let A_m denote the fourth bent joint if one exists; otherwise, set $A_m = A_n$. Keeping A_k and its successors pinned down, rotate the line of links between A_0 and A_i about A_0 so that A_i moves *away* from A_m (see Figure 5.1). Eventually, one of three events must occur:

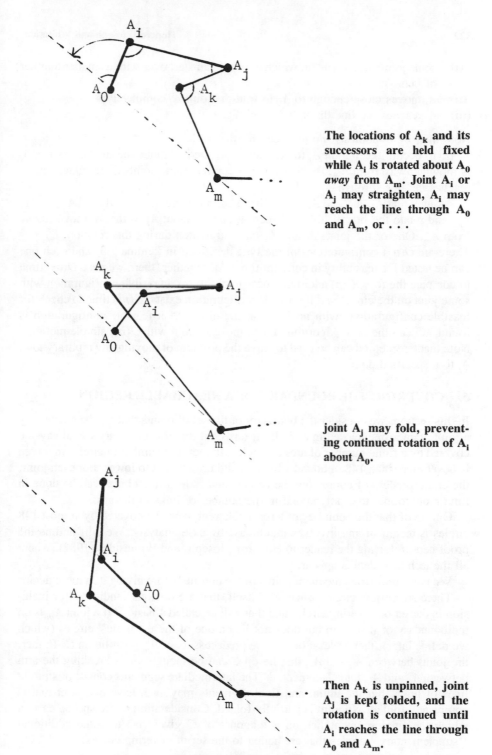

The locations of A_k and its successors are held fixed while A_i is rotated about A_0 *away* from A_m. Joint A_i or A_j may straighten, A_i may reach the line through A_0 and A_m, or . . .

joint A_j may fold, preventing continued rotation of A_i about A_0.

Then A_k is unpinned, joint A_j is kept folded, and the rotation is continued until A_i reaches the line through A_0 and A_m.

Figure 5.1.   Reaching p with at most two bent joints.

(i)   some joint straightens (in which case we can start over with a smaller number
      of bends);

(ii) A_i moves close enough to A_k to fold the joint A_j completely;

(iii) A_i reaches the line through A_0 and A_m.

Note that, by hypothesis, no joint can hit the circle.

If (ii) occurs, keep joint A_j folded, unpin A_k, and continue the rotation. Since A_i
is moving away from A_m, the rotation can continue until joint A_k straightens or A_i
reaches the line through A_0 and A_m.

Assume that A_i, A_0, and A_m are collinear. Pin down A_0, . . . , A_i and A_m, . . . ,
A_n, and rotate the line of links between A_i and A_j about A_i so that A_j moves *away*
from A_m. One of the joints A_i and A_k *must* straighten during this rotation. $\square$

There are $O(n^2)$ configurations of the type described in Lemma 5.3, and each one
can be tested for feasibility in constant time. All together, then, we need $O(n^2)$ time
to compute the R_j's, $O(n)$ additional time to check for a feasible configuration with
some joint on the circle, and if no such configuration exists, $O(n^2)$ time to check for
feasible configurations with no joint on the circle. If a feasible configuration is
found, we can then use Algorithm 4.1 to move A_n to p with $O(n^3)$ simple motions.
Note that our method can be used to solve the problem of moving any arbitrary joint
A_j to a specified point.

6.   COVERING THE BOUNDARY OF A REACHABLE REGION

In this section, we consider the boundary of the set of points that can be reached by
a joint of an arm enclosed in a circle. It turns out that this boundary can always be
covered by a finite number of arcs of circles. In fact, the number of circles involved
is never more than 148, a bound which we did not attempt to lower. For each joint,
the entire process of computing the centers and radii of the circles can be done in
time proportional to a polynomial in the number of links in the arm.

Our proof that the boundary of a reachable region can be covered by at most 148
circles is technical and involves much case by case analysis. We will outline the
proof here, referring the reader to Hopcroft, Joseph, and Whitesides (1982), where
all the technical details appear.

We now summarize the main points of the outline before giving it in more detail.

There are certain circles, such as C itself, that are obvious candidates for inclu-
sion in the set of covering circles and that will be called "basic." If a joint A_m is on
the boundary of its region but does not lie on one of these "basic" circles (which
we define later), then at least one of the predecessors of A_m must lie on C. In fact,
the joints between A_0 and A_m that lie on C can be thought of as breaking the arm
between A_0 and A_m into "segments." The intermediate segments consist of straight
lines of links, but the initial and final segments may each have one joint that is
completely folded. No joint is partially folded. Consideration of the special case in
which the final segment lies on a diagonal of C gives rise to some additional
"supplementary" circles that are added to the set of covering circles.

If A_m lies on the boundary of its region S_m but does not lie on a basic or supplementary circle, then the portion of the arm between A_0 and A_m must lie in one of several possible configurations. In each of these configurations, the number of possible locations for the last joint A_j before A_m that lies on C is small. These possible locations become centers for covering circles of radii determined by the final segment between A_j and A_m. The number of possible locations is small because certain inequalities in the link lengths must hold, and these inequalities can have only a small number of solutions.

Our final observation before giving the detailed outline is that we may assume A_0 is the only joint whose location is fixed. Of course by our definition of "arm," A_0 is the only joint that is fastened to the plane. However, it may be that other joints are effectively fixed for geometric reasons. For example, A_0 may be located on C, and the first link L_1 may have length equal to the diameter d of C so that the location of joint A_1 cannot change. However, it can be shown with the aid of the Normal Form Lemma 4.1 that there is a joint index j such that the location of A_i can change if, and only if, $j < i \leq n$. Furthermore, this index can be found quickly. This result takes care of regions consisting of single points and allows us to assume without loss of generality that A_0 is the only fixed joint.

We now give a detailed outline of the proof of the following theorem.

THEOREM 6.1: For any joint A_m of an n-link arm enclosed in a circle, the boundary of the set S_m of points that A_m can reach can be covered by at most 148 circles. Descriptions of these circles can be computed in $p(n)$ steps, where p is a polynomial.

As a notational matter, we denote a straight line of links between a joint x and a joint y by $[xy]$.

Basic Circles

To define two of the four basic circles that we immediately put into the set of covering circles, recall that in Section 4 we proved that the minimum and maximum distances c_m and d_m that a joint A_m can move off the circle can be computed in $p(n)$ steps, where p is a polynomial in the number of links. Call the two circles centered at the center O of circle C that have radii c_m and d_m basic.

To obtain the other two basic circles, note that summing the lengths of the links preceding A_m gives an upper bound for the maximum distance A_m can move from A_0. If A_m is preceded by a link L_j that is so long that

$$1_j - \sum_{i=1, i \neq j}^{m} 1_i > 0,$$

then this difference gives a positive lower bound for the minimum distance between A_0 and A_m; otherwise, 0 is a bound. Call the circles centered at A_0 with these radii, which are easy to compute, basic also.

Facts About Joints

Before continuing to build up a collection of circles covering the boundary points of S_m, we first need to observe some facts about joints. These are stated in Lemmas 6.1 through 6.3 below.

Consider a joint A_j that does not lie on the circle C. If L_j and L_{j+1} form a $0°(=360°)$ or $180°$ angle, A_j is said to be a *fold* or a *straight joint*, respectively. If A_j does not lie on the circle and is open to any other angle, it is called an *elbow*. (It is important to note that the definition of an elbow requires that the joint not be on C.) The next lemma gives a simple but fundamental observation about elbows. Its simple proof is a consequence of the fact that links in the tail beyond a given joint never constrain its motion along any path that stays within the minimum and maximum distances that the joint can move off the circle C (recall Remarks 4.1 and 4.2).

LEMMA 6.1: Suppose that no joint strictly between A_i and A_k lies on circle C but that some joint A_j between them is an elbow (A_i and A_k may or may not lie on C). Then the location of A_i can be held fixed while A_k is moved to all those points in some open ball centered at A_k that do not violate the minimum and maximum distances that A_k can be located from the circle (see Figures 6.1a and b).

Another basic observation is that a fold can sometimes be turned into an elbow.

LEMMA 6.2: Suppose that u and v are two joints of an arm enclosed in a circle C and that all joints between u and v are straight with one exception, x, which

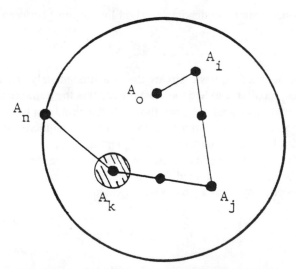

Figure 6.1a. The elbow at A_j enables A_k to reach the points in the shaded area while the location of A_i remains fixed.

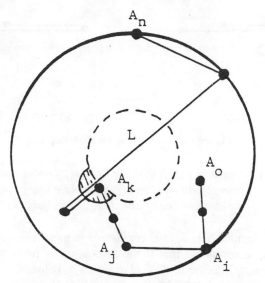

Figure 6.1b.   Link L is so long that A_k cannot
reach any points inside the dashed circle.

is a folded joint not lying on C. If the lines of links $[xu]$ and $[xv]$ from x to u and x to
v are not equal in length and if the longer contains at least two links, then an elbow
can be created at x without changing the locations of u and v. If the lines $[xu]$ and
$[xv]$ (possibly of equal length) *each* contain at least *two* links, then again, x can be
turned into an elbow without moving u and v (see Figure 6.2).

A final basic observation is that an elbow can be created from two folds that are
joined by a straight line of links unless the line consists of a single "long" link.

LEMMA 6.3:   Let u and v be joints of an arm embedded in a circle C. Suppose
all joints strictly between u and v are straight with two exceptions, which are folds.
Then the locations of u and v can be held fixed while the arm is moved to create an
elbow between u and v unless the folds are joined by a single link that is at least as
long as the sum of the lengths of all the other links between u and v.

Segments

If a configuration of the arm places A_m on the boundary of region S_m but not on one
of the four basic circles, then it is not difficult to use Lemmas 6.1–6.3 to prove that
some joint strictly between A_0 and A_m must lie on circle C. In particular, we can find
some *last* joint A_j between A_0 and A_m that is on C. We will say that the links
between A_j and A_m form the *final segment of the configuration before A_m*, or simply,
the *final segment*. Similarly, we will say that the links between A_0 and the *first* joint
beyond A_0 on C form the *initial segment* of the configuration. (Here, A_0 may or may
not be on C.)

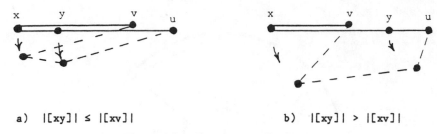

a) |[xy]| ≤ |[xv]| b) |[xy]| > |[xv]|

Figure 6.2.   Creating an elbow at x.

It is clear from Lemmas 6.1–6.3 that *the final segment is made up of either a straight line of links from a joint on C to A_m or a single link from a joint on C to a fold that is followed by a straight line of links to A_m.* In either case, the final segment lies along a line.

Recall that, by Lemma 5.1, $S_j \cap C$ consists of at most two arcs of C. A routine analysis of several cases shows that if the final segment from A_j to A_m lies on a diagonal of C, then R_j', the arc of $S_j \cap C$ to which A_j belongs, consists of a single point. Using this fact, together with Lemmas 6.1–6.3, it is easy to establish the general form of a configuration that places A_m on the boundary of S_m but not on a basic circle.

LEMMA 6.4:   Suppose that an arm has been moved to a configuration that places A_m on the boundary of S_m but not on a basic circle. Let A_i be the first joint beyond A_0 on C, and let A_j be the last joint before A_m on C. Then all joints between A_0 and A_m that do not lie on C are straight, with the possible exceptions of A_{i-1} and A_{j+1}. If A_{i-1} and A_{j+1} are not straight, then they must be folds.

This general form will help us to enumerate the remaining configurations that might have A_m on the boundary of its reachable region. Before we do this, we first observe that we can simplify the enumeration by assuming that the final segment does not lie on a diagonal of C. In order to assume this, however, we must add some more circles to our collection.

Supplementary Circles

If A_m lies on the boundary of S_m but not on a basic circle, and if the final segment from A_j to A_m lies on a diagonal of C, then it can be shown that the initial segment consists of a single link L_1 that is connected directly to the final segment, which begins at $A_j = A_1$. The proof involves the analysis of several cases and uses Lemma 6.4 together with the fact cited previously that R_j' consists of a single point. The important consequence of this new fact is that if the final segment lies on a diagonal of C, then A_m lies on one of at most four "supplementary" circles that we are about to describe. Note that in this situation, there are at most two possible locations for A_1, corresponding to the two possible orientations for L_1 (see Section 4 for the definition of orientation). Then, for a fixed position of A_1, A_m lies on a circle centered at A_1 of radius either $\sum_{j=2}^{m} \mathbf{1}_j$ or, when positive, $\mathbf{1}_2 - \sum_{j=3}^{m} \mathbf{1}_j$. This defines at most four circles, which we call *supplementary* and add to our set.

Enumerating Configurations

From now on, we assume that A_m is neither on a basic circle nor on a supplementary circle so that we need only concern ourselves with situations in which *the final segment before A_m* does not lie on a diagonal of C.

In order to enumerate the possible configurations of the arm between A_0 and A_m, it is useful to establish some forbidden subconfigurations. In Hopcroft, Joseph, and Whitesides (1982) we listed six of these, two of which are shown in Figure 6.3. These subconfigurations are forbidden because they can be moved to form elbows, which are excluded by Lemma 6.4, without changing the location of their endpoints. In Figure 6.3a, the locations of u and x can be held fixed while $[uv]$ is rotated about u. This requires that v move closer to x, which can be accomplished by creating an elbow between v and x. Similarly in Figure 6.3b, the locations of u and y can be held fixed while $[xy]$ is rotated about y. This requires that x move away from u, which can be accomplished by opening the joint at v while simultaneously rotating $[uv]$ about u.

By using the list of six forbidden configurations, we were able to list by careful

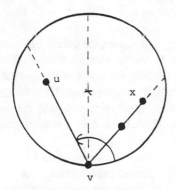
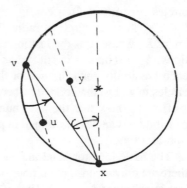

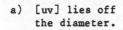
a) [uv] lies off   b) [vx] lies off
   the diameter.   the diameter.

Figure 6.3. Configurations that give rise to elbows. Arrows indicate the angular range for a line of links. A sharp tip indicates that the endpoint of the arc belongs to the range, and a round tip indicates that it does not. No order is implied by the letters at joints: u could come before or after v. There may be additional joints between the ones that appear in the figure. A dashed extension of a link indicates that its endpoint may lie on C.

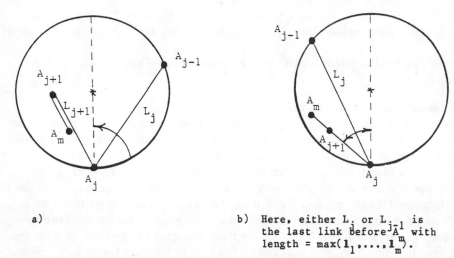

a)

b)   Here, either L_j or L_{j-1} is
     the last link before A_m with
     length = $\max(1_1, \ldots, 1_m)$.

Figure 6.4. Two of the ten possible configurations.

and tedious analysis a set of ten possible configurations for the arm when the final segment contains more than one link. (Two typical ones are shown in Figure 6.4.) Consequently, the boundary of S_m can be covered by a collection of circles consisting of the basic circles (at most four), the supplementary circles (at most four), circles of radius 1_m centered at the endpoints of R_{m-1} (at most four) together with circles covering the ten configurations enumerated. In half of these, as in Figure 6.4a, A_m lies on a circle of radius $1_{j+1} - |[A_{j+1}A_m]|$ centered at the endpoint of an arc of R_j, where A_j is the last joint on C between A_0 and A_m. In the other configurations, A_m lies on a circle of radius $|[A_jA_m]|$, where A_j has the same definition. Thus each possibility for j in each of the configurations gives rise to at most four new circles to add to the collection because R_j has at most four endpoints. Therefore, it suffices to show that the total number of possibilities for A_j is small, and that the possibilities can be determined in polynomial time. This can be done one configuration at a time.

The basic idea for handling each configuration is this. Show that for a fixed m, there are only a constant number (independent of the arm) of possibilities for A_j, the last joint before A_m on C. Then a constant number of circles (at most eight for the worst choice of A_j) can be added to the basic and supplementary circles to form a collection that covers the boundary of S_m. This is because A_m must lie on a circle of radius either $\sum\limits_{k=j+1}^{m} 1_k$ or $1_{j+1} - \sum\limits_{k=j+2}^{m} 1_j$ about A_j, and A_j must lie at one of the endpoints of R_j, of which there are at most four. The possibilities for A_j will be determined by inequalities involving the link lengths that can only be satisfied in a few ways.

Consider, as a simple example, the configuration in Figure 6.4b. There are only two choices for A_j. It could be the higher indexed endpoint of the highest indexed link of longest length, or it could be the next joint after that.

As for the configuration in Figure 6.4a, it can be shown that unless 1_j or $1_{j+1} > r$, then the entire arm could be moved so that the configuration between A_{j-1} and A_{j+1} would go to its mirror image with respect to the initial line determined by A_{j-1} and A_{j+1}, while the configuration of the rest of the arm would be restored. Since this would create elbows, it must be the case that $1_j + 1_{j+1} > r$. Of course $d > 1_{j+1} > |[A_{j+1}A_m]|$. If there are solutions to these inequalities, let z be the largest feasible choice for index $j+1$. Then note that there can be at most three feasible choices for $j+1$ that are smaller than z, giving a total of four choices for A_j.

The idea of moving a subconfiguration to its mirror image to show that certain inequalities must hold is used to handle several of the configurations.

Summing over all ten possible configurations listed in Hopcroft, Joseph, and Whitesides (1982), the total number of choices for A_j is at most 34. Since R_j may have as many as four endpoints, this generates at most 136 circles. There were at most 12 circles initially, so the total number of circles needed is at most 148. This completes the outline of the proof of Theorem 6.1.

The bound of 148 is probably very generous. The important point, though, is that the bound does not depend on the arm.

References

Aho, A., Hopcroft, J., & Ullman, J. (1974). *The design and analysis of computer algorithms*. Reading, MA: Addison-Wesley.

Akritas, A.G., (1980). The fastest exact algorithms for the isolation of the real roots of a polynomial equation. *Computing, 24,* 299–313.

Arnon, D.S. (1981). *Automatic analysis of real algebraic curves* (Technical Report, Computer Science Department, Purdue University). (See also *ACM Sigsam Bulletin, 15* (4), 3–9.)

Arnon, D.S. (1981). *Algorithms for the geometry of semi-algebraic sets* (Technical Report 436, Computer Science Department, University of Wisconsin, Madison).

Atallah, M. (1983). Dynamic computational geometry. *Proceedings of the 24th IEEE Symposium on Foundations of Computer Science* (pp. 92–99). Tucson: IEEE Computer Science Press.

Ben-Or, M., Kozen, D., & Reif, J.H. (1985). Complexity of elementary algebra and geometry. *Journal of Computer and System Sciences.*

Bentley, J.L., & Ottman, T.A., (1979). Algorithms for reporting and counting geometric intersections. *IEEE Transactions on Computers, C-28,* 643–647.

Boyse, J.W., & Gilchrist, J.E. (1982). GMSolid: Interactive modeling for design and analysis of solids. *IEEE Computer Graphics and Applications, 2,* 69–84.

Brady, M., Hollerbach, J.M., Johnson, T.L., Lozano-Perez, T., & Mason, M.T. (Eds.). (1983). *Robot motion: Planning and control*. Cambridge: MIT Press.

Brown, C.M. (1982). PADL-2: A technical summary. *IEEE Computer Graphics and Applications, 2,* 27–40.

Brown, W.S., & Traub, J.F. (1971). On Euclid's algorithm and the theory of subresultants. *Journal of the ACM, 118,* 505–514.

Chazelle, B. (1982). A theorem on polygon cutting with applications. *Proceedings of the 23rd IEEE Symposium on Foundations of Computer Science* (pp. 339–344), Chicago.

Cohen, P.J. (1969). Decision procedures for real and p-adic fields. *Communications on Pure and Applied Mathematics, 22,* 131–151.

Collins, G. E. (1975). Quantifier elimination for real closed fields by cylindrical algebraic decomposition. In: Second GI conference on automata theory and formal languages, *Lecture Notes in Computer Science, 33,* 134–183. Berlin: Springer-Verlag.

Collins, G.E., & Loos, R. (1976). Polynomial real root isolation by differentiation. In *SYMSAC 76 Proceedings, ACM Symposium on Symbolic and Algebraic Computations* (pp. 15–25), Yorktown Heights, NY.

Comba, P.G. (1968). A procedure for detecting intersections of three-dimensional objects. *Journal of the ACM, 15,* 354–366.

Cooke, G.E., & Finney, R.L. (1967). *Homology of cell complexes* (Mathematical notes). Princeton, NJ: Princeton University Press.

Davenport, H. (1971). A combinatorial problem connected with differential equations, II. *Acta Arithmetica, 17,* 363–372.

Davenport, H., & Schinzel, A. (1965). A combinatorial problem connected with differential equations. *American Journal of Mathematics, 87,* 684–694.

Drysdale, R.L. III (1979). *Generalized Voronoi diagrams and geometric searching*. Ph.D. Thesis, Department of Computer Science, STAN-CS-79-705, Stanford University, Stanford, CA.

Drysdale, R.L. III, & Lee, D.T. (1978). Generalized Voronoi diagrams in the plane. *Proceedings of the 16th Annual Allerton Conference on Communications, Control and Computing* (pp. 833–842).

Edelsbrunner, H. (1982). *Intersection problems in computational geometry* (Technical Report F-93). Technical University of Graz, Austria.

Fisher, W.B. et al (1977). *The PADL-1.0/n processor: Overview and system documentation* (PADL System Document-01). Rochester, NY: Production Automation Project, University of Rochester.

Fisher, W.B. et al. (1978). *Part and assembly description languages—II* (Technical Memorandum 20b). Rochester, NY: Production Automation Project, University of Rochester.

Garey, M.R., & Johnson, D.S. (1979). *Computers and intractability: A guide to the theory of NP-completeness.* San Francisco: Freeman.

Hamilton, W.R. (1969). *Elements of quaternions.* New York: Chelsea.

Hartquist, E.E. et al. (1977). *Representations in the PADL-1.0/n processor: Boundary representations and the BFILE/1 system* (PADL System Document-10). Rochester, NY: Production Automation Project, University of Rochester.

Heindel, L.E. (1971). Integer arithmetic algorithms for polynomial real zero determination. *Journal of the ACM, 22,* 533–549.

Hillyard, R. (1982). The Build group of solid modelers. *IEEE Computer Graphics and Applications, 2,* 43–52.

Hironaka, H. (1975). Triangulations of algebraic sets. In *Proceedings, Symposia in Pure Math, 29,* 165–185. Providence, RI: American Mathematical Society.

Hodge, W.V.D., & Pedoe, D. (1952) *Methods of algebraic geometry.* Cambridge, England: Cambridge University Press.

Hopcroft, J.E., Joseph, D.A., & Whitesides, S.H. (1982). *Determining points of a circular region reachable by joints of a robot arm* (Tech. Rep. No. TR 82-516). Computer Science Department Ithaca, NY: Cornell University.

Hopcroft, J.E., Schwartz, J.T., & Sharir, M. (1983). Efficient detection of intersections among spheres. *Robotics Research, 2* (4), 77–80.

Hopcroft, J.E., Schwartz, J.T., & Sharir, M. (1984). *On the complexity of motion planning for multiply independent objects: PSPACE hardness of the warehouseman's problem* (TR-103). New York: Courant Institute of Mathematical Sciences.

Hopcroft, J., & Ullman, J.D. (1979). *Introduction to automata theory, languages, and computation.* Boston: Addison-Wesley.

Ignat'yev, M.B., Kulakov, F.M., & Pukroviskiy, A.M. (1973). *Robot manipulator control algorithms* (Report No. JPRS 59717). Springfield, VA; NTIS.

Keller, O.H. (1974). *Vorlesungen über algebraische geometrie.* Leipzig: Akademische Verlagsgesellschaft, Geest & Portig.

Kahn, P. (1979). Counting types of rigid frameworks, *Invent. Math, 55,* 297–308.

Kempe, A.B. (1976). On a general method of describing plane curves of the nth degree by linkwork. *Proceedings of the London Mathematical Society, 7,* 213–216.

Kirkpatrick, D. (1979). Efficient computation of continuous skeletons. In: *Proceedings of the 20th IEEE Symposium on Foundations of Computer Science* (pp. 18–27), Puerto Rico. IEEE Computer Science Press.

Kung, H.T., & Traub, J.F. (1978). All algebraic functions can be computed fast. *Journal of the ACM, 25,* 245–260.

Lazard, D. (1977). Algebre lineaire sur K $(X_1, \ldots, X_n)$ et elimination. *Bulletin de Societe Mathematique de France, 105,* 165–190.

Lee, D.T., & Drysdale, S. (1981). Generalization of Voronoi diagrams in the plane. *SIAM Journal of Computing, 10* (1), 73–87.

Levin, J. (1976). A parametric algorithm for drawing pictures of solid objects composed of quadric surfaces. *Communications of the ACM, 19,* 555–563.

Lewis, H.R., & Papadimitriou, C.H. (1982). Symmetric space-bounded computation. *Theoretical Computer Science, 19,* 161–187.

Lipson, J. (1976). Newton's method: A great algebraic algorithm. In *SYMSAC 76, Proceedings of the ACM Symposium on Symbolic and Algebraic Computations* (pp. 260–270). Yorktown Heights, NY.

Lockwood, E.H., & Prag, A. (1961). *Book of curves*. Cambridge: Cambridge University Press.

Lozano-Perez, T. (1980). *Automatic planning of manipulator transfer movements* (AI Memo 606). Cambridge: M.I.T. Artificial Intelligence Laboratory.

Lozano-Perez, T., & Wesley, M. (1979). An algorithm for planning collision-free paths among polyhedral obstacles. *Communications of the ACM, 22*, 560–570.

Mahler, K. (1964). An inequality for the discriminant of a polynomial. *Michigan Math Journal, 11*, 257–262.

Marden, M. (1949). The geometry of zeroes of a polynomial in a complex variable. *Mathematical Surveys, 3*. Providence, RI: American Mathematical Society.

McCreight, E.M. (1982). *Priority search trees* (Technical Report CSL-81-5). Palo Alto, CA: Xerox Corporation.

Mignotte, M. (1976). Some problems about polynomials. *SYMSAC 76 Proceedings, 1976 ACM Symposium on Symbolic and Algebraic Computations* (pp. 227–228). Yorktown Heights, NY.

Miller, G., & Reif, J.H. (1985). *Robotic movement planning in the presence of friction is undecidable.*

Moenck, R. (1973). *Studies in fast algebraic algorithms*. Ph.D. dissertation, University of Toronto.

Moravec, H.P. (1980). *Obstacle avoidance and navigation in the real world by a seeing robot rover*. Doctoral dissertation, Stanford University (Report No. AIM-340,).

Nievergelt, J., & Preparata, F.P. (1982). Plane-sweeping algorithms for intersecting geometric figures. *Communications of the ACM, 25*, 739–747.

Nilsson, N.J. (1969). A mobile automaton: An application of artificial intelligence techniques. *Proceedings of the IJCAI-69* (pp. 509–520).

Ocken, S., Schwartz, J.T., & Sharir, M. (1983). Precise implementation of CAD primitives using rational parametrizations of standard surfaces (Technical Report 67, Robotics Activity), Computer Science Department, Courant Institute, New York University.

Ó'Dúnlaing, C., Sharir, M., & Yap, C.K. (1983). Retraction: A new approach to motion planning. In *Proceedings of the Symposium on the Theory of Computing* (pp. 207–220), Boston.

Ó'Dúnlaing, C., Sharir, M., & Yap, C.K. (1984a). *Generalized Voronoi diagrams for moving a ladder: I. Topological analysis*, Computer Science Technical Report No. 139, New York: Courant Institute, New York University.

Ó'Dúnlaing, C., Sharir, M., & Yap, C.K. (1984b). *Generalized Voronoi diagrams for moving a ladder: II. Efficient construction of the diagram*, Computer Science Technical Report No. 140, New York: Courant Institute, New York University.

Paul, R. (1972). *Modelling trajectory calculation and servoing of a computer controlled arm*. Ph.D. Thesis, Stanford University.

Peaucellier, M. (1864). Correspondence. *Nouvelles Annales de Mathematiques, 3*, 414–415.

Preparata, F.P. (1981). A new approach to planar point location. *SIAM Journal of Computing, 10*, 739–747.

Reif, J. (1979). Complexity of the mover's problem and generalizations (pp. 421–427). *Proceedings of the 20th Symposium on the Foundations of Computer Science*.

Reif, J.H., & Storer, J. (1985). *Shortest paths in Euclidean space with polyhedral obstacles*. Center for Research in Computing Technology, Harvard University.

Reif, J.H., & Sharir, M. (1985, October). *Motion planning in the presence of moving obstacles*. Paper presented at the 26th IEEE Symposium on Foundations of Computer Science, Portland, OR.

Requicha, A.A.G. (1977). *Part and assembly description languages—I: Dimension and tolerancing* (Technical Memorandum 19). Rochester, NY: Production Automation Project, University of Rochester.

Requicha, A.A.G., & Voelcker, H.B. (1982). Solid modeling: A historical summary and contemporary assessment. *IEEE Computer Graphics and Applications, 2*, 9–24.

Savitch, W.J. (1970). Relationships between nondeterministic and deterministic tape complexities. *Journal of Computer and Systems Science, 4*, 177–192.

Rump, S. (1976). On the sign of a real algebraic computation. In *Proceedings, 1976 ACM Symposium on Symbolic and Algebraic Computations* (pp. 238–241), Yorktown Heights, NY.

Schwartz, J.T. (1968). *Differential geometry and topology*. New York: Gordon and Breach.

Schwartz, J.T. (1969). *Lectures on differential geometry and topology*. New York: Gordon and Breach.

Schwartz, J.T. (1980). Fast probabilistic algorithms for verification of polynomial identities. *Journal of the ACM, 27*, 701–717.

Schwartz, J.T., & Sharir, M. (1983a). On the "piano movers" problem I. The case of a two-dimensional rigid polygonal body moving amidst polygonal barriers. *Communications on Pure and Aoplied Mathematics, 36*, 345–398.

Schwartz, J.T., & Sharir, M. (1983b). On the "piano movers" problem. II. General techniques for computing topological properties of real algebraic manifolds. *Advances in Applied Mathematics, 4*, 298–351.

Schwartz, J.T., & Sharir, M. (1983c). On the "piano movers" problem: III. Coordinating the motion of several independent bodies: The special cases of circular bodies moving amidst polygonal barriers. *International Journal of Robotics Research, 2*, 46–75.

Schwartz, J.T., & Sharir, M. (1984). On the piano movers' problem: V. The case of a rod moving in three-dimensional space amidst polyhedral obstacles. *Communications on Pure and Applied Mathematics, 37*, 815–848.

Shamos, M.I. (1975). *Computational geometry*. Ph.D. Dissertation, Yale University.

Sharir, M. (1983). *Intersection and closest-pair problems for a set of planar bodies* (Technical Report 56). New York: Computer Science Department, Courant Institute, New York University.

Sharir, M., & Ariel-Sheffi, E. (1984). On the piano movers' problem: IV. Various decomposable two-dimensional motion planning problems. *Communications on Pure and Applied Mathematics, 37*, 479–493.

Sharir, M., & Schorr, A. (1984). On shortest paths in polyhedral spaces. *Proceedings of the 16th ACM symposium on the Theory of Computing* (pp. 144–153), Washington, DC.

Spirakis, P. (1984). *Very fast algorithms for the area of the union of many circles*. Technical report, Computer Science Department, Courant Institute, New York University.

Spirakis, P., & Yap, C. (1985). Strong NP-hardness of moving many discs. *Information Processing Letters*.

Szemeredi, E. (1983). On a problem by Davenport and Schinzel. *Acta Arithmetica, 25*, 213–224.

Tarski, A. (1951). *A decision method for elementary algebra and geometry* (2nd ed.). Berkeley: University of California Press.

Tilove, R.B. (1977). *Representations in the PADL 1.0/n processor: Simple geometric entities* (PADL System Document-02). Rochester, NY: Production Automation Project, University of Rochester.

Udupa, S. (1977). *Collision detection and avoidance in computer-controlled manipulators*. Ph.D. Dissertation, California Institute of Technology, Pasadena.

Van Der Waerden, B.L. (1939). *Einfuhrung in die algebraische geometie*. Berlin: Springer.

Van Der Waerden, B.L. (1960). *Algebra* (5th ed.). Berlin: Springer.

Wesley, M.A. (1980). Construction and use of geometric models. In J. Encarnacao (Ed.), *Computer aided design* (pp. 79–136). New York: Springer-Verlag.

Widdoes, C. (1974). *A heuristic collision avoider for the Stanford robot arm* (CS Memo 227). Stanford, CA: Stanford University.

Yap, C.K. (1983). *Motion coordination for two discs* (Robotics Laboratory Technical Report No. 16). New York: Courant Institute, New York University.

Yap, C.K. (1984). *An O(nlog n) algorithm for the Voronoi diagram of a set of simple curve segments* (Computer Science Report No. 161). New York: Courant Institute, New York University.

Yap, C. K. (1985). *An o(nlog n) algorithm for the Voronoi diagram of a set of curve segments* (Computer Science Report No. 161). New York: Courant Institute, New York University.

Author Index

Italics indicate bibliographic citations

Subject Index

C

cell complex, 61
closest-pair problems, 214 ff
Collins cells, adjacency of, 88
Collins decomposition, 56
computational geometry, pref. 2
connectivity graph, 4, 12, 35
    for moving rod in 3 dimensions, 183
    for moving 'spider', 150
    for several moving circles, 111
critical curves, 2, 9
    example of, 48
    for moving rod or polygon in two
        dimensions, 27
        types of, 14 ff, 28 ff
    for moving 'spider', 147
    for several moving circles, 104, 123
        types of, 106, 125 ff
critical positions,
    for moving rod in 3 dimensions, 163 ff, 172
    types of, 173 ff
crossing rules, 35 ff
    for several moving circles, 113, 133
cylindrical algebraic decomposition, 55

D

decomposition method, for motion
        planning, (i)
displaced corner, 101
displaced wall, 101

F

free position, 4
    for moving 'spider', 143
    for several moving circles, 101, 121, 139

G

generalized mover's problem, 267
generalized Voronoi diagram, (ii), 198
    for circular objects, 218
        efficient construction of, 225

I

incidence function, 63
intersection problems, 214 ff
    for spheres, 240 ff

L

linkages, 282
    for multiplication, 293
    Peaucellier's, 285
    simple, 284

M

Mahler's Lemma, 77
motion planning, (i)
    computational cost of, (i)
    for a disc, 188
    for a ladder,
        by retraction method in two dimensions,
            197 ff
    lower bounds for computational cost of, (iii)
    problems,
        decomposable, 141
        complexity of, 267 ff, 275

N

noncritical regions, 9

P

piano mover's problem, 1

Q

quadric surfaces, 248 ff

R

rational parametrizations, 245
rational surfaces, 246
reachability problem, 300
resultants of polynomials, 73
retraction technique, for motion planning, (ii),
        187, 193
ruler folding problem, 306

S

semi-free position, 4
    for several moving circles, 101, 121
Sturm sequence, 78
subresultants, 73, 84

T

Tarski's algorithm, 52
Tarski sentence, 55